# Socio-metabolic Perspectives on the Sustainability of Local Food Systems

# Human-Environment Interactions

## VOLUME 7



More information about this series at http://www.springer.com/series/8599

Eva Fraňková · Willi Haas
Simron J. Singh
Editors

# Socio-metabolic Perspectives on the Sustainability of Local Food Systems

## Insights for Science, Policy and Practice

Springer

*Editors*
Eva Fraňková
Department of Environmental Studies, Faculty of Social Studies
Masaryk University
Brno
Czech Republic

Willi Haas
Institute of Social Ecology, Faculty for Interdisciplinary Studies
Alpen-Adria University
Vienna
Austria

Simron J. Singh
Associate Professor in the School of Environment Enterprise and Development (SEED)
University of Waterloo
Waterloo, ON
Canada

and

Associate Dean of the Faculty of Environment
University of Waterloo
Waterloo
Canada

Human-Environment Interactions
ISBN 978-3-319-69235-7 ISBN 978-3-319-69236-4 (eBook)
https://doi.org/10.1007/978-3-319-69236-4

Library of Congress Control Number: 2017955653



English proofreading: Simon Botten
Graphics: František Eliáš, http://www.f-e.cz/
Technical support and editing: Nikola Šťastná
Front cover photo: Eva Fraňková

Printed on acid-free paper

This Springer imprint is published by the registered company Springer International Publishing AG part of Springer Nature.
The registered company address is: Gewerbestrasse 11, 6330 Cham, Switzerland

*To all the unsung heroines and heroes,*
*practitioners of "quiet" sustainability*

# Foreword

The food system has been fundamentally transformed during industrialisation and is still in a process of transition. It has shifted from subsistence dominated agriculture, where people largely consume what they produce, towards transfers of large amounts of agricultural products through global processing chains and via international markets. Value added is not created on farms but rather in up- and downstream industries. Consumption patterns and food habits have changed from cereals and other starchy staple crops towards a high share of meat, and from home cooking to processed food and food away from home. The distance between consumers and producers has increased, both in terms of mileage and also culturally. Agricultural production has been transformed from mixed farming and on-farm recycling of biomass and plant nutrients towards highly specialised throughput systems based on fossil-fuel-powered machinery and high inputs of agrochemical products. With the tremendous increases in land and labour productivity of the green revolution, the intensity of human management and control of nature has also intensified, with severe negative side effects on the environment. Agriculture is one of the key drivers of global environmental change. In spite of gains in resource efficiency, the food system contributes a large share to global greenhouse gas emissions and is responsible for biodiversity loss and soil degradation. Overall, the food system must be considered a major driver of humanity reaching planetary boundaries. And, even though food production has increased faster than global population, the Food and Agricultural Organization of the United Nations estimates that one in nine people on the planet still suffers from chronic undernourishment, while at the same time obesity is also on the rise.

The modern food system is in a crisis, but promising new concepts and alternatives to the mainstream are emerging: Organic agriculture is expanding in all parts of the world; a rising number of people care about where and how their food is produced and decide to eat either no or less meat. Alternative concepts like food sovereignty, promoted by NGOs like *Via Campesina*, are gaining popularity in political discourse. While these are positive signals, it may be doubted that we are already observing a transition process towards a new and more sustainable food system. Interdisciplinary research on food systems, providing important knowledge for such a transition, is urgently needed and high on the agenda in sustainability science.

This edited volume presents innovative research on food systems and contributes to science for a more sustainable use of natural resources. It stands out for three main reasons: The book investigates transition processes of food systems with a focus on local scale. While a more sustainable food system is a global challenge, it also has an inevitable local dimension, since it concerns the specific places where people live and work, where they farm the land and consume their meals. Agriculture is always a local process, and closing cycles between production and consumption and reducing the dependence of production systems on external inputs require a local perspective. The chapters in the book present research from diverse local case studies in a larger context, with a broad geographical coverage ranging from Cambodia to Spain. The second important feature of this book is that it emphasises a long-term socio-ecological perspective. Several of the chapters have a historic focus and investigate pre-industrial food systems and their transformation through time. Such a long-term historical perspective is rare in food system studies, although it is key to understanding the emergence of the current food system and ongoing transition processes. The third unique feature is the specific methodological focus. The book introduces a systemic socio-metabolic perspective on food systems. The concept of social metabolism has emerged as a new approach in sustainability science, used to study biophysical aspects of society–nature relations, and it allows the linking of economic activity and ecosystem properties at multiple temporal and spatial scales. It has been applied at different scales, from global to national or urban systems. The application to local food systems presented here is new and promising. The studies assembled in this volume all apply different methods and approaches related to the concept of socioeconomic metabolism. They investigate the changing metabolic characteristics of the food system, how they link to environmental pressures and how we can use this information to create more sustainable food systems.

The book opens new perspectives for interdisciplinary food system studies. It provides an excellent introduction to different approaches and methods to study food systems from a socio-metabolic perspective. The research presented in this book contributes important knowledge on food systems from different temporal and geographic perspectives from historical to current practices, from the Global North and South, and from mainstream to novel food systems. The findings are of high relevance both for sustainability science and for the development of strategies and policies towards more sustainable food systems. I am convinced that this book will not only be read with interest by scholars in the broad and interdisciplinary field of food system science and sustainable development, but also by people engaged in all sorts of initiatives in the search for new pathways to make our food system sustainable. And hopefully it finds its way on the bookshelves of food consumers who care for more sustainable futures for our food system.

Vienna, Austria
June 2017

Fridolin Krausmann

# Acknowledgements

As with many other adventures in life, this book began as a very ordinary, nice, small idea. Our conversations often took place in Klikov, a tiny village close to the Czech–Austrian border, where we taught students how to conduct social metabolism studies at the local level. Then it became “a project”, a new challenge and a commitment that consumed nearly 2 years of our lives. As with most ambitious ideas, and knowing what such a work would entail, we would think three times, and then would say ‘no’. As time passed, from one Klikov seminar to another, we naively began, and then it was just too late to stop. That is not to say we are not grateful for the experience—indeed we are. We profoundly believe that our efforts have paid off well. The project brought us in contact with some incredible scholars and activists working on local food systems, and quite a few of them have entered this volume. A workshop in Vienna brought us all together, the conversations continued, and we continued to learn from each other, reading and commenting on each other’s texts and ideas. At the end, we present you with this result that we find both interesting and important to share. We hope you will feel the same while reading it.

Eva Fraňková is grateful to her partner and friends, because without their support this endeavour would have been impossible. Eva also offers great appreciation to the Department of Environmental Studies at the Faculty of Social Studies, Masaryk University in Brno, Czech Republic. A truly friendly space with plenty of academic freedom, and support at the same time. Willi Haas thanks the Institute of Social Ecology at the Faculty of interdisciplinary studies, Alpen Adria University (Austria), for providing an intellectual and fruitful ground, with the conversation of colleagues who are always prepared to offer their unconditional support. Simron Singh gratefully acknowledges the inspirational conversations with Eva and Willi; it has been a pleasure to work together. He profoundly thanks the Faculty of Environment, University of Waterloo (Canada) for providing a fertile ground for this product, and for supporting travels to Europe. He is particularly thankful to colleagues, Jennifer Clapp, Bruce Frayne, Mary Louise McAllister and Goretty

Dias for the opportunity to converse as well as collaborate in projects and events related to sustainable food systems.

As editors, we are very grateful to all the authors involved in this collection—it was an inspiring and enjoyable experience, and we can only admire all the work that it is only possible to partially introduce in this volume. Thanks for your friendly, effective and fruitful cooperation!

Most of the editorial work on the book was supported by the Czech Science Foundation, grant no. 13-38994P: Quest for sustainable food production: Social and financial metabolism of selected local food systems. Besides all the organisational work during the book preparation process, the support also extends to Chap. 1 (Introduction) and Chap. 12 (Conclusions). Several chapters (see below) were written on the basis of the project 'Sustainable Farm Systems: Long-Term Socio-Ecological Metabolism in Western Agriculture' that was funded by the Social Sciences and Humanities Research Council of Canada (SSHRC 895-2011-1020). Without this funding, the interesting and coherent views expressed in these chapters would not have been feasible.

Other acknowledgements and dedications for each particular chapter are listed below.

**Enric Tello and Manuel González de Molina (Chap. 2):** The research for this chapter was supported by the Canadian Social Sciences and Humanities Research Council, Partnership Grant 895-2011-1020, and the coordinated Spanish Research Projects HAR2015-69620-C2 01 and 02. We would also like to thank Eva Fraňková, Willi Haas and Tiziano Gomiero for their helpful comments, which have helped to improve the chapter, without being responsible for any mistake it may contain.

**Tiziano Gomiero (Chap. 3):** I wish to acknowledge Eva Fraňková and Claudio Cattaneo for their very helpful review of the early drafts. I gratefully acknowledge prof. Mario Giampietro (ICREA professor at ICTA, Universitat Autònoma de Barcelona, Spain) for his comments on social/societal metabolism and MuSIASEM, and prof. Joan Martinez-Alier (Emeritus Professor of Economics and Economic History and researcher at ICTA, Universitat Autònoma de Barcelona, Spain) for discussing some ideas concerning the history of the concept of societal metabolism. Any possible mistake in the paper is attributable solely to the author. I wish to thank Dr. Lucio Marcello, researcher at the Rivers and Lochs Institute, University of the Highlands and Islands, UK, for editing the manuscript. This work has been partially carried out within the framework of the EU FP7 research project 'Global and Local food chain Assessment: A Multidimensional performance-based approach (GLAMUR)' (KBBE.2012.2.5-03—Grant Agreement No.: 311778). The European Union or the European Commission cannot be held responsible for results and opinions expressed in the text.

**Joan Marull and Carme Font (Chap. 4):** This work has been supported by the international Partnership Grant SSHRC-895-2011-1020 on 'Sustainable farm systems: long-term socio-ecological metabolism in western agriculture' funded by the Social Sciences and Humanities Research Council of Canada.

**Roc Padro and colleagues (Chap.5):** We would like to thank Arnim Scheidel and Eva Fraňková for their useful comments on earlier drafts, which improved the final version. This work has been supported by the Spanish research project HAR2012-38920-C02-02, and the international Partnership Grant SSHRC-895-2011-1020 on 'Sustainable farm systems: long-term socio-ecological metabolism in western agriculture' funded by the Social Sciences and Humanities Research Council of Canada. The corresponding author was also granted with two scholarships from the Spanish Ministry of Economy and Competitiveness with references BES-2013-06415 and EEBB-I-17-12073. We also thank the Vallès Occidental County Archive for its assistance.

**Michael Gizicki-Neundlinger and colleagues (Chap.6):** We would like to thank Fridolin Krausmann, Roc Padro i Caminal and Simron Jit Singh for comments on an earlier version of the manuscript, Renate Strempfl for conducting a broad literature survey on local food systems, and the Niederösterreichisches as well as the Oberösterreichisches Landesarchiv and Gutsverwaltungen Grafenegg for generously opening their (private) archives to us. This research was supported by the Canadian Social Sciences and Humanities Research Council, Partnership Grant 895-2011-1020.

**Eva Fraňková and Claudio Cattaneo (Chap.7):** We are thankful for the support of the Czech Science Foundation, Grant no. 13-38994P: Quest for sustainable food production: Social and financial metabolism of selected local food systems. Claudio Cattaneo acknowledges support from the Sustainable Farm Systems project, partnership Grant 895-2011-1020 awarded by the Social Sciences and Humanities Research Council of Canada, from grant no. HAR2015-69620-C2-1-P"¿Sistemas agrarios sustentables? Una interpretación histórica de la agricultura en España desde la perspectiva biofísica" financed by Ministerio de Economía y Competitividad, and from grant no. HAR2016-76814-C21-P (AEI/FEDER, UE) also financed by Ministerio de Economía y Competitividad. The historical cadastral data were provided for free by the Moravian Provincial Archives in Brno (Czech Republic). We are further grateful for the help of Majka Chudíková and Lucka Jahnová with the field data collection, Bára Machová, Hana Bernardová, Kamila Svobodová and Hana Prymusová with translation and interpretation of the historical data, Martin Černý and Nikola Šťastná with data transcription, and Simone Gingrich for her kind support during the analysis. We are also very grateful for comments by Panos Petridis and Michael Gizicki-Neundlinger.

**Arnim Scheidel and colleagues (Chap.8):** The authors are indebted to the villagers for their warm hospitality and great cooperation during the field research. Intellectual and institutional support is acknowledged from The Cambodian Center for Study and Development in Agriculture (CEDAC). Data collection, analysis and writing were funded by the Catalan government grants FI-DGR-2009, BE-DGR-2011 and the Catalan Beatriu de Pinós postdoctoral grant 2014 BP_A 00129, respectively. We are thankful for helpful comments and constructive feedback from Inés Marco Lafuente and Willi Haas. Shortcomings and errors are, of course, our own.

**Panos Petridis and Julia Huber (Chap.9):** We thank Carlota Marañon, Jacqueline Kirby and Mary Pitiakoudi for their invaluable support during fieldwork, and Marina Fischer-Kowalski, Simron Singh and Marian Simon-Rojo for their advice and helpful comments on earlier drafts. Financial support from the Austrian Science Fund (FWF) projects: Susaki (P27951-G7) and CiSciSusaki (F15TCS00022) is gratefully acknowledged.

**Gonzalo Gamboa and colleagues (Chap.10):** This work has been carried out within the EU FP7 research project 'Global and Local food chain Assessment: a Multidimensional performance-based approach (GLAMUR)' (KBBE.2012.2.5-03—Grant Agreement No.: 311778). The European Union or the European Union Commission cannot be held responsible for results and opinions quoted in the text. The research of Gonzalo Gamboa has been funded by the Spanish National project HAR2013-47182-C2-1-P and the Catalan Research group 2014SGR591.

**Marian Simon-Rojo and Barbora Duží (Chap.11):** We would like to thank Franco Llobera, from Madrid Agroecologico, who inspired the project Madrid Agrocomposta and directed our attention to more sustainable urban waste management systems; Pedro Almoguera, master of bio-intensive agriculture who invites us to introduce a social justice perspective (whose soil are we eating?) when thinking about closing nutrient and material loops, and Guillermo Lozano for his support and readiness to discuss and improve the final article. Barbora Duží from the Institute of Geonics, Academy of Sciences of the Czech Republic, would like to thank the long-term conceptual development of research organisation, RVO: 68145535 for its support.

Thank you.
For everything.

# Contents

# About the Editors

**Eva Fraňková** works as Assistant Professor at the Department of Environmental Studies, Faculty of Social Studies, Masaryk University in Brno, Czech Republic. Her long-term research interests and passions include the concept of eco-localisation, sustainable degrowth and various grass-root alternative economic practices including eco-social enterprises, local food initiatives, etc. Recently, she has been involved in the mapping of heterodox economic initiatives in the Czech Republic, and in research on the social metabolism of local food systems. She is also involved in several NGOs—the Association of Local Food Initiatives, the Society and Economy Trust, and NaZemi (On Earth), a global education and Fair Trade organisation in the Czech Republic.

**Willi Haas** is senior researcher and Lecturer at the Institute of Social Ecology in Vienna, an institute of the Faculty of Interdisciplinary Studies at the Alpen-Adria University. In the 90s, he became fascinated by social ecology, the study of society–nature relations across time and space. He is interested in the past transition from agrarian to industrial societies and the insights that can be drawn from these far-reaching changes for the next transition to a post-fossil society. In his view, the question of how to overcome the system inertia and the unsustainable reproduction dynamics of the present fossil fuelled societies is crucial for fostering a sustainable future. He was Chair of Greenpeace CEE for 9 years. Professionally, he was a public official at the Ministry of Social Affairs, Director of the Institute of Applied Ecology (Vienna), Acting Director of the Environmental Monitoring Group (Cape Town) and Researcher at the International Institute of Applied System Analysis (IIASA, based in Austria). At the Institute of Social Ecology, he has headed numerous scientific projects and is the coordinator of the institute's team of thematic research coordinators.

**Simron J. Singh** is Associate Professor in the School of Environment, Enterprise, and Development (SEED), and Associate Dean of the Faculty of Environment, University of Waterloo, Canada. Drawing on the concept of social metabolism, his research focuses on the systemic links between material and energy flows, time use and human well-being. His particular interest lies at the local and subnational scales, as well as small islands. He has conducted social metabolism studies in the Nicobar Islands (India) and Samothraki (Greece), and supervised work on the biomass metabolism of Jamaica (as part of the Canadian project *Hungry Cities*), the Region of Waterloo as well as Canada. As work package leader, he led work on biomass flows and social conflicts in an EU project (EJOLT).

# Abbreviations

| | |
|---|---|
| AFEROI | Agroecological Final Energy Return On Investment |
| AFN | Alternative Food Networks |
| ALEP | Agricultural Labour Energy Productivity |
| ARC2020 | Agricultural and Rural Convention |
| ATT | Agro-ecosystem Total Turnover |
| AWU | Agricultural Working Units |
| BR | Biomass Reused |
| cap | Capita |
| CAP | Common Agricultural Policy |
| CEE | Central and Eastern European (countries) |
| CPR | Common Pool Resource |
| CSA | Community Supported Agriculture |
| DE | Domestic Extraction |
| DEC | Domestic Energy Consumption |
| DM | Dry Matter |
| DMC | Domestic Material Consumption |
| DMI | Direct Material Input |
| E | Energy Storage |
| EBR | Economias BioRegionales |
| ECI | Ecological Connectivity Index |
| EFA | Ecological Functional Areas |
| EFEROI | The External Final EROI |
| EI | External Input |
| ELC | Economic Land Concession |
| ELIA | Energy–Landscape Integrated Analysis |
| EMS | Effective Mesh Size |
| EROI, or EROEI | Energy Return On Investment, or Energy Return on Energy Invested |
| EW-MEFA | Economy-Wide Material and Energy Flow Accounting |
| FAO | Food and Agriculture Organization |

| | |
|---|---|
| FBR | Farmland Biomass Reused |
| FEI | Farmland External Input |
| FEROI | Final Energy Return On Investment |
| FFP | Farmland Final Produce |
| FII | Farmland Internal Input |
| FP | Final Produce |
| FTI | Farmland Total Input |
| FTU | Functional Time Use |
| GCV | Gross Calorific Value |
| GHG | Greenhouse gas |
| GIAHS | Globally Important Agricultural Heritage Systems |
| GIS | Geographic Information Systems |
| GJ | Gigajoule (1 GJ = $10^9$ J = one billion joules) |
| GLAMUR Project | Global and Local food chain Assessment: a MUltidimensional performance-based approach |
| H′ | Shannon–Wiener Index |
| ha | hectare (1 ha = 100 ares = $10^{-2}$ $km^2$ = $10^4$ $m^2$) |
| HANPP | Human Appropriation of Net Primary Production |
| I | Energy Information |
| ICLS | Integrated Crop–Livestock Systems |
| IDC | Intermediate Disturbance–Complexity |
| IFAD | International Fund for Agricultural Development |
| ILA | Impredicative Loop Analysis |
| K | Potassium |
| $km^2$ | Square kilometre (1 $km^2$ = 1km × 1 km) |
| L | Landscape Heterogeneity |
| l | Litre |
| LBR | Livestock Biomass Reused |
| LC | Land Cover |
| LCA | Life-Cycle Analysis |
| Le | Landscape Ecology metric |
| LEI | Livestock External Input |
| LFP | Livestock Final Produce |
| LFS | Local Food Systems |
| LFSC | Local Food Supply Chain |
| LM3 | Local Multiplier |
| LPG | Liquefied Petroleum Gas |
| LPS | Livestock Produce and Services |
| LS | Livestock Services |
| LSU | Livestock Unit (1 LSU = 500 kilogrammes live weight) |
| LTI | Livestock Total Input |
| LTSER | Long-Term Socio-Ecological Research |
| LUC | Land Use Change |
| LW | Livestock Waste |
| $m^2$ | Square metre (1 $m^2$ = 1 × 1 m) |

| | |
|---|---|
| MEFA | Material and Energy Flow Accounting |
| MuSIASEM | Multi-Scale Integrated Analysis of Societal and Ecosystem Metabolism |
| N | Nitrogen |
| NCDs | Noncommunicable diseases |
| NPP | Net Primary Production |
| $NPP_{act}$ | Actual Net Primary Production |
| NPPEROI | Net Primary Production $_{actual}$ Energy Return On Investment |
| P | Phosphorus |
| PCA | Principal Components Analysis |
| PILFS | Pre-Industrial Local Food Systems |
| PK | Peasant Knowledge |
| SDGs | Sustainable Development Goals |
| SFA | Substance Flow Analysis |
| SFS | Sustainable Farm System |
| SM | Social Metabolism |
| SOM | Soil Organic Matter |
| SRI | System of Rice Intensification |
| SSCB | Small-Scale Cooperative Banking |
| STK | Scientific and Technological Knowledge |
| T | Topology |
| t | Tonne |
| THA | Total Human Activity |
| UAB | Universitat Autònoma de Barcelona |
| UB/UhB | Unharvested Biomass |
| UN | United Nations |
| X | Land Matrix |

# Chapter 1
# Introduction: Key Concepts, Debates and Approaches in Analysing the Sustainability of Agri-Food Systems

**Eva Fraňková, Willi Haas and Simron Jit Singh**

**Abstract** The Introduction sets the tone for the book by outlining the main concepts, debates and applications illustrated by the various contributions in this volume. The theme of Local Food Systems (LFS) is a complex one, and therefore a multidisciplinary and interdisciplinary effort, drawing on a myriad of research concepts and frameworks. The chapter begins by embedding the food debate within the broader sustainability discussion. It highlights issues around future demand and supply scenarios, current production and consumption patterns, and the challenges of addressing some of these issues in the context of climate change. The chapter also provides a review of some of the responses so far to counter the current unsustainability of the global agri-food system. Initiatives and concepts such as local food systems, localisation, food security and sovereignty are discussed, drawing on examples from both the Global North and South. Since the volume is about socio-metabolic approaches to agri-food systems, the chapter also introduces the conceptual and methodological underpinnings of this approach, and how this relates to political ecology, social conflicts and environmental justice. The chapter ends with an introduction to the various contributions in this volume that discuss the following cross-cutting issues: the necessity to consider local cases as nested in regional, national and global scales, including the debate on what might be an optimal scale for food regionalisation or sovereignty; the agenda of re-localisation and its political and ideological background in relation to biophysical/

E. Fraňková (✉)
Department of Environmental Studies, Faculty of Social Studies, Masaryk University, Joštova 10, 602 00 Brno, Czech Republic
e-mail: eva.slunicko@centrum.cz

W. Haas
Institute of Social Ecology, Faculty for Interdisciplinary Studies, Alpen-Adria University, Schottenfeldgasse 29, 1070 Vienna, Austria
e-mail: willi.haas@aau.at

S.J. Singh
School of Environment, Enterprise and Development (SEED), University of Waterloo, Environment 3 Building, 200 University Avenue West, Waterloo, ON N2L 3G1, Canada
e-mail: Simron.Singh@uwaterloo.ca

 
E. Fraňková et al. (eds.), *Socio-metabolic Perspectives on the Sustainability of Local Food Systems*, Human-Environment Interactions 7,
https://doi.org/10.1007/978-3-319-69236-4_1

socio-metabolic insights with respect to LFS; the inclusion of trade in the methodological and ideological framework of LFS studies; the multi-functionality of agriculture and various related metrics of efficiency in agriculture (including the energy efficiency/EROI variables); the biophysical performance of other, more sustainability-focused, production regimes; the land-sparing versus land-sharing debate in connection to the biodiversity and landscape multi-functionality of both historical and existing agri-food systems; the role of livestock in various agri-food systems, and the related issue of meat consumption and dietary transition as part of a broader metabolic transition happening in many parts of the world; the issue of labour in terms of efficiency, and also in broader social and economic contexts; the political underpinnings of peasant livelihoods, existing social conflicts and uneven ecological exchange related to food; the issue of democratisation of food systems, access to means of production for fulfilling basic needs, and the agenda of food sovereignty. The contributors to this volume all ask the following two questions: Which local food systems or their particular characteristics can serve as the best practice examples for maintaining and designing more sustainable agri-food systems in the future? Which scientific and policy relevant insights can the socio-metabolic approach offer with respect to studying the sustainability of local food systems?

**Keywords** Sustainability · Agri-food systems · Local food systems Localisation · Food sovereignty · Social metabolism · MEFA · MuSIASEM Ecological economics · Political ecology

## 1.1 The Sustainability of Agri-Food Systems—The Challenge

We all eat.

If we are lucky enough, we eat every day.

Depending on age and gender, our bodies require about 1700 kilocalories per day to maintain their basal metabolism, about 2600 if engaged in sedentary work at the computer, 3200 for keeping up with manual factory work, and 3700 for managing hard physical work like ploughing a field (Shetty 2005). Worldwide, the average of 2903 kcal/cap/day was available in 2014 (compared to 2196 kcal/cap/day in 1961, FAO 2015: 24, 48). However, the real calorific intake of all the 7,511,482,383 human bodies (Worldometers 2017) does differ a lot among the world regions.[1] What's more, there are growing differences within

[1]On the national level, the highest dietary energy supply in 2014 was statistically in Turkey (the average of 3715 kcal/cap/day, in comparison to 2957 kcal/cap/day in 1961), in the US, the calorie supply per capita was 3639 kcal/cap/day in 2011 (but only 2880 kcal/cap/day in 1961). In China the historical shift is much more striking (above-average 3156 kcal/cap/day in 2014 in contrast to only 1439 kcal/cap/day in 1961). From the other side of the spectrum, the people of the Central African Republic had to survive with only 1940 kcal/cap/day in 2014, whereas in 1961 they had

nations as well. Whereas, until recently, the most pressing problem related to food was simply hunger (UN 2015), the more recent term *malnutrition* includes undernutrition (stunting and wasting esp. in the case of children), as well as overweight and obesity that are on the rise worldwide, most rapidly in Asia (IFPRI 2016).[2] All these various human bodies create a living anthropomass of about 300 million tons worldwide (Smil 2013: 226; Walpole et al. 2012: 3),[3] and they all demand their calorific energy to maintain themselves and do their jobs.

Every day.

This simple fact has far reaching consequences for both humans and the natural environment. Humans as one species out of almost nine billion (Sweetlove 2011)[4] currently appropriate about 24% of the potential Net Primary Production (NPP) of all terrestrial ecosystems (Haberl et al. 2007), and agriculture remains the most important driver of land use change in this process (Krausmann et al. 2013). During the last century, we have witnessed unprecedented growth in both global food production and associated environmental, social, and economic problems connected to the increasingly industrialised, globalised and commodified food production system (Weis 2010; Gomiero et al. 2011a; Mayer et al. 2015; Clapp 2017).

Projections for the future foresee a continuation of rising food demand due to population increase and growing demand for meat which requires disproportionately more farm land and higher overall crop production (Walpole et al. 2012; Tilman and Clark 2014). Against this backdrop, some scholars develop scenarios of a 70% or even 100% crop production increase (FAO 2009; Tilman et al. 2011); others criticise the very assumptions of these scenarios and ask whether such a significant growth in food production is at all necessary, and if so, whether it is possible to achieve that without compromising the regeneration capacity of the biosphere. Instead, they stress the need to influence dietary expectations towards less-meaty diets, and to address the still growing inequality in access to existing food, the alarming amounts of food waste, and also the broader political agenda of peasants' livelihoods (Holt-Giménez and Altieri 2013; Edelman et al. 2014). All, however, emphasise the far-reaching environmental consequences caused by current dietary patterns and the dominant agri-food system (Foley et al. 2011; Gomiero

(Footnote 1 continued)

an average of 2256 kcal/cap/day available. Data for 2014 from FAO (2015), data for 2011 and 1961 from Food Security Portal (2017); data from 2011 were used for "the present" where the 2014 FAO information was not available.

[2]Still, the body mass of an average Asian was 57.7 kg in 2005, in contrast to 80.7 kg of a North American (Walpole et al. 2012). These authors estimate that about 15 million tonnes out of the total human biomass was due to overweight, and another 3.5 million tonnes due to obesity, 34% out of it being based in North America where 70% of the population is overweight.

[3]In terms of Carbon weight, the anthropomass in 2000 accounted for approx. 55 million tons of C, whereas domesticated animals account for more than double this (120 million tons of C), most of it being cattle (80 Mt of C) (Smil 2013: 228).

[4]Estimations often go as high as 10 billion species. In any case, these include only eukaryotic organisms (Sweetlove 2011). The microbial diversity is even bigger by order, Earth is predicted to be home to upward of 1 trillion microbial species (Locey and Lennon 2016).

et al. 2011b) as highly disturbing, and call for *some* radical change towards more sustainable agri-food systems (for further debate on what sustainability might actually mean for agri-food systems see Chap. 2 in this volume).

In principle, two distinct strategies on how to achieve sufficient and sustainable diets have been discussed recently in the literature (Phalan et al. 2011; Tcharntke et al. 2012). The *land sparing* strategy is based on increasing agricultural yields, reducing farmland area and actively restoring natural habitats on the land spared, thus counteracting the ongoing loss of biodiversity. Some scholars find that a land-sparing strategy has the technical potential to achieve significant reductions in net GHG emissions from agriculture and land-use change, especially when coupled with demand-side strategies to reduce meat consumption and food waste (Lamb et al. 2016). In contrast, scholars promoting *land sharing* (Fischer et al. 2008; Perfecto et al. 2008; Gabriel et al. 2009) doubt that fragmented land plots set aside for nature conservation provide the high quality habitats required for halting loss of species. They promote extensive land use that reduces the pressure on species in an agroecological matrix at landscape level (for more on the agroecological perspective see Chap. 2, for further argumentation regarding the land sparing vs. land sharing debate, and biodiversity on the landscape level, see Chap. 4 in this volume).

Whatever the strategy, it is clear that when considering the sustainability of agri-food systems, the wider context of food-climate nexus cannot be ignored. Due to climate change, the global agri-food system is expected to supply ever more biomass products for a world population growing in numbers and affluence (Godfray et al. 2010), as well as contributing to climate change mitigation by providing bioenergy and carbon sequestration (Edenhofer et al. 2014; Lamb et al. 2016, for more on multifunctionality, and the energy issues in agriculture, see Chap. 3 in this volume).

In the context of climate mitigation, the spatial scale at which drastic reductions in GHG emissions or even negative emissions (sequestration) can be achieved are decisive for improving agri-food systems. All regions can either meet the emission constraint equally, or through specialisation. For example, certain regions are net-GHG emitters, because, for instance, they host intensive agriculture and provide biomass to other regions, which in turn may be net-sinks for carbon if they dedicate their land to forest regrowth. Biomass trade can balance the GHG emissions of these regions. Given the fundamentality of the required changes (Steffen et al. 2015), local "autarky" strategies can play a significant role. At the same time, regional specialisation is considered a suitable means of increasing land use efficiency (Kastner et al. 2014). Therefore, the study of agri-food systems should consider nested design, which embeds the local and regional scale into the national, regional, and global production-consumption context (see Chap. 5 in this volume for a good example of this multilevel approach, including the issue of biomass trade, and Chap. 10 analysing the sustainability implications of different types of supply chains).

This once again reveals the crucial link between food production and consumption, and the challenge of the ongoing dietary transition in the Global South, along with the predominant western diet in the Global North that entails high levels of meat consumption, far beyond what can be considered as healthy (McMichael et al. 2007, see Chap. 9, on dietary transition on a Greek island, and the virtues of the Mediterranean diet). A vegetarian diet reduces the relative risk for type II diabetes by 40%, cancer by 10% and coronary mortality by 20% compared to the conventional western omnivorous diet. At the same time, the vegetarian diet reduces $CO_2$ emissions by more than two thirds compared with the conventional diet (Tilman and Clark 2014). Consequently, changes in diets are one of the crucial components for achieving improved sustainability (Veeramani 2015).

However, so far it has proven to be quite difficult to intervene in the behaviour of individuals (Cecchini et al. 2010), partly because of the high level of disconnection between humans and agri-food systems, both in terms of production and consumption. Especially in the Western context, consumers are often dependent on transnational food chains managed by anonymous individuals and corporations so distant and disconnected from our direct experience, and often ignorant of the complex processes, from photosynthesis, through Haber-Bosch synthesis, to soda glutamate production (in the case of industrially produced and processed food) and to farming itself. While food *consumption* is still a very intimate experience of direct contact between the bodies of animals and plants providing nutrition, and our own bodies that incorporate the organic matter into their very own structures, food *production* is for most intangible and abstract. For most industrialised societies, the focus is about what we want to buy and eat, but rarely about what we want to grow, gather or hunt. This does not hold true for about half of the world's population (Fischer-Kowalski et al. 2011a) that still live off the land, i.e. experience a direct first-hand contact with food production. This segment of the world's population is, however, gradually declining as they join the growing middle class and associated dietary patterns and consumer behaviour.

At the same time, a growing population from both North and South are increasingly engaged in influencing the way the current dominant agri-food system works, reflected in both activist and academic focus on, broadly speaking, "agri-food alternatives". Concepts such as alternative food networks, local food initiatives and local food systems (Renting et al. 2003; Goodman et al. 2012; Zumkehr and Campbell 2015) have become popular during the last 20 years, encompassing many on the ground initiatives such as community supported agriculture (CSA), community gardens, urban gardening, food coops and the like. Most of these initiatives are based on the idea that our food systems should be somewhat more (re-)localised, i.e. more embedded in local ecosystems and social relations, and shorter in both physical and psychological distance between producers and consumers.[5] Richard Douthwaite, one of the proponents of food (and also energy and money) localisation, calls for the creation of *short circuits,* both in terms of resource flows and human interactions (Douthwaite 1996).

[5]Sometimes shortening the distance to the extent of becoming *prosumers* (*pro*ducers and con*sumers* at the same time).

## 1.2 Local Food Systems, Localisation and Food Sovereignty

In the literature, the term Local Food Systems (LFS) is often used to represent all alternate food production and distribution models to the dominant globalised, industrialised and commodified agri-food system trying to create its more sustainable variants. When defining LFS, physical distance is surely one key ingredient, but still very much dependent on the cultural and geographical context.[6] LFS are also characterised by "the story behind the food" (Martinez et al. 2010: 4) or the food "provenance" (Thompson et al. 2008), referring to a combination of cultural, social and environmental attributes, including commitment to a specific place, to concrete people who produce and consume the food, nurturing of traditional and/or sustainable practices of production and processing, knowledge about these processes and the possibility of direct participation in them (Martinez et al. 2010).

According to proponents of "localisation" (see Fraňková and Johanisová (2012) for a systemic analysis of the concept, and Chap. 7 in this volume for its application on a small-scale organic family farm in the Czech Republic), localised food production is supposed to bring predominantly positive results, combining social, environmental and economic benefits of lower transport dependence resulting in less consumption of fossil fuels, lower $CO_2$ emissions, less waste from packaging, more closed cycles of matter and energy within production systems, and also stronger local economies showing a higher level of local circulation of money, lower dependence on foreign investments, and less dependency on, and more resilience towards, fluctuations of the global economy. Among the potential drawbacks are the possibility of strong community control by a few, and limited choice of local and seasonal goods, but they are mostly seen by the localisation proponents as of lower importance, and as acceptable or—in the case of direct producer-consumer relations—negotiable (Douthwaite 1996; Norberg-Hodge et al. 2002; Desai and Riddlestone 2002).

Such an overtly positive framing of LFS has been criticised by scholars as being too naïve. They argue that the reality of such endeavours, even if positively motivated, can be and indeed sometimes are, socially exclusive, supplying its participants more with a good feeling about themselves than with the required amount of food; critics also doubt its overall sustainability impacts and potential for spreading beyond small niche markets (Hinrichs 2003; Allen 2010; Cadieux and Slocum 2015). Hence, after the "first generation" of academics enthusiastically "discovered" the grass-root local food initiatives in the 1990s and 2000s, the "second" generation are more critical and aware of the complexity in operationalising these concepts. Their scholarly sophistication is reflected in new labels

[6]Whereas in the USA, a 100–400 mile radius might be seen as local, in Europe it is more about 20–100 km (Norberg-Hodge et al. 2002; NEF 2002), and the particular type of goods in question: the supposedly "local" area differs when it comes to eggs, clothes and refrigerators, see NEF (2010) for rough "local" distance indications in the European context.

such as "social or grass-root innovations" and "scaling-up" of alternatives (Seyfang and Smith 2007; Beckie et al. 2012; Zumkehr and Campbell 2015).

Beside the food localisation agenda, the related concept of "food sovereignty", that has been the subject of intense debate in the fields of political ecology and agrarian and peasant studies, should also be acknowledged. This stream of activism and thinking puts a much stronger accent on the political aspects of LFS, including the question of peasant livelihoods, and shows more sensitivity to Global North/South relations. At the same time, the strong point in favour of more localised food systems is also present in the food sovereignty agenda, and thus it can be interpreted as a fresher incarnation of the ideas previously formulated within the localisation movement, mainly in the 1990s. Indeed, in practice many localisation and food sovereignty initiatives overlap and both movements offer cross-fertilisation [see for example, the declaration of Food Sovereignty (Nyéléni 2007) and the recent European CSA declaration (Urgenci 2016)]. Thus, it can be said that the LFS has recently been enriched and received impetus by the food sovereignty agenda to include political issues, and to embody characteristics of a global movement trying to build dialogue between different regions of the world.

Still, when looking for sustainable alternatives to the dominant agri-food system, both academics and activists often focus only on Northern initiatives in the sense of the purposeful, often politically engaged LFS initiatives as described above. Yet, there are several other modes of sustainable agricultural and land use practices that have long existed in terms of cultivation management and production-consumption relationships (Martínez-Torres and Rosset 2014). These are often overlooked by academics as contributing to sustainability. In the context of the Global East, Smith and Jehlička (2013) coined the term "quiet sustainability" for the widespread practices of food self-provisioning and gardening, a practice spread across all ages, education levels and socio-economic statuses, providing food in substantial varieties and amounts (see also Sovová (2015); Smith et al. (2015); Chaps. 8 and 9 in this volume provide other examples of "quiet sustainability": the case of a rural village in Cambodia, and the Greek island of Samothraki respectively).

Besides such prevailing subsistence gardening practices in many Central and Eastern European (CEE) countries, half of the world's population, mainly in the Global South, still remain involved in subsistence farming (Fischer-Kowalski et al. 2011a). This segment of the world's population is often described as backward, poor and underdeveloped (see Escobar (2014) for a critique of the very term development), although by their everyday practices fulfilling, albeit unintentionally, many of the sustainability ideals formulated by the Northern alternative food initiatives. Efforts are underway to change this framing (see e.g. Shrivastava and Kothari (2012)) and to take these "quiet sustainability" practices on board, along with the progressive "innovations" discussed in the literature dominated by western scientists.

Thus, we argue that both the intentional and unintentional initiatives and practices discussed above, based geographically in the North, East and South, can be framed within the localisation perspective and may be labelled not only as Local Food Systems, but at the same time as (re-)Localis*ed* Food Systems, i.e. they do fulfil certain aspects of localisation as defined at the beginning of this section, including the potentially positive implications for their sustainability.

Indeed, contrary to what some of the critics argue, we believe there are several environmental, social, economic and political benefits for upscaling (re-)localisation of agri-food systems in comparison to the globalised agri-food regime (see Chap. 2 for a more detailed elaboration on these reasons). However, we also believe that the respective level of desired localisation should be open, case-specific, and thus explicitly questioned, and therefore not a one-size-fits-all approach. On this, there is a high consensus among virtually all (food) localisation proponents, even if some are vague on how and to what extent international trade should be regulated and organised, and what procedures and principles should be used to decide on this (Fraňková and Johanisová 2012). The issue of scale of (re-)localisation and trade are raised in some of our case studies that look at both historical and current practices (see Chaps. 5 and 6 for a historical account of the Spanish region of Vallés, and several Austrian agrarian villages respectively).

This also explains why the term LFS is kept analytically open in this volume. The contributors do not presuppose any modern or traditional agrarian practices, nor are the cases conditioned by any political or activist orientation. Further, our understanding of the "local" in LFS is intended to mean anything lower-than-national, i.e. on the level of a household, farm, community, village or region. The focus on local scale is also motivated by our understanding that although sustainable food production is a global challenge that indeed needs to be discussed in the global context, it also has an inevitable local dimension. It is at the local level where people live and work, where environmental, economic, social, cultural and institutional issues are interlocked and experienced (that is, where the food is produced, processed, transported, traded, and consumed or wasted).

From the methodological point of view, available LFS studies use predominantly qualitative research methodologies (Seyfang 2007), discussing the biophysical impacts of LFS only to a very limited extent (Martinez et al. 2010; Kneafsey et al. 2013; Matacena 2016). However, well-founded and tested methods to study the biophysical aspects of agriculture in the fields of social ecology, ecological economics and industrial ecology do exist. The conceptual and methodological framework of Social Metabolism (SM) has been applied to study the nexus between energy, material, land, and time use within agri-food systems for decades (Giampietro 2003; Gomiero et al. 2006; González de Molina and Toledo 2014), however, only rarely in the case of LFS.[7]

[7]It is already common to apply the social metabolism approach at national and global scales (Grešlová Kušková 2013; Erb et al. 2016). Existing local studies either focus on describing the transition from traditional to industrial agri-food regimes historically (Haberl and Krausmann 2007; Haas and Krausmann 2015), or focus on contemporary cases in the Global South (Fischer-Kowalski et al. 2011; Scheidel et al 2013; Ringhofer and Fischer-Kowalski 2016; Birke 2014; Singh et al. 2001). However, applications of social metabolism studies at the local level in the Global North are still rather exceptional.

## 1.3 Social Metabolism and Political Ecology

Whether a given local food system is ecologically sustainable or not is a tricky question. This requires a non-reductionist approach to investigate how social and natural systems interact and co-evolve over time and different scales. This interaction entails substantial impacts upon one another, with bidirectional causality (Haberl et al. 2016). There have been efforts to study the biophysical relations between society and nature over the last one-and-a-half centuries. Analogising the metabolism of organisms, Karl Marx first referred to metabolism when he described work as a process in which people regulate, steer and control their metabolism with nature through their own action (Marx 1968, 1867). A century later, these considerations were re-introduced from quite another angle by Ayres and Kneese (1969) that led to what is termed as "industrial metabolism" (referring to the material and energy throughput of industrial societies), and later "social metabolism" (to generally refer to the metabolism of any society at any scale) (for an intellectual history of social metabolism, see Fischer-Kowalski (1998)).

The fundamental premise of "social metabolism", a term that is now widely used, is that social systems need to maintain and reproduce their biophysical stocks (like human and livestock population, and built artefacts) for their survival and evolution. This requires a continuous flow of energy and materials extracted from, and eventually released to, the environment (Ayres and Kneese 1969).[8] Thus, the metabolism of social systems refers to material, substance and energy flows related to socio-economic activities. Maintaining the social metabolism includes the "colonisation of nature" (discussed later in this section), denoted as purposive intervention into natural systems aimed at improving their utility for societal purposes (Haberl et al. 2016).

Socio-metabolic studies range from describing the metabolism of national economies to analysing them across geographical regions over decades or even centuries (e.g. Krausmann et al. 2009; Schaffartzik et al. 2014). Socio-metabolic transitions, especially from hunter-gatherer to agrarian, to industrial societies, and to the so-called Anthropocene, provide a specific analytical perspective to understanding long-term changes in society-nature interactions (Fischer-Kowalski et al. 2014; Fischer-Kowalski and Haberl 2007). To better understand these transitions, a series of local studies have been performed (Fischer-Kowalski et al. 2011a; Birke 2014; Haas and Krausmann 2015).

[8]Ayres and Kneese (1969, p. 283) claimed that the common failure of (currently mainstream) economics results from viewing the production and consumption processes in a manner that is somewhat at variance with the fundamental law of the conservation of mass; similarly, Georgescu-Roegen (1971) argued that mainstream economic theory and modern economies are at variance with thermodynamics and the law of conservation of mass. These insights, recognising the biophysical aspects of all socio-economic processes, established one of the fundamental theoretical underpinnings of the whole field of Ecological Economics (Røpke 2004).

Within social metabolism studies there are different variants such as substance flow analysis (SFA),[9] physical input-output (PIOT), life-cycle analysis (LCA), material and energy flow analysis (MEFA), and multi-scale integrated analysis of societal and ecosystem metabolism (MuSIASEM). They all share many similarities but differ due to their special focus. For example, while the MEFA's original purpose was to add a biophysical dimension to the usual monetary description as national accounts, MuSIASEM is more focused on analysing processes and functions inside a system and at the same time linking this analysis on different scales in an integrated way (land use, human activity, energy and material flows, monetary flows) (Gerber and Scheidel 2018, see also Chaps. 2 and 3 of this volume).

MEFA indicators provide now highly standardised methodological inventory that can be derived to discuss a system's sustainability over time, and to compare with other cases or other levels of scale [for standardised accounting guidelines for countries, see Eurostat (2013), and for local systems, see Singh et al. (2010)].[10] Over the years, MEFA indicators were further developed to be practicable, meaningful and sound. One of the classic indicators is the Domestic Extraction (DE) that accounts for all material appropriated from nature from the social system's domestic territory, be it harvested biomass, extracted crude oil, mined metal ores or minerals. Since livestock, a biophysical stock of society, can partly feed itself, grazing is considered an extraction as well. In addition to domestic extraction, a social system also imports materials from other social systems, while also exporting some. Thus, imports and exports are also basic indicators. The Direct Material Input (DMI) of a social system is the sum of DE and Imports. Domestic Material Consumption (DMC) is calculated by subtracting exports from DMI and reflects the social system's consumption, which goes far beyond the apparent consumption of households since it includes all losses as well as all materials used to extend, maintain and operate the stocks at the system level. These flows can be disaggregated for a wide range of material categories such as biomass, fossil fuel carriers, metal ores, and non-metallic minerals (essentially construction minerals). All flows are reported in material units (e.g., metric tonnes) in fresh weight or dry matter as well as in energetic units (e.g., joules, kilowatt-hours) and per unit of time, usually per year (see Haberl et al. 2004). The resulting energy indicators deviate from standard technical energy accounting, insofar as the former—besides technical energy—also includes all the primary energy required for the endosomatic metabolism of livestock and humans (feed and food).

MuSIASEM studies, on the other hand, have focused on the performance of societies' subsectors (Giampietro et al. 2012) and often on particular resources such as food and agriculture, energy or water (e.g., for agriculture: Ravera et al. 2014;

[9]Substance refers to a single type of matter such as a chemical element, e.g. Cadmium (Brunner and Rechberger 2004).

[10]For nation states such as Japan, as well as in the European Union (EU) and several other countries, Material Flow Accounting (MFA) has become a regular part of public statistical reporting (Fischer-Kowalski et al. 2011b). This allows the provision of reliable annual accounts of material use in physical terms and their comparison across time and with economic accounts.

Scheidel et al. 2014; for energy: Ramos-Martín et al. 2007; Giampietro et al. 2013; for water Madrid-López and Giampietro 2015). But MuSIASEM is also concerned with the nexus between these resources (Giampietro et al. 2014).

While MEFA focus on stocks and flows, MuSIASEM look at *funds and flows*. These are not just differences in words but in concepts. "Funds" refer to agents maintained throughout an economic process such as labour, land or technological capital, and "stocks" to human population, livestock and artefacts. In sum, MEFA is a useful framework for understanding the biophysical dynamics between stocks and flows in quantities, and helps grasp the *figure of a social system's metabolism*, in terms of the tonnes processed (Fishman et al. 2014; Matthews et al. 2000). In contrast, the fund-flow approach conceptualises the *functions of a social system's metabolism*. So funds (e.g., labour) require and produce flows (e.g., food) at a given rate (Gerber and Scheidel 2018, see also Chaps. 2 and 3 of this volume).

What different approaches to "social metabolism" have in common is that they encompass biophysical stocks, flows and/or funds and the mechanisms regulating these flows. Any socio-metabolic analysis starts with the definition of the social system under consideration, such as a household, a business, a village, a local food system, a nation state, or the global community. The first step is to define the social system's boundary, a prerequisite for studying society-nature interactions. Delineating this boundary, and deciding which elements are part of the social system and which belong to the system's environment, is not simply an arbitrary construct of the researcher. According to modern systems theory, a social system constitutes itself through the marking and reproduction of its boundaries in order to fulfil its functions. Accordingly, the researcher needs to understand how the social system itself defines its stocks and flows along ownership boundaries, accepted entitlements, or governance responsibility (see also Netting 1981, 1993; Singh et al. 2010).

In addition to material and energy flow analysis, land use reflects another form of society-nature interaction, which is a consequence of the human colonisation of natural systems. Land use involves the colonisation of ecosystems, organisms and, increasingly, the genomes of crop plants. The human appropriation of net primary production (HANPP) is an indicator of the intensity of land use or the colonisation of ecosystems. HANPP is the quantity of energy appropriated through human interventions into energy flows in ecosystems, that is, net primary production (NPP). Net primary production (NPP) is a measure of the quantity of organic material produced by plants through photosynthesis. NPP is an important process in ecosystems; it is the source of entire food energy for humans and all other heterotrophic food webs and provides the basis for the creation of vegetation cover and soils and their associated carbon stocks. NPP is one of the most important indicators of ecosystem capacity and forms the basis for the existence of all biodiversity (Vitousek et al. 1986; Wright 1990). The difference between the NPP of potential natural vegetation ($NPP_{pot}$, or NPP of the ecosystem with no human influence) and vegetation that is pre-dominant due to a certain land use (e.g. for agriculture, buildings, roads) at a particular point in time ($NPP_{act}$, or actual NPP) is defined as $HANPP_{luc}$ (HANPP resulting from land use). Added to this is the harvest of biomass for human use ($HANPP_{harv}$, or HANPP through harvest), often the ultimate purpose of land use.

A completely different biophysical indicator, yet a key component of a socio-metabolic analysis, is Functional Time Use (FTU), which views human time as a key resource of social systems. The total time available to a social system each day (for example, a village) is the sum of the population multiplied by 24 h. This is the maximum time the village has at their disposal per day, also referred to as the stock of time. With respect to the flows of human time, one fraction of daily time use is expended on certain metabolic functions (such as sleeping or eating) necessary for an individual's basic reproduction, whereas the remainder is used according to socio-cultural norms, economic necessities or simply individual preferences. We distinguish between flows serving four functional subsystems, each of which requires time for reproduction: the person system, the household system, the economic system and the community system [for a discussion on methodology, see Singh et al. (2010)]. Such a systemic analysis offers a perspective on how much human time is available and what it is used for on a system level, thereby helping us to understand the specific opportunities and constraints a society faces in its interaction with the natural environment. At the same time, because the lifetime/labour time ratio is calculated for all the age/sex groups in this system, FTU sheds light on the 'labour burden' or 'time poverty' some of these groups bear with regard to important aspects of social inequality (Ringhofer and Fischer-Kowalski 2016).

These basic components of the socio-metabolic approach (material and energy flow analysis, land use analysis, functional time use analysis) not only provide an important methodological tool, but also generate information that has significant social and political implications, and thus are not only constituent parts of Ecological Economics (see footnote 8) but also contribute significantly to the broad field of Political Ecology. Broadly speaking, political ecology strives to explain the root causes of social vulnerability by studying ecological processes as part of the broader political economy. Watts (2000) defines the role of Political Ecology as being "to understand the complex relations between nature and society through a careful analysis of what one might call the forms of access and control over resources and their implications for environmental health and sustainable livelihoods" (p. 257).

In the context of international trade, the concept of social metabolism and material flow analysis have been used by a few political ecologists to quantify "unequal ecological exchange", that is, the asymmetrical flow of resources from the core (industrialised countries) to the periphery (countries of the Global South) (see for example, Eisenmenger and Giljum 2007; Singh and Eisenmenger 2011; Schaffartzik et al. 2014; Mayer et al. 2017). Joan Martinez-Alier and colleagues have made important contributions to show the links between social metabolism and conflicts. They argue that current patterns of industrial metabolism draw on huge amounts of resources from all over the world, but the costs and benefits of extraction, production, consumption and eventual waste disposal are not equally distributed, often leading to what is termed "ecological distribution conflicts"

(Martinez-Alier 2009; Martinez-Alier et al. 2016). European funded projects[11] have resulted in large collaborative efforts to underline these connections through a number of publications framed around environmental justice (Martinez-Alier et al. 2010a; Healy et al. 2012) and degrowth (Martinez-Alier et al. 2010b). The themes linking social metabolism, political ecology, social conflicts and environmental justice are inextricably woven throughout this book.

The social metabolism approach has thus provided us with valuable insights e.g. on the biomass metabolism occurring at global, national and local scales. For much of the 20th century, biomass dominated total global material extraction for socio-economic activities, and only in the 1990s was it overtaken by construction minerals (Krausmann et al. 2009). Between 1950 and 2010, growth in global biomass extraction was slow, increasing only by a factor of 2.5, unlike other material categories that grew several-fold in the same period. In an absolute sense, biomass production increased from 7 Gigatons (Gt) to 19 Gigatons, or from 2.7 to 3.1 t/cap, indicating a strong relationship with population growth (Steinberger et al. 2010). Much of the growth in biomass production was in primary food crops that increased by a factor of 4, but the related land area increased only by a factor of 1.4, indicating a transformation in agriculture with high levels of inputs of fossil fuels and fertilisers (Mayer et al. 2015).

On the other hand, international trade in biomass increased almost 5-fold in the same period, from 0.3 to 1.4 Gt. Most of the new land dedicated to crop production in the mid-1980s was aimed at export. This period corresponds to what is referred to as the "corporate—or third—food regime" that seeks to address global food security through international trade, neo-liberal policies and market instruments. Those with purchasing power could source fresh fruits, vegetables and animal products from supermarkets (Mayer et al. 2015). The changing dietary patterns of a growing middle class (from plant to animal based diets), and growing industrial use of biomass (e.g. biofuels) continues to provide impetus to its production and trade.

Erb et al. (2009) examined the increasing spatial disconnect between where biomass is being consumed and the environmental impacts on where it is being produced. A closer look at the global biomass metabolism reveals that wealthy industrialised nations of the Global North supply large parts of the Global South with cheap staples (namely cereals and pulses) as well as fodder for meat production, while their imports are comprised mainly of luxury and exotic fruits, vegetables and animal products (Mayer et al. 2015). The unequal distribution of costs and benefits of our current global food production and international trade have provoked resistance movements and related social conflicts worldwide. These trends not only raise questions around food security (i.e. the amount of food available to any given society) but also food sovereignty that emphasises control over food production and access (i.e. who decides to produce what, and how) both

[11]For example CEECEC (http://www.ceecec.net/), EJOLT (http://www.ejolt.org/), ENV-JUSTICE (http://www.envjustice.org/), ACKnowl-EJ (http://acknowlej.org/),

in the Global North and South, giving rise to diverse alternate local food movements as discussed in this chapter (see also McMichael 2009; Clapp 2014).

## 1.4 The Contributions in this Volume

As the title suggests, this volume is an attempt to critically examine the sustainability of a wide range of local food systems through the lens of social metabolism. The chapters in this book cover a range of methods related to social metabolism at different temporal and spatial scales, with a mix of cases from the Global North, transition economies and Global South. The cases in the book also capture the various stages of production, distribution, consumption, and waste related to LFS (for an overview of the particular case studies, see Fig. 1.1). As such, two overarching questions drive our enquiry throughout this book: Which scientific and policy relevant insights can we draw from our LFS cases that might suggest the best practices for maintaining and designing more sustainable agri-food systems in the future? What are the methodological strengths and shortcomings vis-a-vis social metabolism for studying LFS?

Part I, including Chaps. 2, 3 and 4 outline the methodological complexity in studying agroecological systems when considering appropriate space and time dimensions. In Chap. 2, Enrico Tello and Manuel Gonzalez de Molina review several methods for a sustainability assessment of agri-food systems, and offer an agroecological perspective to account for energy and material flows of farm systems, linking them to landscape ecology patterns and processes which sustain farm-associated biodiversity and derived ecosystem services. Tiziano Gomiero, in Chap. 3, explains that food production is both multi-functional (having a dual purpose of maintaining both societal and ecosystem health), as well as multi-scale (connecting soil, farm, landscape, watershed and beyond) that requires consideration of different criteria at different hierarchical levels. This implies a significant complexity of sustainability assessment of agri-food systems; within the chapter, the issue of energy efficiency and energy flow (power) are discussed particularly extensively, in the broader context of biophysical and socio-economic analysis, and the different schools of social metabolism that have already been introduced in this chapter (Sect. 1.3). In the same vein, in Chap. 4, Joan Marull and Carme Font attempt an integrated approach, combining different spatial scales with long-term socio-metabolic balances and changes in the ecological functionality of farm systems. The authors propose an Intermediate Disturbance-Complexity model of agro-ecosystems to assess how different levels of human appropriation of net primary production affect the regional functional landscape structure and biodiversity.

Part II presents seven particular case studies (see Fig. 1.1). Chapters 5, 6 and 7 discuss cases in different periods of time and emphasise the social, economic and community aspects of local food systems. Roc Padró and colleagues (Chap. 5) present the case of Vallès County in Spain in three periods of time (1860, 1956 and 1999) to show the socio-ecological transition from an organic to an industrial

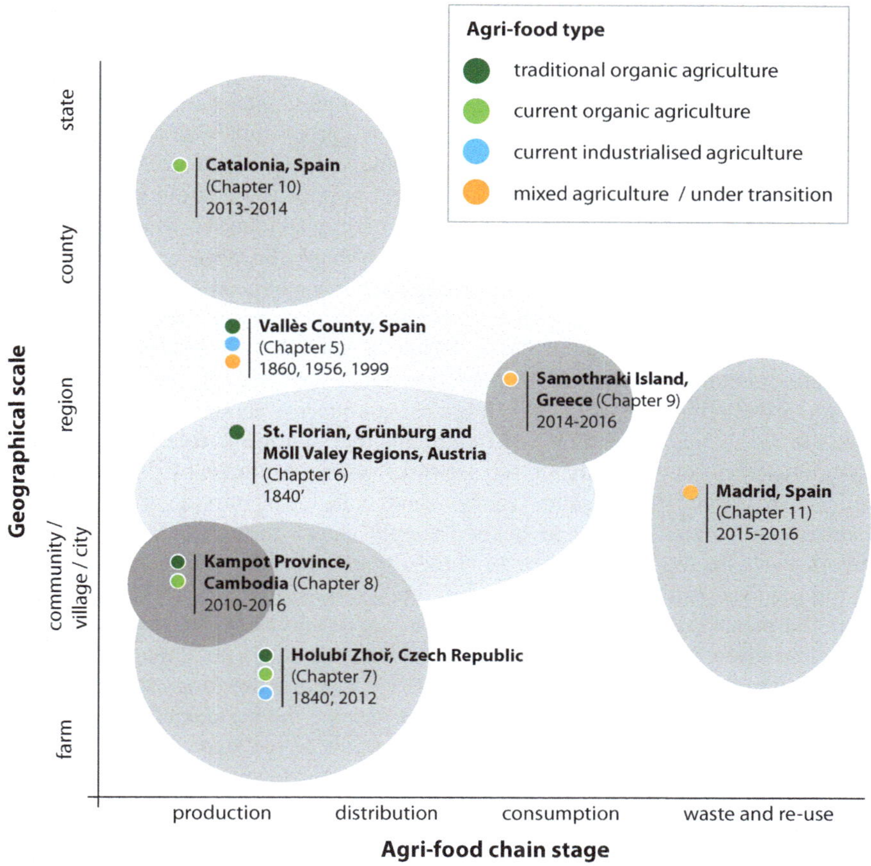


**Fig. 1.1** Overview of case studies in this volume (Chaps. 5–11) according to their main characteristics covered by the respective analysis: geographical location and time period (grey bubble), main stage(s) of agri-food chain (X axis), main geographical scale(s) (Y axis), and main type(s) of the agri-food system (indicated by colours, see the legend for details)

regime. They compare diets and labour productivity, the role of multi-functionality in agro-ecosystems, the loss of landscape diversity and species richness through 150 years of agro history. In Chap. 6, Michael Gizicki-Neundlinger, Simone Gingrich, and Dino Güldner take a Long-Term Socio-Ecological Research (LTSER) perspective to focus on the relationship between soil fertility, food provision and social inequality. Using several village level cases in 19th century Austria, the authors highlight a number of social issues underlying pre-industrial agriculture, and what we might learn from this with respect to the current debate on the sustainability of LFS. Chapter 7 (Eva Fraňková, Claudio Cattaneo) provides a historical comparison of traditional and current forms of organic agriculture in the Czech Republic, focusing mainly on the current case study of an organic family farm to illustrate the concept of food localisation. Here, the nexus of Food-Feed-Fuel-Fibre production is extended to include the aspect of *Finance* that is too often neglected in current socio-metabolic studies.

Chapters 8, 9, 10 and 11 provide other examples and assessments of local initiatives and efforts with the aim of nurturing and promoting local and sustainable food systems. In Chap. 8, Arnim Scheidel and colleagues present a case from Kampot (Cambodia) that highlights the collective efforts of farmers to intensify farming with low labour input, renewable energy, and buffer food prices and shortages through cooperatives and a community bank. Kampot is an example of "leapfrogging" to showcase a model of sustainable livelihoods embedded in local socio-economic dynamics. Chapter 9 (Panos Petridis and Julia Huber) discusses the ongoing transition on the small Greek Island of Samothraki, socio-metabolically, from a traditional agrarian lifestyle to a modern industrial society, and nutritionally, towards a westernisation of diets. While raising concerns about increasing imports from the mainland, the authors also highlight the significant role of agricultural self-provisioning and informal food networks, as an example of "quiet sustainability". In Chap. 10, Gonzalo Gamboa and colleagues provide a comparative assessment of initiatives related to food supply chains in Catalonia (Spain). The authors evaluate the performance of three organic tomato supply chains in the region: the local and global supply chains, and a mixed supply chain based on a set of multidimensional indicators (based on a commodity, environment and livelihoods narrative). The authors show that organic production is not enough to achieve a sustainable food provision system, and the context and the food supply chain in which organic food is commercialised also determine the sustainability of organic food production. Marian Simon-Rojo and Barbora Duží, in Chap. 11, present a case of urban metabolism (Madrid, Spain) where efforts are underway to close the nutrient loops. The civil society platform *Madrid Agroecologico* has initiated a bottom-up approach to reintegrate organic waste into the regional alternative food networks. The project was instrumental in raising awareness on the issue of waste and its potential re-use as compost to amend local soils.

The concluding Chap. 12 summarises both the promising characteristics of the historical and current local food systems, and the pros and cons of the socio-metabolic methodology applied at local level to recognise, preserve and offer more sustainable pathways (along with necessary policy context) for dealing with the challenges in the sustainability of agri-food systems.

## 1.5 Final Remarks

The book is meant as both a scientific and a political project: it's *scientific* in terms of asking critical questions, and discussing the sustainability of local food systems in the context of the existing body of theoretical and empirical knowledge with the aim of providing new data and insights into this debate, including advancement of relevant methodological approaches; it's *political* and value-based in its basic focus on sustainability and local food systems, aiming to bring these into light as potentially beneficial and important, and thus to break the (partly performative) dominance of the industrialised agri-food regime as the only one capable of providing food for the world's population.

Along with Gibson-Graham (2008) we believe that not only activists, but also we as researchers influence and shape reality by making some things and relations visible, and by silencing/ignoring others. We should be aware of this and ethics should inform what we do. According to Gibson-Graham (2008), the dominance of the current global capitalist system (of food production, too) is partly performative, i.e. there are other relations and systems of social and economic organisation than the capitalist, but they are (even by those criticising the capitalist system) framed as "alternative", parallel etc. thus actually confirming the dominant position of capitalism. We as scientists have the power to influence what is seen as "normal", and what is seen at all—in this sense, by this book and the research in it we contribute to making local food systems visible and important as part of the debate on how to feed the world's population, and how to maintain and/or create agro-ecosystems in a more sustainable way.

This book is both academic and activist in nature, thus trying to actively combine these different domains of activities. We tried to make the process of the creation of this book convivial and participative—we are all passionate about our research and the issue this book is about. We see this book as only one part of the process which has to be continuous and democratic—we offer the results of the time and effort which we had the chance to devote for research, but we fully acknowledge that it is only one of many valid ways of creating/gaining knowledge and we do not position ourselves in the roles of those who know and tell others what to do. We try to derive conclusions from our research and experience which can be generalised to some extent (see Chap. 12 for our main conclusions), but we stress the importance of the particular local context and conditions, and the importance of participative processes for exploring and creating more sustainable ways of fulfilling human needs, including the need for food, feed, fibre and fuel provided by agroecosystems.

Thus, we hope to provide rich food for thought on food, and to contribute to a vivid—both academic and political—debate (and action!) on this important topic that affects us all. We would be extremely happy for any feedback, and meanwhile we hope you will thoroughly enjoy reading this book.

## References

Allen, P. (2010). Realizing justice in local food systems. *Cambridge Journal of Regions, Economy and Society, 3*(2), 295–308.

Ayres, R. U., & Kneese, A. V. (1969). Production, consumption and externalities. *American Economic Review, 59*(3), 282–297.

Beckie, M. A., Kennedy, E. H., & Wittman, H. (2012). Scaling up alternative food networks: Farmers' markets and the role of clustering in western Canada. *Agriculture and Human Values, 29*(3), 333–345.

Birke, A. H. (2014). *People, energy, and environment—A socio-ecological analysis in Bashkuriti watershed, Ethiopia—Dissertation*. Klagenfurt, Wien, Graz: Alpen Adria Universitaet.

Brunner, P. H., & Rechberger, H. (2004). Practical handbook of material flow analysis. *The International Journal of Life Cycle Assessment, 9*(5), 337–338.

Cadieux, K. V., & Slocum, R. (2015). What does it mean to do food justice? *Journal of Political Ecology, 22*(1), 1–26.

Cecchini, M., Franco, S., Lauer, J. A., Lee, Y. Y., Guajardo-Barron, V., & Chisholm, D. (2010). Tackling of unhealthy diets, physical inactivity, and obesity: Health effects and cost-effectiveness'. *The Lancet, 376*(9754), 1775–1784.

Clapp, J. (2014). Food security and food sovereignty: Getting past the binary. *Dialogues in Human Geography, 4*(2), 206–211.

Clapp, J. (2017). *Responsibility to the rescue?*. Governing private financial investment in global agriculture: UWSpace. doi:10.1007/s10460-015-9678-8.

De Molina, M. G., & Toledo, V. M. (2014). *The social metabolism: A socio-ecological theory of historical change*. Switzerland: Springer International Publishing. doi:10.1007/978-3-319-06358-4.

Desai, P. & Riddlestone, S. (2002). *Bioregional solutions for living on one planet*. Schumacher Briefing No. 8, Bristol: Schumacher Society, Dartington, Devon: Green Books.

Douthwaite, R. (1996). *Short circuit: Strengthening local economies for security in an unstable world*. Dublin: Lilliput Press. doi:10.1604/9781874675600.

Edelman, M., Weis, T., Baviskar, A., Borras, S. M., Jr., Holt-Giménez, E., Kandiyoti, D., et al. (2014). Introduction: Critical perspectives on food sovereignty. *Journal of Peasant Studies, 41*, 911–931. doi:10.1080/03066150.2014.963568.

Edenhofer, O., Pichs-Madruga, R., Sokona, Y., Farahani, E., Kadner, S., Seyboth, K., et al. (2014). *Climate change 2014: Mitigation of climate change*. Contribution of working group III to the fifth assessment report of the intergovernmental panel on climate change, 5.

Eisenmenger, N., & Giljum, S. (2007). *Evidence from societal metabolism studies for ecological unequal trade* (pp. 288–302). The world system and the earth system: Global socioenvironmental change and sustainability since the Neolithic.

Erb, K. H., Krausmann, F., Lucht, W., & Haberl, H. (2009). Embodied HANPP: Mapping the spatial disconnect between global biomass production and consumption. *Ecological Economics, 69*(2), 328–334.

Erb, K. H., Lauk, C., Kastner, T., Mayer, A., Theurl, M. C., & Haberl, H. (2016). Exploring the biophysical option space for feeding the world without deforestation. *Nature Communications, 7*, 11382. doi:10.1038/ncomms11382.

Escobar, A. (2014). Development, critiques of (Chap. 3). In G. D'Alisa, F. Demaria & G. Kallis. (Eds.), *Degrowth: A vocabulary for a new era*. London: Routledge.

Eurostat. (2013). *Economy-wide material flow accounts (EW-MFA)—Compilation guide 2013*. Eurostat. http://ec.europa.eu/eurostat/documents/1798247/6191533/2013-EW-MFA-Guide-10Sep2013.pdf/. Accessed June 29, 2017.

FAO. (2009). *How to feed the world: High-level expert forum*. Rome: Food and Agriculture Organization of the United Nations. http://www.fao.org/fileadmin/templates/wsfs/docs/Issues_papers/HLEF2050_Global_Agriculture.pdf. Accessed June 7, 2017.

FAO. (2015). FAO statistical pocketbook: Food and agriculture. Rome: Food and Agriculture Organization of the United Nations. http://www.fao.org/3/a-i4691e.pdf. Accessed June 29, 2017.

Fischer, J. et al. (2008). Should agricultural policies encourage land sparing or wildlife friendly farming? *Front. Ecol. Environ* 6. doi:10.1890/070019.

Fischer-Kowalski, M. (1998). Society's metabolism. *Journal of industrial ecology, 2*(1), 61–78.

Fischer-Kowalski, M. & Haberl, H. (Eds.). (2007). *Socioecological transitions and global change: Trajectories of social metabolism and land use*. Cheltenham: Edward Elgar Publishing. doi: 10.4337/9781847209436.

Fischer-Kowalski, M., Krausmann, F., Giljum, S., Lutter, S., Mayer, A., Bringezu, S., et al. (2011a). Methodology and indicators of economy wide material flow accounting. State of the art and reliability across sources. *Journal of Industrial Ecology, 15*(6), 855–876.

Fischer-Kowalski, M., Krausmann F., & Pallua, I. (2014). A sociometabolic reading of the anthropocene: modes of subsistence, population size and human impact on earth. *The Anthropocene Review, 1*(1), doi:10.1177/2053019613518033.

Fischer-Kowalski, M., Singh, S. J., Lauk, C., Remesch, A., Ringhofer, L., & Grünbühel, C. M. (2011b). Sociometabolic transitions in subsistence communities: Boserup revisited in four comparative case studies. *Human Ecology Review, 18*(2), 147–158.

Fishman, T., Schandl, H., Tanikawa, H., Walker, P., & Krausmann, F. (2014). Accounting for the material stock of nations. *Journal of Industrial Ecology, 18*(3), 407–420.

Foley, J. A., Ramankutty, N., Brauman, K. A., Cassidy, E. S., Gerber, J. S., Johnston, M., et al. (2011). Solutions for a cultivated planet. *Nature, 478*(7369), 337–342. doi:10.1038/nature10452.

Food Security Portal. (2017). *Data on Calorie supply per capita, crops equivalent*. Available at http://www.foodsecurityportal.org/api/countries/fao-calorie-supply-p. Accessed June 15, 2017.

Fraňková, E., & Cattaneo, C. (2017). Organic farming in the past and today: Sociometabolic perspective on a Central European case study. *Regional Environmental Change*. doi:10.1007/s10113-016-1099-8.

Fraňková, E., & Johanisová, N. (2012). Economic localization revisited. *Environmental policy and governance* 22(5), 307–321. doi:10.1002/eet.1593 (Wiley & Sons and ERP Environment).

Gabriel, D., Stephen, J. C., Durham, H., Kunin, W. E., Palmer, R. C., Sait, S. M., et al. (2009). *Journal of Applied Ecology, 46,* 323–333. doi:10.1111/j.1365-2664.2009.01624.x.

Georgescu-Roegen, N. (1971). *The entropy law and the economic process*. Cambridge, Massachusetts: Harvard University Press. doi:10.4159/harvard.9780674281653.

Gerber, J.F., & Scheidel, A. (2018). In search of substantive economics: comparing today's two major sociometabolic approaches to the economy—MEFA and MuSIASEM. *Ecological Economics*, 144, 186–194.

Giampietro M. (2003). *Multi-scale integrated analysis of agroecosystems*. Boca Raton: CRC press.

Giampietro M., Aspinall, R., Ramos-Martín, J., & Bukkens, S. (Eds.). (2014). *Resource accounting for sustainability assessment: The nexus between energy, food, water and land use*. London Routledge.

Giampietro M., Mayumi, K., & Şorman, A. (2012). *The metabolic pattern of societies: Where economists fall short*. London: Routledge.

Giampietro M., Mayumi, K., & Şorman, A. (2013). *Energy analysis for a sustainable future: Multi-scale integrated analysis of societal and ecosystem metabolism*. London: Routledge.

Gibson-Graham, J. K. (2008). Diverse economies: Performative practices forother worlds'. *Progress in Human Geography, 32*(5), 613–632.

Godfray, H. C. J., Beddington, J. R., Crute, I. R., Haddad, L., Lawrence, D., Muir, J. F., et al. (2010). Food security: The challenge of feeding 9 billion people. *Science, 327*(5967), 812–818.

Gomiero, T., Giampietro, M., & Mayumi, K. (2006). Facing complexity on agro-ecosystems: A new approach to farming system analysis. *International Journal of Agricultural Resources, Governance and Ecology, 5,* 116–144. doi:10.1504/IJARGE.2006.009160.

Gomiero, T., Pimentel, D., & Paoletti, M. G. (2011a). Is there a need for a more sustainable agriculture? *Critical Reviews in Plant Sciences, 30,* 6–23. doi:10.1080/07352689.2011.553515.

Gomiero, T., Pimentel, D., & Paoletti, M. G. (2011b). Environmental impact of different agricultural management practices: Conventional vs. Organic agriculture. *Critical Reviews in Plant Sciences, 30,* 95–124. doi:10.1080/07352689.2011.554355.

Goodman, D., DuPuis, E. M., & Goodman, M. K. (2012). *Alternative food networks: Knowledge, practice and politics*. London: Routledge.

Grešlová Kušková, P. (2013). A case study of the Czech agriculture since 1918 in a socio-metabolic perspective. From land reform through nationalisation to privatisation. *Land Use Policy, 30,* 592–603. doi:10.1016/j.landusepol.2012.05.009.

Haas, W., & Krausmann, F. (2015). Transition-related changes in the metabolic profile of an Austrian rural village. *Working paper Social Ecology, 153,* 42.

Haberl, H., Erb, H., Krausmann, F., Gaube, V., Bondeau, A. Plutzar, C., et al. (2007). Quantifying and mapping the human appropriation of net primary production in earth's terrestrial ecosystems. *PNAS*, 104(31), 12942–12947. doi:10.1073/pnas.0704243104/.

Haberl, H., Erb, K. H., Fischer-Kowalski, M., Groß, R., Krausmann, F., Plutzar, C., et al. (2016). Introduction. In H. Haberl, M. Fischer-Kowalski, F. Krausmann, et al. (Eds.), *Social Ecology: Society-nature relations across time and space*. London: New York Dordrecht.

Haberl, H., Fischer-Kowalski, M., Krausmann, F., Weisz, H., & Winiwarter, V. (2004). Progress towards sustainability? What the conceptual framework of material and energy flow accounting (MEFA) can offer. *Land Use Policy, 21,* 199–213.

Haberl, H., Krausmann, F. (2007). The local base of the historical agrarian-industrial transition, and the interaction between scales. In: M. Fischer-Kowalski & H. Haberl (Eds.), *Socio-ecological transitions and global change: Trajectories of social metabolism and land use,* (pp. 116–138). Cheltenham, Northampton: Edward Elgar. doi: 10.4337/9781847209436.00012.

Healy, H., Martínez-Alier, J., Temper, L., Walter, M., Gerber, J.-F. (2012). Ecological Economics from the Ground Up. London: Routledge.

Hinrichs, C. C. (2003). The practice and politics of food system localization. *Journal of rural studies, 19*(1), 33–45.

Holt-Giménez, E., & Altieri, M. A. (2013). Agroecology, food sovereignty, and the new Green revolution. *Agroecology and Sustainable Food Systems, 37,* 90–102. doi:10.1080/10440046.2012.716388.

IFPRI. (2016). International Food Policy Research Institute. *Global Nutrition Report 2016: From Promise to Impact: Ending Malnutrition by 2030*. Washington, DC. http://ebrary.ifpri.org/utils/getfile/collection/p15738coll2/id/130354/filename/130565.pdf. Accessed June 16, 2017.

Kastner, T., Erb, K. H., & Haberl, H. (2014). Rapid growth in agricultural trade: Effects on global area efficiency and the role of management. *Environmental Research Letters, 9*(3), 034015.

Kneafsey, M., Venn, L., Schmutz, U., Balázs, B., Trenchard, L., Eyden-Wood, T., et al. (2013). *Short food supply chains and local food systems in the EU. A state of play of their socio-economic characteristics*. JRC Scientific and Policy Reports. Joint Research Centre Institute for Prospective Technological Studies, European Commission.

Krausmann, F., Erb, K. H., Gingrich, S., Haberl, H., Bondeau, A., Gaube, V., et al. (2013). Global human appropriation of net primary production doubled in the 20th century. *Proceedings of the National Academy of Sciences, 110*(25), 10324–10329. doi:10.1073/pnas.1211349110.

Krausmann, F., Gingrich, S., Eisenmenger, N., Erb, K.-H., Haberl, H., & Fischer-Kowalski, M. (2009). Growth in global materials use, GDP and population during the 20th century. *Ecological Economics, 68*(10), 2696–2705.

Lamb, A., Green, R., Bateman, I., Broadmeadow, M., Bruce, T., Burney, J., et al. (2016). The potential for land sparing to offset greenhouse gas emissions from agriculture. *Nature Climate Change, 6*(5), 488–492.

Locey, K. J., & Lennon, J. T. (2016). Scaling laws predict global microbial diversity. *PNAS, 113* (21), 5970–5975. doi:10.1073/pnas.1521291113.

Madrid-López, C., & Giampietro, M. (2015). The water metabolism of socio-ecological systems: Reflections and a conceptual framework. *Journal of Industrial Ecology, 19*(5), 853–865.

Martinez, S., Hand, M. S., Da Pra, M., Pollack, S., Ralston, K., Smith, T., et al. (2010). *Local food systems: Concepts, impacts, and issues*. ERR-97, U.S. Department of Agriculture, Economic Research Service. http://www.ers.usda.gov/Publications/ERR97/ERR97.pdf. Accessed June 12, 2012.

Martinez-Alier, J. (2009). Social metabolism, ecological distribution conflicts, and languages of valuation. *Capitalism Nature Socialism, 20,* 58–87.

Martinez-Alier, J., Kallis, G., Veuthey, S., Walter, M., & Temper, L. (2010a). Social metabolism, ecological distribution conflicts, and valuation languages. *Ecological Economics*. doi:10.1016/j.ecolecon.2010.09.024.

Martinez-Alier, J., Pascual, U., Vivien, F. D., & Zaccai, E. (2010b). Sustainable de-growth: Mapping the context, criticisms and future prospects of an emergent paradigm. *Ecological Economics, 69,* 1741–1747.

Martinez-Alier, J., Temper, L., Walter, M., & Demaria, F. (2016). Social metabolism and ecological distribution conflicts in India and Latin America. In: *Green Economy Reader* (pp. 311–332). Berlin: Springer International Publishing.

Martínez-Torres, M. E., & Rosset, P. M. (2014). Diálogo de saberes in La Vía Campesina: Food sovereignty and agroecology. *Journal of Peasant Studies, 41*(6), 979–997.

Marx, K. (1968 [1867]). *Das Kapital*, Bd. 1, Marx Engels Werke, Bd. 23, Berlin.

Matacena, R. (2016). Linking alternative food networks and urban food policy: A step forward in the transition towards a sustainable and equitable food system? *International Review of Social Research, 6*(1), 49–58.

Matthews, E., Amann, C., Bringezu, S., Hüttler, W., Ottke, C., Rodenburg, E., et al. (2000). *The weight of nations-material outflows from industrial economies.* Washington: World Resources Institute.

Mayer, A., Schaffartzik, A., Haas, W., & Sepulveda, A. R. (2015). Patterns of global biomass trade and the implications for food sovereignty and socio-environmental conflict. *EJOLT* (Report No. 20). doi: 10.13140/2.1.1442.5128.

Mayer, A., Haas, W., & Wiedenhofer, D. (2017). How countries' resource use history matters for human well-being—An investigation of global patterns in cumulative material flows from 1950 to 2010. *Ecological Economics, 134,* 1–10. doi:10.1016/j.ecolecon.2016.11.017.

McMichael, P. (2009). A food regime genealogy. *The Journal of Peasant Studies, 36*(1), 139–169.

McMichael, A. J., Powles, J. W., Butler, C. D., & Uauy, R. (2007). Food, livestock production, energy, climate change, and health. *The lancet, 370*(9594), 1253–1263. doi:10.1016/S0140-6736(07)61256-2.

NEF. (2002). *The money trail. Measuring your impact on the local economy using LM3.* Lonodn: New Economics Foundation.

NEF. (2010). *The great transition.* London: New Economics Foundation.

Netting, R. M. C. (1981). *Balancing on an Alp: Ecological change and continuity in a Swiss mountain community.* CUP Archive.

Netting, R. M. C. (1993). Smallholders, householders: Farm families and the ecology of intensive, sustainable agriculture. Palo Alto: Stanford University Press.

Norberg-Hodge, H., Merrifield, T., & Gorelick, S. (2002). *Bringing the food economy home: The social, ecological and economic benefits of local food.* Dartington, Devon: International Society for Ecology and Culture.

Nyéléni. (2007). *Declaration of Nyéléni.* Declaration of the Forum for Food Sovereignty. https://nyeleni.org/IMG/pdf/DeclNyeleni-en.pdf. Accesed June 7.

Perfecto, I., & Vandermeer, J. (2008). Biodiversity conservation in tropical agroecosystems. *Annals of the New York Academy of Sciences, 1134*(1), 173–200.

Phalan, B., Onial, M., Balmford, A., & Green, R. E. (2011). Reconciling food production and biodiversity conservation: Land sharing and land sparing compared. *Science, 333*(6047), 1289–1291. doi:10.1126/science.1208742.

Ramos-Martín, J., Giampietro, M., & Mayumi, K. (2007). On China's exosomatic energy metabolism: An application of multi-scale integrated analysis of societal metabolism (MSIASM). *Ecological Economics, 63*(1), 174–191.

Ravera, F., Scheidel, A., dell'Angelo, J., Gamboa, G., Serrano, T., Mingorría, S., et al. (2014). Pathways of rural change: An integrated assessment of metabolic patterns in emerging ruralities. *Environment, Development and Sustainability, 16*(4), 811–820.

Renting, H., Marsden, T. K., & Banks, J. (2003). Understanding alternative food networks: Exploring the role of short food supply chains in rural development. *Environment and planning A, 35*(3), 393–411.

Ringhofer, L., & Fischer-Kowalski, M. (2016). Method Précis: functional time use analysis. In: H. Haberl, M. Fischer-Kowalski, F. Krausmann & V. Winiwarter (Eds.), *Social ecology: Society-nature relations across time and space* (pp. 519–522). Berlin: Springer.

Røpke, I. (2004). The early history of modern ecological economics. *Ecological Economics, 50,* 293–314.

Schaffartzik, A., Mayer, A., Gingrich, S., Eisenmenger, N., Loy, C., & Krausmann, F. (2014). The global metabolic transition: Regional patterns and trends of global material flows, 1950–2010. *Global Environmental Change, 26,* 87–97.

Scheidel, A., Farrell, K., Ramos-Martín, J., Giampietro, M., & Mayumi, K. (2014). Land poverty and emerging ruralities in Cambodia: Insights from Kampot province. *Environment, Development and Sustainability, 16*(4), 823–840.

Scheidel, A., Giampietro, M., & Ramos-Martin, J. (2013). Self-sufficiency or surplus: Conflicting local and national rural development goals in Cambodia. *Land use policy, 34,* 342–352. doi:10.1016/j.landusepol.2013.04.009.

Seyfang, G. (2007). Cultivating carrots and community: Local organic food and sustainable consumption. *Environmental Values, 16,* 105–123. doi:10.3197/096327107780160346.

Seyfang, G., & Smith, A. (2007). Grassroots innovations for sustainable development: Towards a new research and policy agenda. *Environmental politics, 16*(4), 584–603.

Shetty, P. (2005). Energy requirements of adults. *Public health nutrition, 8*(7A), 994–1009.

Shrivastava, A., & Kothari, A. (2012). *Churning the earth: The Making of Global India.* New Delhi: Viking/Penguin Global.

Singh, S., & Eisenmenger, N. (2011). How unequal is international trade? An ecological perspective using material flow accounting (MFA). *Journal für Entwicklungspolitik, 26,* 57–88.

Singh, S. J., Grünbühel, C. M., Schandl, H., & Schulz, N. (2001). Social metabolism and labour in a local context: Changing environmental relations on Trinket Island. *Population and Environment, 23,* 71–104.

Singh, S. J., Ringhofer, L., Haas, W., Krausmann, F., & Fischer-Kowalski, M. (2010). *Local Studies Manual: A Researcher's Guide for Investigating the Social Metabolism of Rural Systems.* Social Ecology Working Paper 120.

Smil, V. (2013). *Harvesting the biosphere. What we have taken from nature.* Cambridge: Massachusetts Institute of Technology. doi: 10.1111/j.1728-4457.2013.00617.x.

Smith, J., & Jehlička, P. (2013). Quiet sustainability: Fertile lessons from Europe's productive gardeners. *Journal of Rural Studies, 32,* 148–157.

Smith, J., Kostelecký, T., & Jehlička, P. (2015). Quietly does it: Questioning assumptions about class, sustainability and consumption. *Geoforum, 67,* 223–232.

Sovová, L. (2015). Self-provisioning, sustainability and environmental consciousness in Brno Allotment Gardens. *Sociální studia/Social Studies,* 12(3).

Steffen, W., Richardson, K., Rockström, J., Cornell, S. E., Fetzer, I., Bennett, E. M., et al. (2015). Planetary boundaries: Guiding human development on a changing planet. *Science, 347*(6223), 1259855.

Steinberger, J. K., Krausmann, F., & Eisenmenger, N. (2010). Global patterns of materials use: A socioeconomic and geophysical analysis. *Ecological Economics, 69*(5), 1148–1158.

Sweetlove, L. (2011) Number of species on Earth tagged at 8.7 million. *Nature.* doi: 10.1038/news.2011.498.

Thompson, E., Jr., A. M. Harper, and S. Kraus. (2008). *Think Globally—Eat Locally: San Francisco Foodshed Assessment, American Farmland Trust.* http://162.242.222.244/programs/states/ca/Feature%20Stories/documents/ThinkGloballyEatLocally-FinalReport8-23-08.pdf. Accessed June 28, 2017.

Tilman, D., & Clark, M. (2014). Global diets link environmental sustainability and human health. *Nature, 515*(7528), 518–522.

Tilman, D., Balzer, C., Hill, J., & Befort, B. L. (2011). Global food demand and the sustainable intensification of agriculture. *Proceedings of the National Academy of Sciences, 108*(50), 20260–20264. doi:10.1073/pnas.1116437108.

Tscharntke, T., Clough, Y., Wanger, T. C., Jackson, L., Motzke, I., Perfecto, I., et al. (2012). Global food security, biodiversity conservation and the future of agricultural intensification. *Biological Conservation, 151*(1), 53–59. doi:10.1016/j.biocon.2012.01.068.

UN. (2015). The Millennium Development Goals Report 2015. New York: United Nations. http://www.un.org/millenniumgoals/2015_MDG_Report/pdf/MDG%202015%20rev%20(July%201).pdf. Accessed 7 June 2017.
Urgenci. (2016). *Overview of Community Supported Agriculture in Europe*. European CSA Research Group. http://urgenci.net/wp-content/uploads/2016/05/Overview-of-Community-Supported-Agriculture-in-Europe.pdf. Accessed 1 June 2017.
Veeramani. A. (2015). *Carbon Footprinting Dietary Choices in Ontario: A life cycle approach to assessing sustainable, healthy & socially acceptable diets* (Master's thesis, University of Waterloo).
Vitousek, P. M., Ehrlich, P. R., Ehrlich, A. H., & Matson, P. A. (1986). Human appropriation of the products of photosynthesis. *BioScience, 36*, 368–373.
Walpole, S. C., Prieto-Merino, D., Edwards, P., Cleland, J., Stevens, G., & Roberts, I. (2012). The weight of nations: An estimation of adult human biomass. *BMC public health, 12*(1), 439.
Watts, M. J. (2000). Political ecology. In E. Sheppard & T. Barnes (Eds.), *A Companion to Economic Geography* (pp. 257–274). Oxford: Blackwell.
Weis, T. (2010). The accelerating biophysical contradictions of industrial capitalist agriculture. *Journal of Agrarian Change, 10*, 315–341. doi:10.1111/j.1471-0366.2010.00273.x.
Worldometers. (2017). *Current world population*. Data from 14.6.2017 12:38 CET. http://www.worldometers.info/world-population/. Accessed June 14, 2017.
Wright, D. H. (1990). Human impacts on energy flow through natural ecosystems, and implications for species endangerment. *Ambio, 19*, 189–194.
Zumkehr, A., & Campbell, J. E. (2015). The potential for local croplands to meet US food demand. *Frontiers in Ecology and the Environment, 13*(5), 244–248.

## Author Biographies

**Eva Fraňková** works as an Assistant Professor at the Department of Environmental Studies, Faculty of Social Studies, Masaryk University in Brno, Czech Republic. Her long-term research interests and passions include the concept of eco-localisation, sustainable degrowth and various grass-root alternative economic practices including eco-social enterprises, local food initiatives etc. Recently she has been involved in the mapping of heterodox economic initiatives in the Czech Republic, and in research on the social metabolism of local food systems. She is also involved in several NGOs – the Association of Local Food Initiatives, the Society and Economy Trust, and NaZemi (OnEarth), a global education and Fair Trade organisation in the Czech Republic.

**Willi Haas** is senior researcher and lecturer at the Institute of Social Ecology in Vienna, an institute of the Faculty of Interdisciplinary Studies at the Alpen-Adria University. In the 90s, he became fascinated by social ecology, the study of society-nature relations across time and space. He is interested in the past transition from agrarian to industrial societies and the insights that can be drawn from these far-reaching changes for the next transition to a post- fossil society. In his view the question of how to overcome the system inertia and the unsustainable reproduction dynamics of the present fossil fuelled societies is crucial for fostering a sustainable future. He was chair of Greenpeace CEE for 9 years. Professionally, he was a public official at the Ministry of Social Affairs, director of the Institute of Applied Ecology (Vienna), acting director of the Environmental Monitoring Group (Cape Town) and researcher at the International Institute of Applied System Analysis (IIASA, based in Austria). At the Institute of Social Ecology he has headed numerous scientific projects and is the coordinator of the institute's team of thematic research coordinators.

**Simron Jit Singh** is Associate Professor in the School of Environment, Enterprise, and Development (SEED), and Associate Dean of the Faculty of Environment, University of Waterloo, Canada. Drawing on the concept of social metabolism, his research focuses on the systemic links between material and energy flows, time use and human wellbeing. His particular interest lies at the local and sub-national scales, as well as small islands. He has conducted social metabolism studies in the Nicobar Islands (India), and Samothraki (Greece), and supervised work on the biomass metabolism of Jamaica (as part of the Canadian project Hungry Cities), the Region of Waterloo as well as Canada. As work-package leader, he led work on biomass flows and social conflicts in an EU project (EJOLT).

# Part I
# Key Concepts and Approaches

# Chapter 2
# Methodological Challenges and General Criteria for Assessing and Designing Local Sustainable Agri-Food Systems: A Socio-Ecological Approach at Landscape Level

**Enric Tello and Manuel González de Molina**

**Abstract** Agri-food systems are sustainable when they can meet human needs while maintaining the basic funds and ecosystem services of agoecosystems and cultural landscapes in both a reproducible way and a healthy ecological state, at local, regional and global scales. This axiological definition involves a large research agenda to explore the operative criteria and indicators needed to know how to achieve this goal. It has to be a transdisciplinary research, capable of linking some of the already existing methodologies, like Life-Cycle Analysis (LCA), Material and Energy Flow Accounting of Social Metabolism (MEFA), Multi-Scale Integrated Analysis of Societal and Ecosystem Metabolism (MuSIASEM), Agroecology, Landscape Ecology, Political Ecology and valuation of site-specific Biocultural Heritages of Peasant Knowledge. We will examine these approaches and the accounting methods, highlighting their strengths and weaknesses, so as to combine them in innovative ways within a common framework focused on the interactive relations among societal and ecological metabolisms. To do so in a non-eclectic manner requires an agroecological perspective when accounting energy and material flows of farm systems, linking them with landscape ecology patterns and processes which sustain farm-associated biodiversity and derived ecosystem services, and adopting at all times an environmental history standpoint.

E. Tello (✉)
Department of Economic History, Faculty of Economics and Business, Institutions, Policy and World Economy, University of Barcelona, Diagonal Avenue 690, 08034 Barcelona, Spain
e-mail: tello@ub.edu

M. González de Molina
Agro-Ecosystems History Laboratory, Pablo de Olavide Univeristy, Building 2, Carretera de Utrera Km 1, 41013 Seville, Spain
e-mail: mgonnav@upo.es

 
E. Fraňková et al. (eds.), *Socio-metabolic Perspectives on the Sustainability of Local Food Systems*, Human-Environment Interactions 7,
https://doi.org/10.1007/978-3-319-69236-4_2

**Keywords** Agri-food systems · Sustainability · Social metabolism Agroecology · Life-cycle analysis · MuSIASEM · Energy-landscape integrated analysis · Peasant knowledge

## 2.1 Introduction

The present global food regime is unsustainable. In response, all over the world a growing network of local social movements has emerged—i.e. consumer awareness organizations, urban gardeners, rural platforms, transition towns and citizens' initiatives, slow food and zero-km promoters, organic farmers, small-size cooperatives, Via Campesina and other peasant unions. They are striving to open a way for healthier, nearer, fairer, and more ecologically sound agri-food systems. While about 805 million people are still chronically undernourished (FAO et al. 2014), an unhealthy diet stuffed with meat, fat and sugar has been increasingly globalized, turning overweight and obesity into an epidemic among the youth (World Health Organization 2000a; Wang and Lobstein 2006; Swinburn et al. 2011; Kleinert and Horton 2015; NCD-RisC 2016). Inequality in access to food deprives from a sufficient diet an unacceptable share of Humankind living in the Global South, and even a shocking proportion in the Global North—15%, more than 40 million people in the United States (see the Website www.feedingamerica.org)—, while the nutrient and energy content of the whole amount of edible biomass actually produced ends up being concentrated in the unsafe diet consumed by affluent populations of the Global North and South (World Health Organization 2000b; Rosset 2011; FAO et al. 2014). Furthermore, throughout the long current agri-food chains an incredible amount of this food is simply wasted (Parfitt et al. 2010; Gustavsson et al. 2011). At the same time, human population is about to reach nine billion people and a foreseeable shortage of cheap fossil fuels is going to jeopardize the yields of intensive conventional agriculture (Arizpe et al. 2011; Markussen and Østergärd 2013).

Throughout the second half of the twentieth century, the industrialization of agriculture, together with the globalization of diets and transport chains of food products and inputs, have increased energy consumption of agri-food systems one order of magnitude—e.g., in Spain, from about 180,000 TJ in 1960 to 1,850,000 TJ in 2010, rendering the agri-food sector a relevant contributor of one fifth of the total primary energy consumed and the ensuing greenhouse gas emissions in this country (Aguilera et al. 2014). In turn, the ongoing climate change is challenging the resilience capacity of food production, and the loss of species richness is threatening vital ecosystem services (Schröter et al. 2005; Godfray et al. 2010; de Groot et al. 2010; Cardinale et al. 2012). A major driver of this biodiversity depletion is the expansion of cropland and livestock breeding under the prevailing agri-food systems which were adopted after the Green Revolution (Matson et al. 1997; Tilman 1999; Patel 2012; Pingali 2012). Nearly half of the global usable land, and the more productive, is already allocated to intensive farming and grazing whereas only 6–

12% is under any nature protection scheme (Bengston et al. 2003; Tscharntke et al. 2012a, b). All these unsustainable trends—i.e. climate change, biodiversity loss, soil degradation, water shortage and pollution, malnutrition, human deprivation—are interconnected within an imperial food regime driven by big corporations aimed only at the short-term financial accumulation of private profit, and are endangering food security worldwide in the long run (van der Ploeg 2010; McMichael 2013; Sage 2013; Gomiero 2016).

Awareness that society urgently needs to find another food regime is spreading among scientists, doctors and paediatricians, international agencies, and even among some politicians, technicians and planners. In this global context, a core issue that needs to be explored is the capacity of more sustainable ways of self-reliant organic farming to provide a healthy and sufficient diet for the world population, while reducing damaging environmental impacts of agricultural practices, taking into account the socio-economic consequences of their alleged limits in terms of land and labour productivity (Gomiero et al. 2008, 2011; Ponisio et al. 2015; Perignon et al. 2016; Jordan 2016; see also the United Nations Webpage on the Right to Food, http://www.ohchr.org/EN/Issues/Food/Pages/FoodIndex.aspx; the Webpages Feed the World: http://feedtheworld.org/; and Food First: https://foodfirst.org/).

A key point in this complex issue is that the ecological impacts of farming, animal husbandry and forestry depend on the type of management carried out. Industrial or organic farming involve very different levels of energy and material socio-metabolic turnover on the land, and may either entail a decrease or an increase in landscape biological diversity (Altieri 1999; Moguel and Toledo 1999; McNeely and Scheer 2003; Swift et al. 2004; Tscharntke et al. 2005; Gomiero et al. 2011; Cardinale et al. 2012; Erb 2012; see also the IFOAM Webpage: https://www.ifoam.bio/). In a world where sustainable farm systems are increasingly seen as providers of supporting, habitat and regulating ecosystem services, as well as producers of food, feed, fibre and fuel (Tress et al. 2001; Millennium Ecosystem Assessment 2005; Agnoletti 2006, 2014; Agnoletti and Rotherham 2015; see also the TEEB Webpage: http://www.teebweb.org/resources/ecosystem-services/), more research is needed to better understand the relationship between ecological disturbances exerted by different types of farm, forests and livestock management, and biodiversity levels kept in cultural landscapes (Tilman et al. 2002; Benton et al. 2003).

The '*what*' behind this overriding question is closely related with the '*who*' issue—i.e. the agency of change. If we start by focusing our scrutiny on those who are working with nature at the very base of agri-food chains, this once again brings to the fore the role of the 'awkward' class of peasants (Shanin 1972, 1987; McMichael 2008, 2016; Bernstein 2016; Friedman 2016). The shifting mood towards more realistic approaches to global food sovereignty is entailing a greater recognition of the role that peasant family farms and communities play as true global *sustainers* of Humankind (Toledo 2002; Altieri 2002, 2009; Holt-Giménez et al. 2009; Martínez-Torres and Rosset 2010; Holt-Giménez and Shattuck 2011; Martínez Alier 2011; Altieri and Toledo 2011; Altieri et al. 2012; Rosset and Martínez-Torres

2012; Van der Ploeg 2013; Rosset 2013). This is not only true in biophysical terms. In many different parts of the world, farmers have inherited a set of site-specific biocultural knowledge of how to develop and reproduce complex farm systems which are fairly adapted to local conditions.

Thanks to the multi-purpose character of their integrated use of natural resources, peasant knowledge has allowed many rural communities to manage their environments sustainably while meeting subsistence needs without depending to a large extent on external technological inputs (Gliesmann 2000; Toledo et al. 2007; Gómez-Baggethun et al. 2013; González de Molina and Toledo 2014).The Food and Agriculture Organization of the United Nations (FAO) has launched the initiative of creating a catalogue of Globally Important Agricultural Heritage Systems (GIAHS), defined as 'remarkable land use systems and landscapes which are rich in globally significant biological diversity evolving from the co-adaptation of a community with its environment and its needs and aspirations for sustainable development' (http://www.fao.org/giahs/giahs/en/; Koohafkan and Altieri 2010). This requires the opening of a dialogue on sustainable agroecological innovation between different types of knowledge and actors at stake (Toledo et al. 2002, 2003; Barrera-Bassols and Toledo 2005; Toledo and Barrera-Bassols 2008; IAASTD 2009; Martínez-Torres and Rosset 2014).

The study of social metabolism is not only about biophysical fluxes being interchanged between societal and natural systems. It also includes the consideration of information that drives these flows and loops (Odum 1988) through a knowledge-based land-use pattern that gives rise to distinctive cultural landscapes (Marull et al. 2015, 2016a, b). Furthermore, the way peasants and farmers work with nature to produce the food that society requires, while trying to keep their agroecosystems in a reproducible state by preserving their funds, is also embedded in a set of social relationships and ownership rules that establishes their actual access to or deprivation of natural resources (Georgescu-Roegen 1977; Odum 2001, 2007; Martínez Alier 2011; Foster and Holleman 2014). Lessening the dependence of peasant communities on external inputs, by enhancing the integrated multiple use of local resources based on their own site-specific knowledge, may involve a real chance to empower themselves, and to democratise rural-urban interchanges (Pimbert 2009; Borras et al. 2015; Claeys 2015).

However, how can these agroecological landscapes, endowed with a rich biocultural heritage, be identified, understood and valued as an important resource for a sustainable future? Why, where and when does a farming pattern of complex agro-forest mosaics become suitable to host a rich farm-associated biodiversity and derived ecosystem services? And how does the maintenance of these healthy agroecological landscapes relate with the food diets and consumption baskets, which are culturally defined and sold throughout a long food chain that ends up in cities? Such a complex and dynamic research subject needs to be addressed by means of an innovative transdisciplinary approach, capable of bridging social and natural sciences in order to develop ground-breaking methods, indicators and criteria seen from a long-term perspective. This poses a big challenge to policy makers, stakeholders of civil society, and to scientific research as well, given that all pre-existing barriers that separate the nature conservation policies from farming, forestry or animal husbandry have to be overcome.

In a world full of anthropogenic ecological impacts, and where growing food production may either endanger or improve all sorts of ecosystem services (Millennium Ecosystem Assessment 2005; de Groot et al. 2010), this research focus is also connected with the ongoing debate within biological conservation between the so-called ‘land sparing’ and ‘land sharing’ approaches (Fischer et al. 2008; Perfecto and Vandermeer 2010). The land-sparing strategy is based on increasing agricultural intensification in some areas, with the aim of allocating others to nature conservation (Green et al. 2005; Matson and Vitousek 2006). The land-sharing approach argues that a set of natural reserves kept isolated is not enough for safeguarding biodiversity, and promotes a wildlife-friendly farming that can provide a complex agroecological matrix to connect those natural sites to each other, and jointly maintain high species richness at landscape level (Bengston et al. 2003; Perfecto and Vandermeer 2010; Tscharntke et al. 2012a, b). Behind this scientific controversy, and the important dilemmas for policy-making that entails, there exist many contrasting bio-geographical features on Earth, different human settlement patterns and socio-ecological histories, as well as diverse intellectual traditions (McDonnell and Pickett 1993; Farina 1997; Fischer et al. 2008). Even those who try to combine both perspectives recognise that more research is needed to understand how biodiversity is kept in diverse landscape patterns that give rise to different ecological processes (Phalan et al. 2011; Gomiero et al. 2011; Gomiero 2015).

Our main point is that this ground-breaking research agenda needs to be addressed adopting a socio-metabolic approach, focused on an agroecological standpoint on how farming biophysical flows are continuously interacting with land-use patterns and landscape ecological functioning, and with the societal structure and functioning, seeing this coevolution from a long-term historical perspective and a multidimensional and multi-scalar view (Giampietro 2003; Fischer-Kowalski and Haberl 2007; Giampietro et al. 2012, 2013; González de Molina and Toledo 2014). The energy and material flow analysis of agroecosystems brings to light the interaction and co-evolutionary dynamics set forth by farm systems in a land matrix, and how this anthropogenic disturbance may positively or negatively affect the species richness hosted in cultural landscapes (Guzmán and González de Molina 2015, 2017; Tello et al. 2016; Galán et al. 2016).

After framing the subject of sustainability of food-producing agroecosystems in Sect. 2.2, in Sect. 2.3 we present five interlinked answers to the question of why sustainable agri-food chains needs to be re-localised: closing nutrient cycles, keeping landscape mosaics well-endowed with biodiversity, improving farming energy efficiency, putting peasant knowledge to work again for agroecology innovations, and making economically fairer and more democratic urban-rural relationships feasible. Section 2.4 considers the multi-scalar hierarchical character of rural-urban interchanges and global trade throughout food chains up to final consumption, and argues that there is a maximum scope for sustainable trade when the closure at landscape scale of biophysical flows needed to reproduce vital agroecological funds is taken into consideration. Section 2.5 examines the strengths and weaknesses of the existing methodological entryways to analyse agri-food systems, and proposes to combine them in a non-eclectic way through an agroecological fund-flow integrated assessment of farm systems able to consider landscape ecology patterns and processes from a socio-metabolic standpoint. Section 2.6 concludes by suggesting the aim of creating agroecological networks of healthy biocultural sites as a social innovation approach to devise more sustainable agri-food chains, kept by people for people, in order to make a living while they stay working with Nature.

## 2.2 Sustainability of What? Framing the Subject

The global agri-food system may become sustainable if it is able to meet the food needs of all human beings in a fair manner through the biomass produce obtained from cropland, pastureland and woodland, while at the same time keeping all agroecosystems on Earth in a reproducible and healthy ecological state, so as to ensure that they will be able to keep satisfying the needs of future generations (Young and Burton 1992; Neher 1992; Costanza and Patten 1995). This is an ethical definition that postulates equality among all existing human beings, as well as between human generations throughout time—an extended notion of justice that aims at avoiding the dictatorship of present over future (Georgescu-Roegen 1977). The socio-ecological interactions of human society with Nature are implicitly involved in this definition, and they entail a complex set of variables at stake—from the size of human population, the proportion of meat in consumption baskets, or unequal access to food on the demand side, to the different types of farming, agroecological landscapes and food chains on the supply side. This multidimensional and multi-scalar complexity arises when the means to achieve this ethical goal are taken into account, which requires adopting a fund-flow socio-metabolic approach to agroecosystem functioning and socio-ecological performance of food chains (Georgescu-Roegen 1971; Giampietro et al. 2013; González de Molina and Toledo 2014).

In brief, agri-food sustainability means meeting human intake of an adequate diet fairly distributed to all, while the funds of agroecosystems, providing the whole set of biomass flows needed for that purpose, are kept in a healthy ecological state. This is only an axiological definition, easy to write in order to communicate what a sustainable agri-food system aims to achieve. Yet, if besides saying what sustainability means, we want to explore how to achieve this goal, more operative criteria are needed. Searching for them, and for derived indicators, is a complex issue that has to tackle the hierarchical, multidimensional and multi-scalar character of production-consumption chains interwoven with the ecosphere by the human techno-sphere (Commoner 1990). To begin with, this means specifying the spatiotemporal boundaries of any sustainability assessment (Costanza and Patten 1995): Of what system? How long? Considering which dimensions? These questions can only be answered, providing clear criteria and precise indicators, after a long research and practical experimentation which have still to be done. Our chapter aims only at exploring the main focus of this research, to suggest some very general criteria as working hypotheses, and propose some useful methodological approaches. It is mainly addressed at the landscape scale, where the basic funds are located together with the labouring people who manage them, given the multiple possible synergies that can be set up at this spatial level among farms, rural communities, urban consumers and land-use planners in order to carry out the significant rearrangement of agriculture, livestock breeding and forest management that agri-food sustainability requires (Sayer et al. 2013).

Understandably, beyond some very basic consensus on a few operative principles like minimizing resource use, preventing pollution and turning to renewable energy sources, the research developed in sustainability assessment has led to a large list of more specific goals such as recycling, reusing, repairing, remanufacturing, cleaner production, responsible care, supply chain management, and so on and so forth (Glavič and Lukman 2008). In turn, this has led to a closer scrutiny of production chains, initially 'from the cradle to the grave' and then 'from cradle to cradle'. This means adopting a supply chain-wide approach to sustainability assessment (Schöggl et al. 2016). Yet, most of the existing literature on these sustainability assessments is about industrial ecology rather than farm systems. Even after having been industrialized, agriculture is not like any other production sector that transforms raw material into final consumption goods.

First of all, agroecosystems are a part of living nature, which, despite being transformed by farmers' labour and knowledge in order to supply human food, continue to have their own ecological functions. Secondly, agroecosystems are the kind of human intervention on natural systems with by far the widest territorial scope on Earth. To give just one example, sealed urban-industrial areas occupy 3% of the European continent, while agricultural lands (34%) and forests (38%) make up 72% of the land covered by agro-forest mosaics (the other quarter are wetlands, water bodies, areas of bare rock and other land covers; see http://www.eea.europa.eu/data-and-maps/data/data-viewers/land-accounts). Furthermore, when they are kept at an appropriate intermediate level, farming and forestry ecological disturbances may keep and sometimes even somehow improve the ecosystem functioning of a landscape by increasing its capacity to host biodiversity. None of this can be said of any industrial site even if they act as eco-industrial parks, or of a metropolitan region despite their citizens and town councils being highly committed to sustainability. Hence, a sustainability assessment of the agroecosystems that lay at the very base of food chains is a very special case, where farmers' and consumers' socio-economic viability has to be achieved by keeping in a long-lasting, reproducible way their basic living funds, such as soil and plants, and domestic animals, leaving room for many non-domesticated species that provide key supporting, habitat and regulating ecosystem services (de Groot et al. 2010; Mitchell et al. 2013; González de Molina and Toledo 2014). It entails a new transdisciplinary approach to agriculture, able to go beyond the field and farm-gate scales, so as to assess at landscape level the role played by the heterogeneity of land-use patterns, and the ensuing ecological processes and species richness (Barrett 1992; Gliesmann 2000).

These distinctive features of agroecosystems—i.e. the continuous interaction with living nature, and their unparalleled spatial extent as biophysical structures of human society (Haberl et al. 2004)—entail that the issue of scale becomes particularly relevant. Economists talk about diseconomies of scale when a growing activity increases instead of reduces the cost per unit of output.[1] From a sustainability standpoint, the scale problem should not only be seen as a microeconomic question of optimality, but as a macroeconomic-ecological issue at a global extent as well: the impact of the whole economic biophysical move upon a finite biosphere (Daly 2005). An apparent example of economic activity outweighing these global limits is the level attained by anthropogenic greenhouse gas emissions, and the ongoing climate change. There is no doubt that food production, delivery, freezing, cooking and consumption must replace fossil fuels with renewable energy sources, reduce wastes by reusing and recycling materials, and become carbon-neutral, as in any other production-consumption chain of a sustainable circular economy (Haas et al. 2016). Indeed, agriculture is not only able to achieve this goal, but could even become the most important carbon sink for climate change mitigation—e.g. by enriching soils with more organic matter and biochar (Freibauer et al. 2004; Atkinson et al. 2010; Spokas 2010; Schmidt et al. 2011; Aguilera et al. 2013).

## 2.3 Why Making Agri-Food Systems Sustainable Requires Re-Localisation

The problem of sustainable scale becomes clearer when we go down the agri-food chain, from urban consumption towards a myriad of agroecosystems which remain at the very base of food regimes in rural areas. Food sovereignty movements claim that achieving sustainable agri-food systems necessarily means reallocating current food chains in a much narrower scope (Aubry and Kebir 2013). Why do they need to be re-localised? The answer lies in a combination of agroecological, environmental, social and political ecology reasons (Kneafsey 2010). We are going to group these reasons around five basic criteria (Table 2.1).

The first of these criteria lies in agroecology, and comes from the '*dis-ecologies*' of scale that our present globalized food chains entail for nutrient soil replenishment. In the same vein as economists talk about economies and diseconomies of scale, from a circular bioeconomic standpoint we can talk about *scale dis-ecologies* when the widening of the spatial scope of biophysical flows entail an increase in the

[1]Diseconomies of scale are a pervasive situation in agriculture, mainly due to the relevance of 'landesque capital' (Sen 1959) that is always site-specific and untransferable—i.e. long-lasting investments in the land such as stone terraces, hedgerows, deep furrows to prevent waterlogging, woody crop plantations such as vineyards or tree groves, organic matter replenishment and limestone amendment for acid soil through liming, to name but a few. This, in turn, also involves problems of labour control and discipline for capitalist investors (usually referred to as 'moral hazard' in Economics).

**Table 2.1** Five reasons why more sustainable and fairer agri-food chains have to be re-localised

| Criteria/reasons | Current fossil-fuelled industrial agri-food systems | More sustainable agri-food systems re-localized |
|---|---|---|
| Agroecological functioning of farm systems, and nutrient cycling | Linearity and globalization of biophysical flows through a massive use of external inputs prevent the closure of nutrient cycles (*dis-ecologies* of scale) | Enhancing internal loops in complex agroecosystems, and shorter food chains that allow for compost reuse, make closure of nutrient cycles feasible |
| Landscape ecology patterns and processes, and species richness | Monocultures simplify land cover patterns and reduce landscape complexity, damaging habitat differentiation and biodiversity | Self-reliant agroecosystems, and closure of internal loops, give rise to heterogeneous landscapes with greater farm-associated biodiversity |
| Energy efficiency of farming and food systems, and GHG emissions derived | Substituting internal biophysical loops with industrial external inputs has led to lower Energy Returns on Investment (EROI), greater fossil fuel consumption, and higher GHG emissions | Lessening dependence on external industrial inputs, by replacing them with internal renewable reuses, may increase EROIs, reduce GHG emissions and increase carbon sequestration in agroecosystems |
| Role of site-specific knowledge of farm communities | Peasant Knowledge (PK) no longer matters when techno-economic decisions are taken faraway, by big corporations in industrial urban centres | PK interacting with agroecology innovators can matter again when decision-making processes become more evenly distributed throughout re-localised agri-food chains |
| Democratization of market power along new eco-agri-food value chains | Dependence on external inputs, and PK contempt, diminishes farmers' control over value chains in the marketplace | Re-localisation empowers farm rural communities and their PK, giving a chance of economic democracy in fairer food chains |

*Source* Our own, from the references given in the text

environmental load per unit of product—always taking into account the different dimensions involved and the multi-scale character of this complex issue (Jungbluth and Demmeler 2005). A relevant example is when the material turnover moved by agri-food systems prevents soil nutrient replenishment through organic means (Billen et al. 2012; see Chap. 11 in this book by Simon-Rojo and Duží). Hence, the first and foremost reason for re-localizing a sustainable agri-food system is to keep soil fertility through an adequate flow of organic matter that replenishes the nutrients extracted, ensures an adequate agglomerate structure, enables a good water retention capacity, and helps to maintain the soil free of chemical pollutants and able to reproduce all the life it contains—a complex issue that involves an entire research field in itself (Karlen et al. 1997; Magdoff and Weil 2004; Engel-Di Mauro 2014).

Soil nutrient replenishment can only be achieved in self-reliant agroecosystems whose main biomass flows are closed at local scale. As said in Chap. 5 of this book, once livestock bioconversion made in industrial feedlots has been disconnected from the local farmland production, the energy and economic cost of redirecting the dung to faraway suppliers of imported feed becomes unfeasible (Marco et al. 2017). Furthermore, sustaining soil fertility goes beyond only keeping some feasible ways to return N-P-K from the grave of food consumption to the cradle of plant nutrition. Closing nutrient cycles through organic matter moved at local scale, and keeping landscape biological richness, are two agroecological dimensions tightly linked to one another, either belowground or aboveground. On the one hand, self-reliant agroecosystems mainly oriented to meet the intake needs of local population have to produce a diverse food basket, which translates into land cover agro-diversity; on the other, the closure of internal loops through organic procedures requires a multiple-integrated management of land and livestock at farm and farming community level that give rise to complex landscape mosaics (Krausmann 2004; Tello et al. 2012; Gingrich et al. 2015; see Chaps. 6 by Neundlinger et al., 5 by Padró et al., and 7 by Fraňková and Cattaneo in this book). This leads us to the second criterion of why more sustainable agri-food chains need to be locally and bio-regionally re-scaled: landscape ecology. A more heterogeneous farming land cover leads to a greater habitat differentiation able to host biodiversity (Altieri 1999; Gliessman 2000; Marull et al. 2016a).

The underlying mechanism is a basic principle of the functioning of all living systems on Earth, capable of maintaining a dynamic stability far from thermodynamic equilibrium. Thanks to these internal cycles, they can increase their own complexity and energy storage capacity, improve energy throughput and decrease entropy dissipation, opening ascendancy pathways to develop (Ulanowicz 1986; Ho and Ulanowicz 2005; Morowitz and Smith 2007; Guzmán-Casado and González de Molina 2017). In the same vein, sustainable agroecosystems have to close a set of vital biophysical loops at landscape level in order to keep in a reproducible and ecologically healthy state their basic funds, such as soil fertility, agro-diversity and biodiversity. In turn, to allow that the by-products of a loop can become resources for another (Ho 2013), this integrated and multi-purpose management of the agroecosystem funds gives way to a spatial heterogeneity of land covers. Therefore, the agricultural landscape mosaics, well-endowed with biodiversity, become a spatial expression of a complex land-use management aimed at the internal closure of biomass flows that reproduce agroecological funds (Guzmán and González de Molina 2015; Marull et al. 2016b). Conversely, when industrialized farm systems start replacing internal reuses and biomass flows that are kept unharvested with external inputs, the ensuing linearity of energy flows lessens agroecosystem complexity, reduces the diversity of cultural landscapes, and jeopardises biodiversity. Here, we are again facing a problem of *dis-ecologies* of scale—as shown in several studies and meta-analysis that demonstrate how self-reliant organic farming, that generates complex agroecological landscapes, contributes positively to the organic matter content of soils, agro-diversity and biodiversity (Mondelaers et al. 2009; Winqvist et al. 2011).

Considering the importance of internal flows in agroecosystems leads us, in turn, to the third biophysical criterion of its scalar limits: energy efficiency, and derived GHG emissions. It is well-known that throughout the industrialization of farm systems following the Green Revolution, the marketable output obtained per unit of labour and land invested has grown, but only at the cost of sharply reducing the Energy Return On Investment (EROI) —even though recent improvements in the energy efficiency of producing and operating fertilizers and machinery have somewhat counteracted this poor energy performance (Schroll 1994; Pelletier et al. 2011; Pracha and Volk 2011; Hamilton et al. 2013; Aguilera et al. 2015). The EROI decrease of industrial farming has been mainly due, as expected, to the big increase in the consumption of External Inputs (EI) coming directly from fossil fuels, or indirectly through feed imports for livestock fattening in industrial feedlots (Marco et al. 2017). This also explains why organic farming can again raise energy yields of farming by a self-reliant closure of internal biomass flows, once the critical transition period has been overcome (Dalgaard et al. 2001; Tripp 2008). However, most of the energy analyses of farm systems only focus on the external part of the story, neglecting the internal flows of Biomass Reused (BR), and the Unharvested Biomass (UhB) that is left for all non-domesticated species, vegetal and animal. Under this one-dimensional input-output approach, which only takes into account what enters into and exits from the system boundaries of a farming production unit, or the whole agricultural sector, the agroecosystem functioning is concealed inside a black box (Tello et al. 2016).

To overcome this shortfall, a wider approach is required, like the one developed by the international Sustainable Farm System (SFS) research project (Guzmán and González de Molina 2015, 2017; Tello et al. 2016; Galán et al. 2016). After applying this SFS energy accounting to a large dataset of case studies in Europe and America along the socio-ecological transition from past organic farm systems in the 19th century to current industrial ones, the results confirm the expected long-term trend of decreasing returns to EI (*External Final EROI*, i.e. $EFEROI = \frac{Final\,Product}{External\,Inputs}$) due to the huge amount of industrial inputs consumed that led to a high external dependence of farm functioning. Despite some differences derived from diversity of natural resource endowments, combined with differences in regional specializations, these results also reveal to some extent a trend towards lower reinvestments of BR per unit of output extracted from agroecosystems (thus increasing Internal Final EROI, i.e. $IFEROI = \frac{Final\,Product}{Biomass\,Reuse}$). These trends mean a growing land-use exploitation while at the same time disregarding the effort needed to keep the internal renewal of the underlying funds, giving rise to an environmental degradation of agroecosystem functioning and widespread pollution.

This holds true even when the amount of BR has scarcely been reduced or has even increased, leading to more or less stable or decreasing IFEROI values in industrialized agriculture (Gingrich et al. 2017 forthcoming). The main reason for this is the dietary shift towards greater meat consumption, and the ensuing spread of livestock fattening in industrial feedlots (Soto et al. 2016). While only some 10% of the grain harvested worldwide was diverted towards animal feed in 1900, it rose to

20% in 1950, and attained 45% of the total grain cropped at the end of the 20th century (Smil 2000; Fischer-Kowalski and Haberl 2007). When animal feed monocultures are coupled with a linear feed-meat bioconversion that becomes very energy inefficient and polluting, this type of BR can no longer have the capacity to keep the agroecosystems' complexity that animal husbandry provides in a tightly integrated mixed organic farming (Krausmann 2004; see also Chap. 5 by Padró et al. in this book). Something similar has occurred with the increased flows of UhB, when they only take place in steep lands, spontaneously reforested following their abandonment after the mechanization of agriculture. In this case, the ensuing land cover polarization between flat land intensively cropped with industrial monocultures, and steep lands abandoned, prevents these growing fractions of the actual Net Primary Production (NPP), left free for non-domesticated species, from resulting in a growing biodiversity, due to the lack of habitat differentiation in overly homogenous landscapes (Tello et al. 2014; Marull et al. 2014; Otero et al. 2015). In order to go deeper into these different complexity degrees in the relationship between energy flows and land uses, we have to go beyond energy balances, applying an Energy-Landscape Integrated Analysis (Marull et al. 2016a, b, and Chap. 4 in this book by Marull et al.).

These SFS results bring to light that, following industrialization, agricultural systems have shifted towards substituting former internal reinvestments of Biomass Reuses (BR) and Unharvested Biomass (UhB) by a much greater amount of External Inputs. The trend towards decreasing $\frac{BR+UhB}{EI}$ ratios largely explains why the outcome has been a very low energy performance, ensuing industrial farming. The strategy of merely replacing internal renewable loops by external industrial inputs coming from fossil fuels has ended up being a trap for farm energy efficiency (Tello et al. 2016). An agroecological fund-flow analysis helps to better explain the reason behind this outcome (Georgescu-Roegen 1971). Lower amounts of Biomass Reused and Unharvested Biomass per unit of Final Product extracted from the land also involves lessening a much needed investment to keep in a good ecological state the underlying funds of the agroecosystems: soil fertility, complex landscape mosaics, and biodiversity. The ensuing reduction of Final EROI makes apparent how damaging for energy performance the attempt to substitute with industrial EI internal local ecosystem services has been (Giampietro 1997). This substitution, massively carried out during the Green Revolution, has proven to be a serious mistake with deleterious environmental effects. An alternative agroecological way to attain true improvements of farming energy efficiency is to invest in the betterment of agroecosystem funds, and the derived ecosystem services, in order to jointly raise IFEROI and EFEROI in more integrated multi-purpose farming. It follows that closer eco-agri-food chains, with a more complex and integrative agroecosystem management, aimed at lowering the dependence on external inputs and taking advantage of the multi-use of diversified practices, can improve the energy efficiency of farm units and farming communities, while reducing the energy embodied in food consumption baskets and the derived greenhouse gas emissions (Guzmán and Alonso 2008; Snyder et al. 2009; Alonso and Guzmán 2010;

Infante-Amate et al. 2013; Aguilera et al. 2015; Ponisio et al. 2015; Theurl 2016; Jordan 2016).

Therefore *dis-ecologies* of scale matter when it comes to improving energy efficiency and reducing the impact of agri-food chains on climate change. If the sustainable enhancement of energy returns of farm systems has to rely on keeping a multiple-integrated renewal of agroecosystem funds, it follows that there exists a close synergy between this aim and the other two reasons given above to explain why the scale of a more sustainable agroecosystem functioning has to be re-localised—i.e. closing nutrients cycling and keeping complex landscape mosaics well-endowed with biodiversity. Even more, if the only actual way to attain higher energy yields of agroecosystems is by means of discovering wise combinations of EI, BR and UhB, this undoubtedly brings to the fore the key role performed by the information that spatially shapes the biophysical flows and loops of agroecosystems. Without information, any energy conversion only increases entropy, not life (Ulanowicz 2001; Morowitz 2002; Ulanowicz et al. 2009). The question is of what information we are talking about, of the place where it is located along the agri-food chain, and of who rules it.

This leads us directly towards the fourth reason for keeping a local restrained scale in farm systems operation: the irreplaceable importance of the site-specific Peasant Knowledge (PK) developed by rural communities through history, and the role it may play again in a more sustainable and innovative food regime (Berkes et al. 2000; Barrera-Bassols and Toledo 2005; Toledo and Barrera-Bassols 2008; Reyes-García et al. 2007; Gómez-Baggethun et al. 2013). No one can deny that theoretical scientific knowledge, and derived technological applications, have largely improved labour performance and living standards, and will continue to do so. But PK is another sort of site-specific know-how, gained and developed through experience working with Nature. It is the knowledge of places, of what thrives better here or there, in order to adapt farming practices to different types of soils, climates and environments—e.g. multiple species management, crop rotations and associations, succession management, landscape patchiness maintenance, livestock breeding integration with cropland tillage and forestry, among others. It is because of this agroecological suitability, and not for being 'traditional', that this peasant knowledge has presently become a valuable resource for more sustainable agri-food systems (IAASTD 2009; Koohafkan and Altieri 2010; Altieri et al. 2012).

The technocratic endeavour to void and put an end to this PK, which ruled farming over millennia, has been another major mistake of the Green Revolution. After WWII, the new agro-industrial technical package was systematically deployed by state-led demonstration farms, agronomists, engineers, experimental stations, some US Foundations, and big agribusiness salesmen in a way that largely dismissed all sorts of PK. It was scorned as a set of unscientific beliefs devoid of practicality, a true epitome of backwardness. Indeed, PK was not backward at all, but simply too awkward for the generation of technicians, businessmen and politicians who launched the Green Revolution. It is no surprise that, along with the increase of marketed produce, the outcomes have been a homogenization of farmland, a decrease in fund-flow complexity and energy efficiency of agroecosystems, a pervasive pollution, and a loss of farm-associated biodiversity—as shown by many widespread evidences, like the collapse of Europe's farmland bird populations in the last thirty years (Donald et al. 2001; Inger et al. 2015).

It is not that the whole agri-food chain was losing information as such, but that a deep shift in the epistemic character and location of knowledge took place. While cultural landscapes have actually lost a great deal of spatial information imprinted on them by peasants' labour and knowledge (Marull et al. 2016a, b, and Chap. 4 in this book by Marull et al.), new sorts of technological and business information has been concentrated in the urban headquarters of big corporations, distribution operators, financial future markets, and university labs. As a result of the socio-metabolic production-consumption disconnect of biomass flows that are moved along globalized agri-food chains (Erb et al. 2009b), knowledge has been alienated from its wide spatial base and from the people who actually work with Nature, while it has been appropriated upstream in increasingly fewer hands. By alienation, we refer to the growing inequality in the share of all the information which runs a system that every participant have at their own disposal—as might be assessed by Theil or Gini indices, provided we could measure this. It is similar to what in Economics is called information asymmetry, except that economists tend to limit its scope to the kind of information they consider directly affecting the outcome of a market deal (Greenwald and Stiglitz 1986). Another aspect is the diversity of places along agri-food chains where each of these different types of information is important. Local and regional information, encoded by PK at peasant community level, no longer matter when the main decisions are taken at higher hierarchical levels, and when individual farms are left devoid of instruments for collective memory and action (Krausmann 2004). Conversely, it might again matter if decision-making processes become more evenly distributed throughout re-localised agri-food chains.

This biocultural heritage includes a long-lasting process of seed and breed selection that has given rise to precious richness of germplasm as a common-pool resource of Humankind (Sand 2004; Timmermann and Robaey 2016), which is currently being recovered and reused by innovative organic farming. Even where these PKs have been largely eroded by long-lasting industrial farming, anthropologists and historians can do a great job recovering them from oral, written, biological and archaeological sources (Agnoletti and Emanueli 2016). Its double character, biological and cultural, is also apparent in many farming landscape components such as stone terraced soils, small-scale water catchments and irrigation or drainage facilities, wooded pastures, tree orchards, vineyards and vegetable gardens (Agnoletti and Rotherham 2015). A common feature of all of them is the careful adaptation to particular places. The site-specific character of the PK embodied in cultural landscapes and foodscapes is something that scientists and technologists will never be able to replace. By its own theoretical nature, scientific and technological knowledge (STK) is inherently devoid of this concreteness of place.

Agroecological innovation can take this PK as a body of information on how to combine all these site-specific capacities, and limitations of land, plants and livestock endowments, in an integrated and multi-purpose organic management able to take advantage of as many synergies as possible (Edwards et al. 1993; Davidson and Howarth 2007; Ponisio et al. 2015). No doubt, a scientifically-based agroecological innovation has to critically examine, select and adapt this biocultural heritage to current contexts and new challenges through an open dialogue drawing on different types of knowledge and expertise (Toledo et al. 2002, 2003; Toledo and Barrera-Bassols 2008). Accordingly, STK can build upon it and go beyond, but should never try to substitute PK—this would be like admitting that the attempt to substitute the living functioning of agroecological funds by external industrial inputs has proven to be a trap. Surely, going back in time is impossible, and discussing whether such an attempt might be worth making or not is a waste of time. Instead of trying to reproduce traditional PK as such, or to just keep it like in a living museum, agroecological innovation has to learn from what worked in the past in order to develop new sustainable solutions for our completely new current situations.

Despite the dismissal it has suffered in recent generations, PK needs to be put to work again through a wider participatory dialogue between different branches and types of knowledge, as claimed by the proponents of a post-normal science focused on sustainability options (Ravetz and Funtowicz 2015). Indeed, PK has never been a fixed set of ideas and prescriptions, routinely transmitted intact over generations, but an ever-changing adaptive socio-cultural response to shifting socio-ecological conditions (Gómez-Baggethun and Reyes-García 2013; Gómez-Baggethun et al. 2013; Tello et al. 2017). Hence, PK can change again and needs to do so by regaining its necessary place in more sustainable food systems that become bio-physically and culturally re-localised. In short, new PK also means new peasantries and peasant discoveries (van der Ploeg 2009; Ravera et al. 2014). Sustainability science can help this endeavour, by pointing out how public policies, private investments and consumption patterns can be reshaped in order to help PK to better respond to the new societal and environmental challenges, in a new governance that help peasants and rural communities to empower themselves (McMichael 2011; González de Molina 2013).

No doubt, the harsh contempt of their useful knowledge has entailed a significant disempowerment for peasant farms and rural communities, either in the political arena or the marketplace, in favour of big agribusiness and state political rulers placed in large global cities (Martínez Alier 2002). It is not surprising that the greater land yields and labour productivity brought about by the Green Revolution has gone hand in hand with its major socio-economic curse: a steady downturn in the net added value retained by farmers, after selling their crops and buying their external inputs. For the same reason, and with very few exceptions, large agribusinesses have withdrawn from food production to concentrate on the higher levels of the agri-food chain, where there are better opportunities for profit such as the production and sale of agrochemical inputs and patented germplasm, distribution logistics, and large supermarkets (Shanin 1987; Friedmann 1987; IFAD 2016). This leads us to the fifth and last reason why more sustainable agri-food chains need to be re-localised: peasant empowerment and economic democratization of rural-urban exchanges are necessary dimensions of fairer and more sustainable food regimes (Thompson and Scoones 2009; Holt-Giménez 2011; McMichael 2013; McKeon 2015).

What from a biophysical standpoint is seen as a problem of *dis-socio-ecologies* of scale reappears again here, as a socio-cultural and economic problem of *dis-empowerment* that needs to be addressed from a political ecology point of view. The current food regime has to be changed from a placeless 'Food from Nowhere' to a 'Food from Somewhere', as is being discussed among scholars in Sociology, Anthropology and Political Ecology (McMichael 2003, 2009, 2011, 2013; Campbell 2009). Fairer prices for food producers, and better farm incomes, are unavoidable conditions for this change towards more sustainable agri-food systems. These can be better obtained within closer socio-economic circuits (Douthwaite 2003; Mooney 2004). Interestingly, Nicholas Georgescu-Roegen defined a peasant community as a set of social relationships in which everyone is someone through mutual recognition (Georgescu-Roegen 1976). It is well-known that iterated relationships, when kept face-to-face, give rise to cooperation among agents, through a tit-for-tat strategic behaviour that appears either in natural or social contexts (Axelrod 1997, 2006; Riolo et al. 2001; Boyd et al. 2003; Bowles and Gintis 2011). It follows that participatory economic democracy also faces problems of scale, as shown by the evidence that cooperation is totally prevented in global economic circuits managed by large corporations, given the unavoidable alienation that is set up among the producers and consumers that remain isolated at both ends of agri-food chains. A more democratic economic functioning and fairer valuation of food produce are only feasible if biophysical flows moving in agroecosystems, cultural information governing farm labour, and economic exchanges throughout agri-food chains become re-scaled into a closer social fabric (Berkes et al. 2003). In brief, re-localisation is needed to give economic democracy a chance (Commoner 1990; Dahl 1985; Shiva 2005; Johanisova and Wolf 2012; Cattaneo et al. 2012).

The five interlinked reasons listed above provide general criteria about why the current agri-food system needs to be reshaped and re-localised. But if so, in which particular scope, or in which specific set of nested scales, should food chains be re-localised? What would the adoption of these local or bio-regional scales mean for many other dimensions involved in the production and consumption of food? To what extent is this endeavour compatible with food trade, and with which sort and scope of trade? What about dietary patterns, draught power, labour productivity, forestry use, food prices, retailing, public policies, and so on? What kind of society will have such a sustainable agri-food system? Exploring all these questions sets a wide research agenda that involves a complex multidimensional, multi-criterial and multi-scalar assessment, performed using a transdisciplinary 'nexus approach' (see Chap. 3 by Gomiero in this book). A starting hypothesis to begin this 'nexus' research and development, which is still to be done, is that landscape scale becomes a crucial space where it can be better devised and implemented. Proponents of this 'landscape focus' point out that the multi-criterial concepts, tools, indicators and decision-making processes required to confront competing land uses (such as the trade-offs between production and conservation goals), and to devise alternative resource allocations, can be better found, carried out, negotiated, and tested by trial and error at this scale by different stakeholders sharing common sustainability aims (Sayer et al. 2013).

Beyond the general criteria explained so far, we will have no more concrete and reliable answers unless a large transdisciplinary research agenda is undertaken. When it comes to deliberating, approving and enforcing public policies, in the same vein as socio-ecological sustainability problems are simultaneously posed at several dimensions and nested scales, political measures aimed at solving them also have to be addressed at multi-levels by adopting an integrative approach based on federal subsidiarity, combined with the strong participative democracy of empowered local communities (Graham 2008; González de Molina 2015). Income distribution is a particularly relevant example of the multi-scalar character of this complex set of issues. Fairer prices and incomes for sustainable farmers require a consumers' willingness to pay for better food, produced in more sustainable landscapes. Due to the endo-somatic character of food metabolism, people and families can be easily aware of the importance food quality has for their own health. Environmental education can help to enlarge this embodied cognition towards a wider awareness of the importance that the quality of food production also has for the agroecosystem's health. In turn, food consumer awareness can help to establish a fairer and more sustainable rural-urban deal set in a nearer scope, where face-to-face relationships can be endowed with trust. People have to see with their own eyes, either as consumers, citizens or taxpayers that 'your landscape mirrors what you eat'—as said in Chap. 5 of this book by Padró et al. And only in these shortened supply chains can the prices of organic food become affordable enough for consumers, while providing farm producers with fair incomes and a decent livelihood (Renting et al. 2003; Douthwaite 2003).

Furthermore, all this is necessary but not enough. Beyond being willing to pay fair prices for healthy food that sustains environmental quality and ecosystem services, people have to have a sufficient income to do so. While the share of family income devoted to buying food strongly decreased in the Global North, total income has been distributed in an ever more unequal manner within nations over the last thirty years (Atkinson et al. 2011; Stiglitz 2012, 2015; Piketty and Saez 2013). At the same time, the purchase capacity between countries has converged somewhat (Milanovic 2011), allowing large populations in fast growing economies of the Global South to start going up the food ladder towards increasingly globalized, unhealthy diets. All these trends are making any socio-ecological transition towards more sustainable agri-food chains increasingly difficult (Smil 2003). Healthier and more sustainable agroecosystems cannot be widely spread in rural areas without fairer income distribution at a societal and global scale. Re-localising biophysical flows, and setting fairer deals in the political economy of food production-consumption chains, also means changing other important economic and political dimensions at higher hierarchic levels.

## 2.4 The Multi-scale Character of Food Chains, and the Issue of Trade

Before going into methodological details, we ought to pay attention to the issue of trade. The five reasons explained in the previous section make apparent the unsustainability of our current globalised agri-food regime (van der Ploeg 2010; McMichael 2013). This regime relies on non-renewable polluting fossil fuels (Alonso and Guzmán 2010; Infante-Amate and González de Molina 2013; Sage 2013; Perignon et al. 2016), degrades people's and agroecosystems' health, and leaves many food producers in poverty while profits are concentrated in a handful of agribusinesses. Still, one thing is a global trade ruled by these large corporations and financial markets, a very recent phenomenon in historical terms, and another is trade as such that has coexisted with farming and food consumption chains over many centuries.

From the core idea, explained in the previous section, on the need to close a basic amount of biophysical flows at landscape level in order to reproduce the agroecosystem funds, we can raise a set of questions which help to address the issue of trade from a socio-metabolic standpoint. After having closed these vital internal loops, required for a sustainable reproduction of farm systems, and once the food needs of local population are fairly met, is there any surplus available to be exported? If so, is there a demand for it? And where is it located? Can the material and energy expenditure required for transporting and delivering that surplus to those faraway consumers, and the corresponding greenhouse gas emissions, be kept at a level low enough to be globally sustainable?

Putting trade under the perspective of agroecosystems' renewal immediately raises the notion that it may only be sustainable up to a certain scope, which also helps to explain why its socio-ecological effects are twofold. Trading among bio-regions endowed with different but complementary climate and soil capabilities may offer an opportunity to take advantage of their possible complementarities; however, since natural resource endowments and bio-regional productivities are not the only factor influencing specialisation patterns, and since relative—not absolute—advantages used to be decisive in market decisions, socio-ecologically 'perverse' allocative effects are also a possible outcome of trade (van den Bergh and Verbruggen 1999; Andersson and Lindroth 2001). Throughout history, market-driven intensification and specialization of farming has sometimes allowed the complementation of bioregional climate and soil diversities—see, e.g. Chap. 5 by Padró et al. in this book. However, transforming staple food into a commodity that is going to be massively traded far away may also give way to unequal ecological exchanges in terms of biomass and sink-capacities involved (Muradian and Martinez-Alier 2001; Muradian et al. 2002)—an environmental load displacement through trade also largely found through history (Hornborg et al. 2007).

There are several dimensions at stake that need to be taken into account. To begin with, trade may foster exporting monocultures and lead to a large increase in the energy expenditure of transport (Fischer-Kowalski et al. 2004). This would raise the overall energy expenditure of agri-food chains, and lead to lesser energy efficiency and greater GHG emissions per unit of final product consumed through agri-food systems, unless true economies and ecologies of scale were actually able to counter this trend (Schlich and Fleissner 2003, 2005a, b; Jungbluth and Demmeler 2005). For instance, energy consumption per tonne-kilometre carried out by vessels along maritime routes (0.17 MJ) is of one order of magnitude lower than by terrestrial roads (1.78 MJ), and two orders of magnitude lower than by airplane (16.33 MJ) (Aguilera et al. 2014: 39). This means that for some specific cases, and to some extent, a fair long-distance maritime trade may be more sustainable than that supplied from some closer terrestrial locations—a possibility that opens a way to thinking about hybrid agri-food chains that combine many short-distance suppliers with a few ones coming from greater distances.

However, the expansion of cash-crops through monocultures would also tend to disconnect livestock breeding from an integrated management of cropland, pastureland and woodland. If so, the loss of a multiple-use of agroecological funds would make it harder to obtain enough quantities of manure and other organic fertilizers to keep soil fertility. Soil degradation would go hand in hand with the vanishing of former agro-forestry mosaics needed to maintain biodiversity, thus undermining the biotic regulation of agroecosystems (Guzmán and González de Molina 2009; Marull et al. 2010; Guzmán et al. 2011). As a result, trade may become a pathway for environmental load displacement. Instead of a fatal destiny, all these socio-ecological imbalances, ensuing agricultural specialization in cash-crops, might also be counteracted, at least to some extent, by the agroecological ingenuity of farming PK. Provided that peasant family units have fair and sufficient access to natural resources, they have many times proven to act as true shock-absorbers against a wide set of socio-ecological pressures, and repairers of many agroecological imbalances of the so-called 'socio-metabolic rift' (Schneider and McMichael 2010; González de Molina and Toledo 2014). This also partly explains why the issue of trade sustainability is twofold. The political strength of colonial and imperial dominions, which subsumed the 'peasant natural economy' below them, and were able to set favourable price differentials through their market power, also help to understand the other side of the coin (Luxemburg 1964; Mies 1986; Mies and Shiva 1993; van der Ploeg 2010; Tadei 2014).

A starting point for a theoretical foundation of this ambivalent character of the socio-ecological impacts of trade was pointed out as early as 1902 by the Austrian physicist and chemist Leopold Pfaundler (1839–1920), in his attempt to assess Earth's maximum capacity to sustain human needs (Martínez Alier 1987). He argued that any reasonable estimate of this carrying capacity would depend on whether we were to aggregate the local capacities of each small territory taken one by one, or were allowed to consider the Earth as a single global territory. In the former case, the accountancy would have to consider a great deal of limiting factors, which would vary locally, without being allowed to supplement or complement any of these by trade. This would definitely lead to a very low global carrying capacity. But the opposite extreme would entail assuming that any local resource would be globally available from any point on Earth to another through trade, without taking into consideration any ecological restriction, or energy cost. This latter assumption would undoubtedly lead to an incredibly high global carrying capacity.

Pfaundler suggested that both extremes were implausible, and any reasonable answer should lie somewhere in between, once we admit that the possibility of trade exists, but that it also entails strong ecological impacts. He pointed out that trade involves energy-consuming transport and produces environmental disturbances, which become particularly acute in terrestrial ecosystems, and this sets a limit to a sustainable biophysical move. Pfaundler's reasoning entails that the key issue for a sustainability assessment of trade is, once more, scale. Up to a limited spatial scope, an excessive degree of commercial specialization and the ensuing globalization of biophysical flows would entail that ecological impacts outweigh any socio-ecological advantage of trade. Pfaundler's biophysical approach to trade can be related with the well-known socio-economic distinction set forth by Karl Polanyi between markets as places where local goods and useful information are interchanged among relatively equal members of a community, and the alienated market economy that arises when basic sustainability funds like land and labour become commodified (Polanyi 2001).

Agroecology and Landscape Ecology can go beyond these very general starting points, by providing more precise criteria and useful indicators to set a distinction between low-scale, socially embedded, sustainable-oriented ways of agri-food fair trade vs. a financially-driven, ecologically unequal and unsustainable global trade. As expected of a man who studied with Justus von Liebig at the University of Munich, before succeeding Ludwig Boltzmann as professor of physics at the University of Graz, Leopold Pfaundler built his argument on Liebig's approach to limiting factors (Taylor 1934). This single-minded factorial approach to Liebig's Law of Minimum has been long criticized in Agroecology (Vandermeer 2000), and has also been overcome in Ecology by adopting a more synergic and dynamic approach to nutrient availability (Davidson and Howarth 2007). A long-term field experiment in natural grasslands on the Great Plains of the United States demonstrated that plots with larger species richness outperformed the NPP attained in the best monocultures even when chemical Nitrogen (N) was added to correct for this limiting factor. This happened thanks to niche complementarity in nutrient uptake by different plant species, and their joint capacity to reduce N leaching (Tilman et al. 1996, 2001). According to this and other results, biodiversity appears to be the best adaptive way of natural systems to respond to the synergic and dynamic set of capabilities and limitations laid down by resource availability in each environment—an approach that has been generalized to include all sorts of terrestrial, freshwater and marine ecosystems (Elser et al. 2007). Hence, a biomimetic agroecological strategy is to emulate this synergistic adaptive response to natural resource endowment through biodiversity, by reinforcing agro-diversity and the complexity of cultural landscapes. We came, then, to our starting point: the internal closure of basic biophysical loops of agroecosystems is the key. An agroecological synergistic approach to natural resource endowment provides a clearer criterion to go beyond Pfaundler's preliminary exploration of the issue of trade from a bioeconomic standpoint.

We came to the conclusion that autarky is neither a necessary condition for eco-agri-food sustainability, nor a guarantee of attaining it. Indeed, while a self-preserving autarky has sometimes been a way to achieve a lasting food provision for a human population able to keep its limited resource base in a reproducible state, a self-consuming autarky can also erode this resource base and lead towards ecological collapse. Yet, adopting an 'island logic' as a conceptual starting point can be of help in devising feasible socio-ecological transitions towards sustainable agri-food chains in a bottom-up way—i.e. as a large number of self-preserving re-localised food systems that can then explore which sustainable trade possibilities are within their reach, instead of trying to balance the resource flows of a globally-interconnected agri-food system in a top-down manner (Busch and Sakhel 2016). Although we do not know beforehand what the maximum scope of such a sustainable level of trade will be, by adopting a bottom-up agroecological criterion we will focus our assessment on the point where crop specialization may start threatening the closure of biophysical flows needed to keep high levels of soil fertility, landscape complexity, and the biodiversity endowment of agroecosystems in order to ensure the provision of all sorts of ecosystem services over time.

Following this general rule, we can better understand why the relationship between trade and a sustainable agroecological functioning cannot be viewed in black and white terms. Depending on the type and extent of the markets being considered, a fair and sustainable level of trade may help basically self-reliant farm systems to take advantage of different bioregional resource endowments, provided that the renewable capacity of agroecosystem funds is kept in a healthy state; or, conversely, trade can foster cash-crop monocultures sold in unfair fully-globalized markets at the expense of an acute local ecological degradation in the exporting territory. Many current approaches to the actual relationship between human development, agroecosystem functioning and fair trade in many countries of the world have drawn similar conclusions (Shiva 2000; Shiva and Gitanjali 2002). While a network of local and regional markets may be an important tool to empower local sustainable rural communities, a full dependency on globalized markets ruled by big corporations may often become a trap that leads to the environmental load displacement of an ecological unequal exchange (Muradian and Martinez-Alier 2001; Muradian et al. 2002; Hornborg et al. 2007), and sets a barrier for endogenous processes of sustainable human development (Chang 2002; Reinert 2004; Hornborg et al. 2007). In short, what we need is a paradigm shift from a market-driven to a sustainability-driven trade.

## 2.5 Strengths and Weaknesses of Existing Tools in Analysing Agri-Food Systems

Understanding the multi-scalar, multidimensional, and multi-criterial set of agroecological and societal relationships at stake in food regimes poses a big methodological problem for sustainability research. We are going to argue that the search for adequate criteria and indicators to devise new sorts of sustainable agri-food systems, eco-landscape efficiently adapted to very different bio-regions of the Earth, has to work with several existing methodologies and combine them in a consistent manner within a solid theoretical background: Life-Cycling Analysis (LCA), Material and Energy Flow Accounting of Social Metabolism (MEFA), and Multi-Scale Integrated Analysis of Societal and Ecosystem Metabolism (MuSIASEM), combined with Agroecology, Landscape Ecology, Political Ecology, and the valuation of biocultural heritages of site-specific PK.

Life-Cycle Assessment aims at tracing back 'from the cradle to the grave' the environmental load of consumer food baskets, or of some of its components, up to the production process and inputs used (Dutilh and Kramer 2000; Erb et al. 2009a; Infante-Amate and González de Molina 2013). This involves compiling the energy and materials used, and the ecological impacts following the whole chain of producing, elaborating, packaging, freezing, transporting, delivering, retailing and cooking, along with the disposal of all wastes and residues generated along production-consumption chains (Jones 2001; Roy et al. 2009). LCAs have revealed many hidden sides to agri-food systems, and helped to raise public concern, to reinforce consumer awareness, and to develop sustainable public policies. A salient example is the highlighting of the importance of reducing meat consumption and changing dietary patterns in any alternative scenario modelled using the results obtained with LCA of agri-food chains (Scarborough et al. 2012; see Chap. 9 by Petridis and Huber in this book). However, despite having been technically standardized by ISO14040 and other references that can be used for entrepreneurial purposes (ISO 1997; Audsley 1997), the main unsolved problem is the lack of a clear methodology to deal with truncation and allocation decisions with a full consensus among scientific researchers (Reap et al. 2008a, b; Theurl 2016; Theurl and Schaffartzik 2016; Haas et al. 2016).

Truncation and allocation problems arise from decisions about what is included and excluded in LCA when setting system boundaries, defining functional units, dealing with the multi-purpose use of resources and technical farm implements, or using different indicators and units of measurement (e.g. energy content, dry weight, nitrogen content, monetary values, greenhouse gas emissions, etc.). The scale of analysis varies a lot, which greatly affects results and interpretation. At nation-wide scale, LCA has to deal with several double-counting problems when data is aggregated along agri-food chains, whereas at local scale there is a lack of statistical records that obliges either the extrapolation of available averages or generation of data through survey and sampling. Exclusion of economic and social impacts limits the comprehensiveness of this tool and the choice of alternative scenarios. Different researchers use different criteria, and adopt many ad hoc decisions. As a result, interpretations appear to be too conditioned by the different criteria adopted when setting the four steps of LCA: impacts selected, indicators adopted, modelling performed with these indicators, and assigning results to impact categories. All of these methodological problems hamper comparability, and offer opportunities for spurious LCAs biased in favour of partial or commercial interests. Further improvement requires standardized LCA inventory and impact databases. In turn, indicator selection requires agreement on underlying models, which have to rely on some coherent theoretical fundamentals (Reap et al. 2008a, b). In short, LCA still lacks theoretical and methodological consistency. This is why we stress that environmental assessment of agri-food chains needs to be based on the fundamentals provided by agroecology and landscape ecology, and by adopting a socio-metabolic approach.

A salient feature of the Material and Energy Flow Analysis (MEFA) is to have established a clearer accounting method focused in societal boundaries within common spatiotemporal frames, overcoming to some extent the main flaws of LCA (Hinterberger et al. 2003; Krausmann et al. 2009). Either in its Economy-Wide macroeconomic use (EW-MEFA), or applied to a sectoral accounting such as agriculture (Haberl and Weisz 2007; Fischer-Kowalski et al. 2011a; Infante-Amate et al. 2015; Soto et al. 2016), this material and energy bookkeeping has been devised from the onset to attain a rigour at least comparable to GDP national accounts, and it is currently being incorporated into the systems of national accounts by Eurostat and the United Nations (EUROSTAT 2001; UNEP 2016). It is also widely applied in many community-based studies performed at local scale (Singh et al. 2010; Fischer-Kowalski et al. 2011b). This does not mean that MEFA has no unsolved difficulties, particularly when it comes to link network analysis performed in Ecology with MEFA accounting in Ecological Economics (Suh 2005). The most important pending task is to develop a MEFA specifically adapted to analyse agroecosystem functioning (González de Molina and Toledo 2014) and food systems. Currently, MEFA allows the performance of cross-section comparisons and provides time series that bring to light the main trends of societal metabolism (Haberl et al. 2016). These biophysical and energy datasets can be linked with the monetary value added flows of GDP to calculate intensity variations over time, or between countries and regions (OECD 2008; see Eurostat: http://ec.europa.eu/eurostat/web/environment/methodology). Input-output tables can be used to disaggregate EW-MEFA flows among economic sectors, and relink them with sectoral value added flows, labour data, GHG emissions, pollutants emitted, land use, water consumed and foreign trade flows—see, e.g. the WIOD database (http://www.wiod.org/home) and the EXIOBASE Multi-Regional Environmentally Extended Input-Output Tables with Externalities (http://www.exiobase.eu/).

Yet the main virtue of MEFA accounting turns out to be its main limitation when it comes to being applied to agri-food systems. Precisely because it is an input-output way of accounting biophysical and energy flows driven by economic activity, it conceals in a black box the internal flows that take place within agroecosystems (Guzmán and González de Molina 2015, 2017; Tello et al. 2016). As explained, this is a feature of many energy balances of agricultural systems when they only take into account the energy content of those biophysical flows that leave, as products, the agricultural cell of the input-output table, so as to be compared with the energy content of the inputs that come into it from outside. A shortcoming of these accountancies is the weak connection allowed with land-use patterns and landscape ecological functioning—an important one indeed, that prevents the bringing to light of how farming and food consumption affect landscape biodiversity and ecosystem services at landscape scale.

The Human Appropriation of Net Primary Production (HANPP) is the only component of the MEFA toolkit that is aimed at providing a bridge to link socio-metabolic flows with ecological patterns and process that take place on the land (Krausmann et al. 2013). However, the actual appropriation of the effective photosynthetic NPP carried out in a territory, which is the starting point of any socio-metabolic analysis of farm systems (González de Molina and Toledo 2014; Guzmán et al. 2014) is one thing; another is the way in which HANPP is calculated with the aim of including the impact of anthropic land-use changes previously made in this territory by taking as reference an alleged 'natural' state (Smil 2012). In this latter HANPP accountancy that stems from Vitousek et al. (1986), a potential Net Primary Production ($NPP_{pot}$) without human intervention has to be assessed relying on potential vegetation models so as to then set a difference with the actual NPP ($NPP_{act}$) that effectively takes place. This assumption relies on the idea of a single potential climax vegetation that has been long discussed and abandoned in Ecology and Biogeography (Kent 2012), and leads to many problematic HANPP results such as the anthropogenic increase of $NPP_{act}$ in relation to the alleged values of $NPP_{pot}$ resulting from the irrigation of arid zones, or a large use of chemical fertilizers, without taking into account the negative environmental impacts they may entail.

From an agroecological point of view, this renders HANPP results rather unreliable, particularly when zooming down from a broad regional scope towards smaller scales of observation at landscape level. In practice, for most biomes of the world the current HANPP accountancy means setting a score through which all the rest of land covers are compared with forestland—a procedure that conceals the positive ecological role that cultural agro-forest mosaic may play in cultural landscapes. This assumption also tends to take for granted that the forest transitions following rural abandonment always involve positive impacts for the environment, hiding the negative ones underway (Tello et al. 2014; Marull et al. 2014; Otero et al. 2015). However, when performed at regional and global scales, HANPP may sometimes provide a useful measure to account for human ecological disturbances (Marull et al. 2015, 2016a; see also Chap. 4 in this book); and when HANPP values have been spatially correlated with empirical observations of different taxonomic groups in Austria, the results have shown that species richness peaks at levels of 50–60% of HANPP. This result can be taken as a rough corroboration that the biodiversity actually existing in this part of Central Europe coexists with the intermediate levels of ecological disturbance typically carried out in agro-forest mosaics by farming and forestry (Wrbka et al. 2004).

Hence, while HANPP can sometimes be a useful indicator for studies performed at global or bioregional scopes, it becomes inadequate at landscape and local scales. When it comes to studying the actual agroecological patterns and processes taking place at landscape level, in order to assess farm-associated biodiversity and ecosystem services, other methods and indicators are required. Of course these limitations are only referred to the HANPP accounting of a potential NPP ($NPP_{pot}$) and have nothing to do with using the actual values of NPP of a study area. Therefore, we need to find other ways of landscape ecology modelling, able to bring to light how the production of food, fuel and *fibres* in agroecosystems interacts with ecological functioning of cultural landscapes (Guzmán et al. 2014). The Energy-Landscape Integrated Analysis (ELIA) currently developed by Joan Marull et al. (2016b; see also Chap. 4 in this book) is aimed at meeting this challenge by using a graph modelling of energy turnover in agroecosystems that captures the interlinked complexity and information embedded in the pattern of flows that become imprinted in cultural landscapes.

Another problem is the multi-scalar character of these patterns and processes of agroecological functioning, and more so their linkage with many other dimensions interacting at both sides of Societal and Ecological metabolisms. To address this large analytical and methodological challenge, Mario Giampietro et al. (2003, 2012, 2013) have developed a nexus proposal of Multi-Scale Integrated Analysis of Societal and Ecosystem Metabolism (MuSIASEM). MuSiASEM reveals how the different domains at stake in Societal and Ecological metabolic interactions constrain each other, making apparent that everything changes everything all the time. This is a very important aspect, to be taken into account when seeking changes in any complex dynamic system (see Chaps. 3 by Gomiero, and 10 by Gamboa et al.). Hence MuSIASEM is a very useful tool for avoiding undue extrapolations when we are dealing with multidimensional and multi-scalar dynamic issues. In the end, though, the assembling of dimensions taken into consideration, and the interactions among them that it helps to play, can only be as good as the indicators and models with which MuSIASEM works.

Therefore, we come to the conclusion that in order to assess agri-food systems, and devise plausible scenarios of feasible changes, we need to combine all these approaches and methods (LCA, MEFA, HANPP, ELIA, MuSIASEM). We can choose one or another, and combine several of them, depending on the question raised and the scale of the approach adopted, but always adopting the viewpoint of Social Metabolism, that is the most promising analytic approach to bring to light the agroecosystem functioning that sustains the reproduction of farm and agri-food chains over time (González de Molina and Toledo 2014). The study of agri-food systems from a supply chain-wide standpoint (Lauk and Lutz 2016) inherently requires a multidimensional scrutiny, and a basic methodological aim is to overcome reductionists views by adopting a 'nexus approach' among different dimensions involved at landscape level (see Chap. 3 by Gomiero). To do so, a solid theoretical background is essential. Agroecology, Landscape Ecology and Political Ecology can jointly provide the theoretical foundations needed to perform the coherent fund-flow socio-metabolic analysis of agri-food systems we are searching for. The chapters included in this book, and many references given in them, are good examples of this.

## 2.6 Concluding Remarks: Searching for AgroEcological Sites Endowed with Rich Biocultural Heritage

Throughout the last century, increasing market integration led most farmers of Global North nations to intensify industrial farming on flat and fertile lands, while steeper lands were considered marginal, abandoned and reforested. Together with urban sprawl, this polarization between agricultural intensification and abandonment caused loss of land cover diversity. This land-use transformation of rural areas in many rich regions of the world has entailed a worrisome loss of peasant site-specific knowledge and farm practices which were responsible for shaping complex landscape mosaics, and of creating crop varieties and livestock breeds—true biocultural foodscapes, indeed. They accumulated an age-old heritage made up of local biological traits (e.g. species richness, land cover diversity, plant varieties and seeds, animal breeds), as well as cultural legacies (e.g. farming know-how, names of places and paths, local identities, traditions), or both (e.g. healthy diets). The loss of this biocultural heritage, along with the landscapes where they had a place, also poses problems to regional human development in rural areas, causing loss of rural employment opportunities and deepening rural marginalisation (Agnoletti 2006, 2014).

Understanding and inventorying these biocultural heritages involves a challenging research agenda for sustainability science. It has to identify key features of sustainable rural livelihoods, contextualizing them in their landscapes and foodscapes, so as to value them as resources for the future. The interdisciplinary character of this socio-ecological research is crucial to the aim of offering useful results for sustainable policy making, and the communities' involvement in it. To give just one relevant example, there is a growing consensus that biodiversity conservation requires going beyond natural protected sites so as to adopt a new land-sharing approach that highlights the ecological connectivity role performed by new organic types of wildlife-friendly farming, forestry and grazing. Enhancing biodiversity ecosystem services is, in turn, closely linked to the search for more sustainable farm systems, able to provide locally healthy food products while maintaining genetic diversity, pest and disease control, soil fertility, pollination, cleaning of water and runoff stabilization, together with other ecosystem services, such as community identities, aesthetic and recreational values, as well as meaningful local job opportunities.

Maintaining these vital ecosystem services of organic farm systems critically depends on two types of increasingly scarce resources: complex landscapes, and people working on them so as to host a great deal of biological and cultural diversity. They jointly involve an endangered biocultural heritage that deserves to be understood, catalogued and preserved (Koohafkan and Altieri 2010; Agnoletti and Emanueli 2016). To this end, history matters (see Chaps. 5 by Padró et al., 6 by Neundlinger et al., and 7 by Fraňková and Cattaneo in this book). A sustainable land-use planning, aimed at conserving agro-forest landscapes endowed with a rich biodiversity, requires knowledge about the farm practices and cultures that developed this biocultural heritage in the past. Community and NGO initiatives are increasingly claiming to recover the local knowledge tied to site-specific foodscapes (e.g. Slow Food and Zero-Km movements, conservation of local crop varieties and breeds, organic farming, community-based rural development projects). In the Global South, instead, a wide array of complex agroecosystems, ruled by a large diversity of PK, are still in place in many rural areas of fast growing economies, but under increasing threat of suffering land grabbing by foreign investors (Borras et al. 2012), or being forced to give up their traditional ways of farming and adopt the disappointing promises of the Green Revolution (Borras et al. 2015; IFAD 2016; see Chap. 8 in this book by Scheidel et al.). A global leapfrogging convergence towards innovative forms of organic farming can open the way to more sustainable foodscapes—as explained in Chaps. 6 and 8 of this book.

To implement better policies addressed toward that aim, urban or rural planners and environmental agencies lack clear indicators and criteria. The industrial transformation of agroecosystems has been so strong for such a long time in affluent nations of the Global North, the prevailing industrialized farm systems are so homogeneous and away from soil and climate bioregional conditions, unhealthy diets are becoming so globalized and uniform, and rural populations have been so marginalised for such a long time, that the same biocultural heritage increasingly praised and vindicated by the FAO vanishes before our eyes. This crossroads urges a multidisciplinary historical study of the long-term evolution of agroecological landscapes, along with an agroecology critical recovery of farm knowledge, labour practices, seed varieties and local cultures that created and maintained the wide range of traditional organic farming that existed everywhere before the massive diffusion of the Green Revolution, and which still exists in many parts of the world.

The study of these examples of biocultural heritage, slowly accumulated by trial-and-error over generations, can only be carried out by multidisciplinary teams able to bring together natural and social sciences with a shared environmental history standpoint (González de Molina and Toledo 2014). They need to be studied as site-specific sets of farm managements, peasants' know-how, tastes, narratives or identities, as well as biophysical landscapes, seeds and plant varieties, livestock breeds, terraces, irrigation systems and other forms of 'landesque capital' (Sen 1959). A main outcome will be to provide environmental agencies, land-use planners, peasant unions, organic farmers and other non-governmental stakeholders of civil society with a clear set of criteria, indicators and datasets that may help them to envision and design a network of local agroecological sites of a rich biocultural heritage. They will be based on the interplay of farmer's activity and knowledge with landscape diversity and species richness, by putting at the forefront the Social Metabolism of Agri-Food Systems and stressing its historical socio-ecological character, far from either purely naturalistic, socioeconomic or folkloristic approaches. The common aim is feeding people by turning farming and cultural landscapes agroecological. Instead of going to 'conservation', this endeavour is addressed to change.

## References


Agnoletti, M. (Ed.). (2006). *The conservation of cultural landscapes*. Wallingford: CABI Pub.

Agnoletti, M. (2014). Rural landscapes, nature conservation and culture. Some notes on research trends and management approaches from a (southern) European perspective. *Landscape and Urban Planning, 126,* 66–73. doi:10.1016/j.landurbplan.2014.02.012.

Agnoletti, M., & Emanueli, F. (Eds.). (2016). *Biocultural diversity in Europe*. New York: Springer.

Agnoletti, M., & Rotherham, I. A. (2015). Landscape and biocultural diversity. *Biodiversity and Conservation, 24,* 3155–3165. doi:10.1007/s10531-015-1003-8.

Aguilera, E., Guzmán, G.I., Infante-Amate, J., Soto, D., García-Ruiz, R., Herrera, A., et al. (2015). *Embodied energy in agricultural inputs. Incorporating a historical perspective*. Working Papers of the Spanish Society for Agricultural History DT-SEHA 1507. Available at: https://ideas.repec.org/s/seh/wpaper.html. Accessed September 3, 2016.

Aguilera, E., Infante-Amate, J., & González de Molina, M. (2014). *La gran transformación del sector agroalimentario español. Un análisis desde la perspectiva energética (1960–2010)*. Working Papers of the Spanish Society for Agricultural History DT-SEHA 1403. Available at: https://ideas.repec.org/s/seh/wpaper.html. Accessed March 9, 2016.

Aguilera, E., Lassaletta, L., Gattinger, A., & Gimeno, B. S. (2013). Managing soil carbon for climate change mitigation and adaptation in Mediterranean cropping systems: A meta-analysis. *Agriculture, Ecosystems & Environment, 168,* 25–36. doi:10.1016/j.agee.2013.02.003.

Alonso, A. M., & Guzmán, G. I. (2010). Comparison of the efficiency and use of energy in organic and conventional farming in Spanish agricultural systems. *Journal of Sustainable Agriculture, 34,* 312–338. doi:10.1080/10440041003613362.

Altieri, M. (1999). The ecological role of biodiversity in agroecosystems. *Agriculture, Ecosystems & Environment, 74,* 19–31. doi:10.1016/S0167-8809(99)00028-6.

Altieri, M. (2002). Agroecology: The science of natural resource management for poor farmers in marginal environments. *Agriculture, Ecosystems & Environment, 93*(1–3), 1–24. doi:10.1016/S0167-8809(02)00085-3.

Altieri, M. (2009). Agroecology, small farms, and food sovereignty. *Monthly Review, 61*(3), 102–113. Available at: http://monthlyreview.org/2009/07/01/agroecology-small-farms-and-food-sovereignty/.

Altieri, M. A., Funes-Monzote, F. R., & Petersen, P. (2012). Agroecologically efficient agricultural systems for smallholder farmers: Contributions to food sovereignty. *Agronomy for Sustainable Development, 32,* 1–13. doi:10.1007/s13593-011-0065-6.

Altieri, M., & Toledo, V. (2011). The agroecological revolution in Latin America: rescuing nature, ensuring food sovereignty and empowering peasants. *Journal of Peasant Studies, 38*(3), 587–612. doi:10.1080/03066150.2011.582947.

Andersson, J. O., & Lindroth, M. (2001). Ecologically unsustainable trade. *Ecological Economics, 37,* 113–122. doi:10.1016/S0921-8009(00)00272-X.

Arizpe, N., Giampietro, M., & Ramos-Martin, J. (2011). Food security and fossil fuel dependence: An international comparison of the use of fossil energy in agriculture (1991–2003). *Critical Reviews in Plant Sciences, 30,* 45–63. doi:10.1080/07352689.2011.554352.

Atkinson, A. B., Piketty, T., & Saez, E. (2011). Top incomes in the long run of history. *Journal of economic literature, 49,* 3–71. doi:10.1257/jel.49.1.3.

Atkinson, C. J., Fitzgerald, J. D., & Hipps, N. A. (2010). Potential mechanisms for achieving agricultural benefits from biochar application to temperate soils: A review. *Plant and Soil, 337,* 1–18. doi:10.1007/s11104-010-0464-5.

Aubry, C., & Kebir, L. (2013). Shortening food supply chains: A means for maintaining agriculture close to urban areas? The case of the French metropolitan area of Paris. *Food Policy, 41,* 85–93. doi:10.1016/j.foodpol.2013.04.006.

Audsley, E. (Ed.). (1997). Harmonisation of environmental life cycle assessment for agriculture. EU Concerted Action report Final Report. Concerted Action AIR3-CT94-2028. Available at: https://www.researchgate.net/publication/258966401_EU_Concerted_Action_report_HARMONISATION_OF_ENVIRONMENTAL_LIFE_CYCLE_ASSESSMENT_FOR_AGRICULTURE_Final_Report_Concerted_Action_AIR3-CT94-2028. Accessed September 5, 2016.

Axelrod, R. (1997). *The complexity of cooperation. Agent-based models of competition and collaboration*. Princeton: Princeton Univeristy Press.

Axelrod, R. (2006). *The evolution of cooperation*. Cambridge, MA: Basic Books.

Barrera-Bassols, N., & Toledo, V. (2005). Ethnoecology of the Yucatec Maya: Symbolism, knowledge and management of natural resources. *JLAG, 4,* 9–41. doi:10.1353/lag.2005.0021.

Barrett, G. W. (1992). Landscape ecology. Designing sustainable agricultural landscapes. *Journal of Sustainable Agriculture, 2,* 83–103. doi:10.1300/J064v02n03_07.

Bengston,.J, Angelstam, P., Elmqvist, T., Emanuelsson, U., Folke, C., Ihse, M., et al. (2003). Reserves, resilience and dynamic landscapes. *Ambio, 32*, 389–396. doi:10.1579/0044-7447-32.6.389.

Benton, T. G., Vickery, J. A., & Wilson, J. D. (2003). Farmland biodiversity: Is habitat heterogeneity the key? *Trends in Ecology & Evolution, 18,* 182–188. doi:10.1016/S0169-5347(03)00011-9.

Berkes, F., Colding, J., & Folke C. (2000). Rediscovery of traditional ecological knowledge as adaptive management. *Ecological Applications, 10,* 1251–1262. doi:10.1890/1051-0761(2000)010[1251:ROTEKA]2.0.CO;2.

Berkes, J., Folke, C., & Colding, F. (2003). *Navigating social-ecological systems building resilience for complexity and change*. Cambridge: Cambridge University Press.

Bernstein, H. (2016). Agrarian political economy and modern world capitalism: The contributions of food regime analysis. *The Journal of Peasant Studies, 43,* 611–647. doi:10.1080/03066150.2015.1101456.

Billen, G., Garnier, J., Thieu, V., Silvestre, M., Barles, S., & Chatzimpiros, P. (2012). Localising the nitrogen imprint of the Paris food supply: The potential of organic farming and changes in human diet. *Biogeosciences, 9,* 607–616. doi:10.5194/bg-9-607-2012.

Borras, S. M., Jr., Franco, J. C., & Suárez, S. M. (2015). Land and food sovereignty. *Third World Quarterly, 36,* 600–617. doi:10.1080/01436597.2015.1029225.

Borras, S. M., Kay, C., Gómez, S., & Wilkinson, J. (2012). Land grabbing and global capitalist accumulation: Key features in Latin America. *Canadian Journal of Development Studies/Revue Canadienne D'études Du Développement, 33*(4), 402–416. doi:10.1080/02255189.2012.745394.

Boyd, R., Gintis, H., Bowles, S., & Richerson, P. J. (2003). The evolution of altruistic punishment. *Proceedings of the National Academy of Sciences of the United States of America, 100,* 3531–3535. doi:10.1073/pnas.0630443100.

Bowles, S., & Gintis, H. (2011). *A cooperative species: Human reciprocity and its evolution*. Princeton: Princeton Univeristy Press.

Busch, T., & Sakhel, A. (2016). The island logic. Scaling up the concept of self-preserving Autarky. *Journal of Industrial Ecology, 20*(5), 1008–1009. doi:10.1111/jiec.12452.

Campbell, H. (2009). Breaking new ground in food regime theory: Corporate environmentalism, ecological feedbacks and the 'food from somewhere' regime? *Agriculture and Human Values, 2009*(26), 309–319. doi:10.1007/s10460-009-9215-8.

Cardinale, B. J., Duffy, J. E., Gonzalez, A., Hooper, D. U., Perrings, C., Venail, P., et al. (2012). Biodiversity loss and its impact on humanity. *Nature, 486,* 59–67. doi:10.1038/nature11148.

Cattaneo, C., D'Alisaa, G., Kallis, G., & Zografos, C. (2012). Degrowth futures and democracy. *Futures, 44,* 515–523. doi:10.1016/j.futures.2012.03.012.

Chang, H. J. (2002). *Kicking away the ladder. Development strategy in historical perspective*. London: Anthem Press.

Commoner, B. (1990). *Making peace with the planet*. New York: Pantheon Books.

Costanza, R., Bernard, C., & Patten, B. C. (1995). Defining and predicting sustainability. *Ecological Economics, 15*, 193–196. doi:10.1016/0921-8009(95)00048-8.

Claeys, P. (2015). *Human rights and the food sovereignty movement: Reclaiming control*. New York: Routledge.
Dahl, R. A. (1985). *A preface to economic democracy*. Berkeley: The Univeristy of California Press.
Dalgaard, T., Halberg, N., & Porter, J. R. (2001). A model for fossil energy use in Danish agriculture used to compare organic and conventional farming. *Agriculture, Ecosystems & Environment, 87,* 51–65. doi:10.1016/S0167-8809(00)00297-8.
Daly, H. (2005). Economics in a full world. *Scientific American, 293*(100), 107. doi:10.1038/scientificamerican0905-100.
Davidson, E. A., & Howarth, R. W. (2007). Environmental science: Nutrients in synergy. *Nature, 449,* 1000–1001. doi:10.1038/4491000a.
De Groot R., Fisher, B., Christie, M., Aronson, J., Braat, L, Gowdy, J., et al. (2010). The economics of ecosystems and biodiversity: The ecological and economic foundations. Available at: http://www.teebweb.org/our-publications/. Accessed January 1, 2017.
González de Molina, M. (2013). Agroecology and politics. How to get sustainability? About the necessity for a political agroecology. *Agroecology and Sustainable Food Systems, 37,* 45–59. doi:10.1080/10440046.2012.705810.
González de Molina, M., & Toledo, V. M. (2014). *The social metabolism: A socio-ecological theory of historical change*. New York: Springer.
González de Molina, M. (2015). Agroecology and politics: On the importance of public policies in Europe. *Law and Agroecology. A transdisciplinary dialogue* (pp. 395–410). Berlin: Springer.
Donald, P. F., Green, R. E., & Heath, M. F. (2001). Agricultural intensification and the collapse of Europe's farmland bird populations. *Proceedings of the Royal Society of London, Series B: Biological Sciences, 268,* 25–29. doi:10.1098/rspb.2000.1325.
Douthwaite, R. (2003). *Short circuit: Strengthening local economies in an unstable world*. Dublin: Liliput Press.
Dutilh, C. E., & Kramer, K. J. (2000). Energy consumption in the food chain. Comparing alternative options in food production and consumption. *Ambio, 29,* 98–101. doi:10.1579/0044-7447-29.2.98.
Edwards, C. A., Grove, R. R., Harwood, C. J., & Colfer, P. (1993). The role of agroecology and integrated farming systems in agricultural sustainability. *Agriculture, Ecosystems & Environment, 46,* 99–121. doi:10.1016/0167-8809(93)90017-J.
Elser, J. J., Bracken, M. E. S., Cleland, E. E., Gruner, D. S., Harpole, W. S., Hillebrand, H., et al. (2007). Global analysis of nitrogen and phosphorus limitation of primary producers in freshwater, marine and terrestrial ecosystems. *Ecology Letters, 10,* 1135–1142. doi:10.1111/j.1461-0248.2007.01113.x.
Engel-Di Mauro, S. (2014). *Ecology, Soils and the left: An ecosocial approach*. New York: Palgrave Macmillan.
Erb, K. H. (2012). How a socio-ecological metabolism approach can help to advance our understanding of changes in land-use intensity. *Ecological Economics, 76,* 8–14. doi:10.1016/j.ecolecon.2012.02.005.
Erb, K.-H., Haberl, H., Krausmann, F., Lauk, C., Plutzar, C., Steinberger, J., et al. (2009a). Eating the planet: Feeding and fuelling the world sustainably, fairly and humanely—a scoping study. London: Compassion in World Farming, River Court/Friends of the Earth. Available at: https://www.foe.co.uk/sites/default/files/downloads/eating_planet_report.pdf. Accessed August 19, 2016.
Erb, K.-H., Krausmann, F., Lucht, W., & Haberl, H. (2009b). Embodied HANPP: Mapping the spatial disconnect between global biomass production and consumption. *Ecological Economics, 69,* 328–334. doi:10.1016/j.ecolecon.2009.06.025.
EUROSTAT. (2001). *Economy-wide material flow accounts and derived indicators. A methodological guide*. Luxembourg: Office for Official Publications of the European Communities.
FAO, IFAD, & WFP. (2014). *The State of food insecurity in the world. Strengthening the enabling environment for food security and nutrition.* Rome: FAO.

Farina, A. (1997). Landscape structure and breeding bird distribution in a sub-Mediterranean agro-ecosystem. *Landscape Ecology, 12,* 365–378. doi:10.1023/A:1007934518160.

Fischer, J., Brosi, B., Daily, G. C., Ehrlich, P. R., Goldman, R., Goldstein, J., et al. (2008). Should agricultural policies encourage land sparing or wildlife-friendly farming? *Frontiers in Ecology and the Environment, 6,* 380–385. doi:10.1890/070019.

Fischer-Kowalski, M., Krausmann, F., & Smetschka, B. (2004). Modelling scenarios of transport across history from a socio-metabolic perspective. *Review (Fernand Braudel Center), 27,* 307–342. doi:http://www.jstor.org/stable/40241610.

Fischer-Kowalski, M., & Haberl, H. (2007). *Socioecological transitions and global change: Trajectories of social metabolism and land use.* Cheltenham: Edward Elgar.

Fischer-Kowalski, M., Krausmann, F., Giljum, S., Lutter, S., Mayer, A., Bringezu, S., et al. (2011a). Methodology and indicators of economy-wide material flow accounting state of the art and reliability across sources. *Journal of Industrial Ecology, 15,* 855–876. doi:10.1111/j.1530-9290.2011.00366.x.

Fischer-Kowalski, M., Singh, S. J., Lauk, C., Remesch, A., Ringhofer, L., & Grünbühel, C. M. (2011b). Sociometabolic transitions in subsistence communities: Boserup revisited in four comparative case studies. *Human Ecology Review, 18*(2), 147–158.

Foster, J. B., & Holleman, H. (2014). The theory of unequal ecological exchange: A Marx-Odum dialectic. *Journal of Peasant Studies, 41,* 199–233. doi:10.1080/03066150.2014.889687.

Freibauer, W., Rounsevell, M. D. A., Smith, P., & Verhagen, J. (2004). Carbon sequestration in the agricultural soils of Europe. *Geoderma, 122,* 1–23. doi:10.1016/j.geoderma.2004.01.021.

Friedmann, H. (1987). International regimes of food and agriculture since 1870. In T. Shanin (Ed.), *Peasants and peasant societies* (pp. 258–276). Oxford: Basil Blackwel.

Friedmann, H. (2016). Commentary: Food regime analysis and agrarian question: Widening the conversation. *Journal of Peasant Studies, 43,* 671–692. doi:10.1080/03066150.2016.1146254.

Galán, E., Padró, R., Marco, I., Tello, E., Cunfer, G., Guzmán, G. I., et al. (2016). Widening the analysis of Energy Return on Investment (EROI) in agro-ecosystems: Socio-ecological transitions to industrialized farm systems (The Vallès County, Catalonia, c.1860 and 1999). *Ecological Modelling, 336,* 13–25. doi:10.1016/j.ecolmodel.2016.05.012.

Georgescu-Roegen, N. (1971). *The entropy law and the economic process.* Cambridge, London: Harvard University Press.

Georgescu-Roegen, N. (1976). The institutional aspects of peasant communities: An analytical view. In N. Georgescu-Roegen (Eds.), *Energy and economic myths. Institutional and analytical economic essays* (pp. 199–234). New York: Pergamon.

Georgescu-Roegen, N. (1977). Inequality, limits and growth from a bioeconomic viewpoint. *Review of Social Economy, 35,* 361–175. doi:10.1080/00346767700000041.

Giampietro, M. (1997). Socioeconomic constraints to farming with biodiversity. *Agriculture, Ecosystems & Environment, 62,* 145–167. doi:10.1016/S0167-8809(96)01137-1.

Giampietro, M. (2003). *Multi-scale integrated analysis of agroecosystems.* Florida: CRC Press LLC.

Giampietro, M., Mayumi, K., & Sorman, A. H. (2012). *The metabolic pattern of societies: Where economists fall short.* London: Routledge.

Giampietro, M., Mayumi, K., & Sorman, A. H. (2013). *Energy analysis for sustainable future: Multi-scale integrated analysis of societal and ecosystem metabolism.* London: Routledge.

Gingrich, S., Haidvogl, G., Krausmann, F., Preis, S., & Garcia-Ruiz, R. (2015). Providing food while sustaining soil fertility in two pre-industrial Alpine agroecosystems. *Human Ecology: an Interdisciplinary Journal, 43,* 395–410. doi:10.1007/s10745-015-9754-0.

Glavič, P., & Lukman, R. (2008). Review of sustainability terms and their definitions. *Journal of Cleaner Production, 15,* 1875–1885. doi:10.1016/j.jclepro.2006.12.006.

Gliessman, S. R. (2000). *Agroecology. Ecological processes in sustainable agriculture.* Boca Raton: Lewis Publishers/CRC Press.

Godfray, H. C. J., Beddington, J. R., Crute, I. R., Haddad, L., Lawrence, D., Muir, J. F., et al. (2010). Food security: The challenge of feeding 9 billion people. *Science, 327,* 812–818. doi:10.1126/science.1185383.

Gómez-Baggethun, E., Corbera, E., & Reyes-García, V. (2013). Traditional ecological knowledge and global environmental change: Research findings and policy implications. *Ecology and Society, 18*(4). doi:10.5751/ES-06288-180472.

Gómez-Baggethun, E., & Reyes-García, V. (2013). Reinterpreting change in traditional ecological knowledge. *Human Ecology: An Interdisciplinary Journal, 41*. doi:10.1007/s10745-013-9577-9.

Gomiero, T. (2016). Soil degradation, land scarcity and food security: Reviewing a complex challenge. *Sustainability, 8,* 281. doi:10.3390/su8030281.

Gomiero, T. (2015). Effects of agricultural activities on biodiversity and ecosystems: Organic versus conventional farming. In G. M. Robinson & D. A. Carson (Eds.), *Handbook on the globalization of agriculture* (pp. 77–105). Cheltenham: Edward Elgar.

Gomiero, T., Paoletti, M. G., & Pimentel, D. (2008). Energy and environmental issues in organic and conventional agriculture. *Critical Reviews in Plant Sciences, 27,* 239–254. doi:10.1080/07352680802225456.

Gomiero, T., Pimentel, D., & Paoletti, M.G. (2011). Environmental impact of different agricultural management practices: Conventional vs. organic agriculture. *Critical Reviews in Plant Sciences, 30,* 95–12. doi:10.1080/07352689.2011.554355.

Graham, M. (2008). Nesting, subsidiarity, and community-based environmental governance beyond the local scale. *International Journal of the Commons, 2,* 75–97. doi:10.18352/ijc.50.

Green, R. E., Cornell, S. J., Scharlemann, J. P. W., & Balmford, A. (2005). Farming and the Fate of Wild Nature. *Science, 307,* 550–551. doi:10.1126/science.1106049.

Greenwald, B. C., & Stiglitz, J. E. (1986). Externalities in economies with imperfect information and incomplete markets. *The Quarterly Journal of Economics, 101,* 229–264. doi:10.2307/1891114.

Gustavsson, J., Cederberg, C., Sonesson, U., van Otterdijk, R., Meybeck, A. (2011). Global food loses and food waste. Extent, causes and prevention. Rome: FAO. Available at: http://www.fao.org/docrep/014/mb060e/mb060e.pdf. (A full version of this report is available at https://www.researchgate.net/publication/267919405. Accessed November 13, 2016.

Guzmán, G. I., & Alonso, A. M. (2008). A comparison of energy use in conventional and organic olive oil production in Spain. *Agricultural Systems, 98,* 167–176. doi:10.1016/j.agsy.2008.06.004.

Guzmán, G. I., & González de Molina, M. (2009). Preindustrial agriculture versus organic agriculture: The land cost of sustainability. *Land Use Policy, 26,* 502–510. doi:10.1016/j.landusepol.2008.07.004.

Guzmán, G. I., González de Molina, M., & Alonso, A. M. (2011). The land cost of agrarian sustainability. An assessment. *Land Use Policy, 28,* 825–835. doi: 10.1016/j.landusepol.2011.01.010.

Guzmán, G., Aguilera, E., Soto, D., Cid, A., Infante, J., Garcia-Ruiz, R., et al. (2014). Methodology and conversion factors to estimate the net primary productivity of historical and contemporary agroecosystems (I). Working Papers of the Spanish Society for Agricultural History DT-SEHA 1407. Available at: https://ideas.repec.org/s/seh/wpaper.html. Accessed September 3, 2016.

Guzmán, G. I., & González de Molina, M. (2015). Energy efficiency in agrarian systems from an agroecological perspective. *Agroecology and Sustainable Food Systems, 39,* 924–952. doi:10.1080/21683565.2015.1053587.

Guzmán-Casado, G. I., & González de Molina, M. (2017). *Energy in agroecosystems. A tool for assessing sustainability.* Boca Ratón (FL): CRC Press.

Haas, W., Krausmann, F. Wiedenhofer, D., & Heinz, M. (2016). How circular is the global economy? A sociometabolic analysis. In H. Haberl, M. Fischer-Kowalski, F. Krausmann, V., Winiwarter (Eds.), *Social ecology: Society-nature relations across time and space* (pp. 259–276). New York: Springer. doi:10.1007/978-3-319-33326-7_11.

Haberl, H., Fischer-Kowalski, M., Krausmann, F., Weisz, H., & Winiwarter, V. (2004). Progress towards sustainability? What the conceptual framework of material and energy flow accounting (MEFA) can offer. *Land Use Policy, 21,* 199–213. doi:10.1016/j.landusepol.2003.10.013.

Haberl, H., Fischer-Kowalski, M., Krausmann, F., & Winiwarter, V. (Eds.). (2016). *Social Ecology. Society-Nature Relations across Time and Space*. Berlin: Springer.

Haberl, H., & Weisz, H. (2007). *The potential use of the Materials and Energy Flow Analysis (MEFA) framework to evaluate the environmental costs of agricultural production systems and possible applications to aquaculture*. Rome: FAO. Fisheries and Aquaculture Department eng. http://agris.fao.org/agris-search/search.do?recordID=XF2016008927. Accessed August 18, 2016.

Hamilton, A., Balogh, S., Maxwell, A., & Hall, C. (2013). Efficiency of edible agriculture in Canada and the U.S. over the past three and four decades. *Energies, 6*, 1764–1793. doi:10.3390/en6031764.

Hinterberger, F., Giljum, S., & Hammer, M. (2003). Material Flow Accounting and Analysis (MFA). A Valuable Tool for Analyses of Society-Nature Interrelationships. Internet Encyclopedia of Ecological Economics. Avilable at: http://alt.seri.at/en/publications/online-publications/2009/09/22/material-flow-accounting-and-analysis-mfa-a-valuable-tool-for-analyses-of-society-nature-interrelationships/. Accessed August 18, 2016.

Ho, M. W. (2013). Circular thermodynamics of organisms and sustainable systems. *Systems, 1*, 30–49. doi:10.3390/systems1030030.

Ho, M. W., & Ulanowicz, R. U. (2005). Sustainable systems as organisms? *BioSystems, 82*, 39–51. doi:10.1016/j.biosystems.2005.05.009.

Holt-Giménez, E. (Ed.). (2011). *Food movements unite! Strategies to transform our food system*. Oakland: Food First Books.

Holt-Giménez, E., Patel, R., & Shattuck, A. (Eds.). (2009). *Food rebellions: Crisis and the hunger for justice*. Boston: Pambazuka Press.

Holt-Giménez, E., & Shattuck, A. (2011). Food crises, food regimes and food movements: rumblings of reform or tides of transformation? *The Journal of Peasant Studies, 38*(1), 109–144. doi:10.1080/03066150.2010.538578.

Hornborg, A., McNeill, J. R., & Martínez Alier, J. (Eds.). (2007). *Rethinking environmental history. World-system history and global environmental change*. New York: Altamira Press.

IAASTD (International Assessment of Agricultural Knowledge, Science and Technology for Development). (2009). *Agriculture at a crossroads*. In International Assessment of Agricultural Knowledge, Science and Technology for Development Global Report, Island Press, Washington, D.C.

IFAD. (2016). *Rural development report 2016. Fostering inclusive rural transformation*. Rome: International Fund for Agricultural Development.

Infante-Amate, J., & González de Molina, M. (2013). 'Sustainable de-growth' in agriculture and food: An agro-ecological perspective on Spain's agri-food system (year 2000). *Journal of Cleaner Production, 38*, 27–35. doi:10.1016/j.jclepro.2011.03.018.

Infante-Amate, J., Soto, D., Aguilera, E., García Ruiz, R., Guzmán, G., Cid, A., et al. (2015). The Spanish transition to industrial metabolism long-term material flow analysis (1860–2010). *Journal of Industrial Ecology, 19*, 866–876. doi:10.1111/jiec.12261.

Inger, R., Gregory, R., Duffy, J. P., Stott, I., Vorisek, P., & Gaston, K. J. (2015). Common European birds are declining rapidly while less abundant species' numbers are rising. *Ecology Letters, 18*, 28–36. doi:10.1111/ele.12387.

ISO. (1997). Environmental management life cycle assessment principles and framework. The International Organization for Standardization, Geneva. ISO14040. Available at: http://www.iso.org/iso/catalogue_detail?csnumber=37456. Accessed August 18, 2016.

Johanisova, N., & Wolf, S. (2012). Economic democracy: A path for the future? *Futures, 44*, 562–570. doi:10.1016/j.futures.2012.03.017Jungbluth.

Jones, A. (2001). *Eating oil. Food supply in a changing climate*. London: Sustain & Elm Farm Research Centre. Availabe at: https://ia800501.us.archive.org/24/items/Eating_Oil-Food_Supply_in_a_Changing_Climate/Eating_Oil-Food_Supply_in_a_Changing_Climate.pdf. Accessed August 19, 2016.

Jungbluth, N., & Demmeler, M. (2005). The ecology of scale: Assessment of regional energy turnover and comparison with global food. In E. Schlich & U. Fleissner (Eds.), *The International Journal of Life Cycle Assessment, 10*, 168–170. doi:10.1065/lca2004.11.191.

Jordan, C. F. (2016). The farm as a thermodynamic system: Implications of the maximum power principle. *BioPhysical Economics and Resource Quality (On-line Preview)*, 1–9. doi:10.1007/s41247-016-0010-z.

Karlen, D. L., Mausbach, M. J., Doran, J. W., Cline, R. G., Harris, R. F., & Schuman, G. E. (1997). Soil quality: A concept, definition, and framework for evaluation (A guest editorial). *Soil Science Society of America Journal, 61*, 4–10. doi:10.2136/sssaj1997.03615995006100010001x.

Kent, M. (2012). *Vegetation description and data analysis*. Oxford: Wiley.

Kleinert, S., & Horton, R. (2015). Rethinking and reframing obesity. *Lancet, 385*(9985), 2326–2328. doi:10.1016/S0140-6736(15)60163-5.

Kneafsey, M. (2010). The region in food—important or irrelevant? *Cambridge Journal of Regions, Economy and Society, 3*, 177–190. doi:10.1093/cjres/rsq012.

Koohafkan, P., & Altieri, M. A. (2010). *Conserving our world's agricultural heritage. Globally Important Agricultural Heritage Systems (GIAHS)*. Rome: Food and Agriculture Organization of the United Nations.

Krausmann, F. (2004). Milk, manure, and muscle power. Livestock and the transformation of preindustrial agriculture in Central Europe. *Human Ecology: An Interdisciplinary Journal, 32*, 735–772. doi:10.1007/s10745-004-6834-y.

Krausmann, F., Erb, K. E., Gingrich, S., Haberl, H., Bondeau, A., Gaube, V., et al. (2013). Global human appropriation of net primary production doubled in the 20th century. *Proceedings of the National Academy of Sciences of the United States of America, 110*, 10324–10329. doi:10.1073/pnas.1211349110/-/DCSupplemental.

Krausmann, F., Gingrich, S., Eisenmenger, N., Erb, K.-H., Haberl, H., & Fischer-Kowalski, M. (2009). Growth in global materials use, GDP and population during the 20th century. *Ecological Economics, 68*, 2696–2705. doi:10.1016/j.ecolecon.2009.05.007.

Lauk, C., & Lutz, J. (2016). The future is made. Imagining feasible food and farming futures in an unpredictable world. In J. Niewöhner, A. Bruns, P., Hostert, T., Krueger, J. Ø. Nielsen, H. Haberl, C., Lauk, J., Lutz, D., Müller (Eds.), *Land use competition. Ecological, economic and social perspectives* (pp. 233–246). Cham (ZG): Springer International Publishing Switzerland.

Luxemburg, R. (1964). *The accumulation of capital*. New York: Monthly Review Press.

Magdoff, F., & Weil, R. E. (2004). *Soil organic matter in sustainable agriculture*. Boca Raton: CRC Press.

Marco, I., Padró, R., Cattaneo, C., Caravaca, J., & Tello, E. (2017, forthcoming). From vineyards to feedlots: A fund-flow scanning of socio-metabolic transitions in the Vallès County (Catalonia) (1860–1956–1999). *Regional Environmental Change* (published on-line first). doi:10.1007/s10113-017-1172-y

Markussen, M. V., & Østergård, H. (2013). Energy analysis of the Danish Foos Production System: Food-EROI and fossil fuel dependency. *Energies, 6*, 4170–4186. doi:10.3390/en6084170.

Martínez Alier, J. (1987). *Ecological economics: Energy, environment and society*. Oxford: Blackwell.

Martínez Alier, J. (2002). *The environmentalism of the poor. A study of economic conflicts and valuation*. Cheltenham: Edward Elgar.

Martínez Alier, J. (2011). The EROI of agriculture and its use by the Via Campesina. *The Journal of Peasant Studies, 38*, 145–160. doi:10.1080/03066150.2010.538582.

Martínez-Torres, M. E., & Rosset, Peter M. (2010). La Vía Campesina: The birth and evolution of a transnational social movement. *The Journal of Peasant Studies, 37*(1), 149–175. doi:10.1080/03066150903498804.

Martínez-Torres, M. E., & Rosset, Peter M. (2014). Diálogo de saberes in La Vía Campesina: food sovereignty and agroecology. *The Journal of Peasant Studies, 41*(6), 979–997. doi:10.1080/03066150.2013.872632.

Marull, J., Font, C., Padró, R., Tello, E., & Panazzolo, A. (2016a). Energy-Landscape Integrated Analysis: A proposal for measuring complexity in internal agroecosystem processes (Barcelona Metropolitan Area, 1860–2000). *Ecological Indicators, 66,* 30–46. doi:10.1016/j.ecolind.2016.01.015.

Marull, J., Font, C., Tello, E., Fullana, N., Domene, E., Pons, M., et al. (2016b). Towards an energy–landscape integrated analysis? Exploring the links between socio-metabolic disturbance and landscape ecology performance (Mallorca, Spain, 1956–2011). *Landscape Ecology, 31,* 317–336. doi:10.1007/s10980-015-0245-x.

Marull, J., Pino, J., Tello, E., & Cordobilla, M. J. (2010). Social metabolism, landscape change and land-use planning in the Barcelona Metropolitan Region. *Land Use Policy, 27,* 497–510. doi:10.1016/j.landusepol.2009.07.004.

Marull, J., Tello, E., Fullana, N., Murray, I., Jover, G., Font, C., et al. (2015). Long-term biocultural heritage: Exploring the intermediate disturbance hypothesis in agro-ecological landscapes (Mallorca, c. 1850–2012). *Biological Conservation, 24,* 3217–3251. doi:10.1007/s10531-015-0955-z.

Marull, J., Tello, E., Wilcox, P., Coll, F., Pons, M., Warde, P., et al. (2014). Recovering the landscape history behind a Mediterranean edge environment (The Congost Valley, Catalonia, 1854–2005): The importance of agroforestry systems in biological conservation. *Applied Geography, 54,* 1–17. doi:10.1016/j.apgeog.2014.06.030.

Matson, P. A., Parton, W. J., Power, A. G., & Swift, M. J. (1997). Agricultural intensification and ecosystem properties. *Science, 277,* 504–509. doi:10.1126/science.277.5325.504.

Matson, P. A., & Vitousek, P. M. (2006). Agricultural Intensification: Will land spared from farming be land spred for nature? *Conservation Biology, 20,* 709–710. doi:10.1111/j.1523-1739.2006.00442.x.

McDonnell, M. J., & Pickett, S. T. A. (Eds.). (1993). *Humans as components of ecosystems. The ecology of subtle human effects and populated areas*. New York: Springer.

McKeon, N. (2015). *Food security governance: Empowering communities, regulating corporations*. New York: Routledge.

McMichael, P. (2003). Food security and social reproduction: Issues and contradictions. In I. Bakker & S. Gillpp (Eds.), *Power, production and social reproduction* (pp. 169–189). New York: Palgrave MacMillan.

McMichael, P. (2008). Peasants make their own history, but not just as they please. *Journal of Agrarian Change, 8,* 205–228. doi:10.1111/j.1471-0366.2008.00168.x.

McMichael, P. (2009). A food regime genealogy. *The Journal of Peasant Studies, 36,* 139–169. doi:10.1080/03066150902820354.

McMichael, P. (2011). Food system sustainability: Questions of environmental governance in the new world (dis)order. *Global Environmental Change, 21,* 804–812. doi:10.1016/j.gloenvcha.2011.03.016.

McMichael, P. (2013). *Food regimes and agrarian questions*. Halifax: Fernwood.

McMichael, P. (2016). Commentary: Food regime for thought. *The Journal of Peasant Studies, 43,* 648–670. doi:10.1080/03066150.2016.1143816.

McNeely, J. A., & Scheer, S. (2003). *Ecoagriculture. Strategies to feed the world and save biodiveristy*. Washington: Island Press/Future Harvest/UICN.

Mies, M. (1986). *Patriarchy and accumulation at world scale*. New York: Zed Books.

Mies, M., & Shiva, V. (1993). *Ecofeminism*. New York: Seed Books.

Milanovic, B. (2011). *The haves and have nots. A brief and idiosyncratic history of global inequality*. Philadelphia: Basic Books.

Millennium Ecosystem Assessment (MA). (2005). *Ecosystems and human well-being: A framework for assessment*. Washington, D.C.: Island Press.

Mitchell, M. G. E., Benett, E. M., & González, A. (2013). Linking landscape connectivity and ecosystem service provision: Current knowledge and research gaps. *Ecosystems, 16,* 894–908. doi:10.1007/s10021-013-9647-2.

Moguel, P., & Toledo, V. (1999). Biodiversity conservation in traditional coffee systems of Mexico. *Conservation Biology 13*(1), 11–21.

Mondelaers, K., Aertsens, J., & Van Huylenbroeck, G. (2009). A meta-analysis of the differences in environmental impacts between organic and conventional farming. *British Food Journal, 111,* 1098–1119. doi:10.1108/00070700910992925.

Mooney, P. H. (2004). Democratizing rural economy: Institutional friction, sustainable struggle and the cooperative movement. *Rural Sociology, 69,* 76–98. doi:10.1526/003601104322919919.

Morowitz, H. J. (2002). *The emergence of everything: How the world became complex.* Oxford: Oxford University Press.

Morowitz, H. J., & Smith, E. (2007). Energy flow and the organization of life. *Complexity, 13,* 51–59. doi:10.1002/cplx.20191.

Muradian, R., O'Connor, M., & Martinez-Alier, J. (2002). Embodied pollution in trade: Estimating the 'environmental load displacement' of industrialised countries. *Ecological Economics 41,* 51–67. doi:10.1016/S0921-8009(01)00281-6.

Muradian, R., & Martinez-Alier, J. (2001). Trade and the environment: From a 'Southern' perspective. *Ecological Economics, 36,* 281–297. doi:10.1016/S0921-8009(00)00229-9.

NCD Risk Factor Collaboration (NCD-RisC). (2016). Trends in adult body-mass index in 200 countries from 1975 to 2014: a pooled analysis of 1698 population-based measurement studies with 19·2 million participants. *Lancet, 387*(10026), 1377–1396. doi:10.1016/S0140-6736(16)30054-X.

Neher, D. (1992). Ecological sustainability in agricultural systems: Definition and measurement. *Journal of Sustainable Agriculture, 2,* 51–61. doi:10.1300/J064v02n03_05.

Odum, H. T. (1988). Self-organization, transformity, and information. *Science, 242,* 1132–1138.

Odum, H. T. (2001). *A prosperous way down.* Boulder: University Press of Colorado.

Odum, H. T. (2007). *Environment, power and society for the twenty-first century: The hierchy of energy.* New York: Columbia University Press.

OECD. (2008). Measuring Materal Flows and Resource Productivity. Paris: OECD. https://www.oecd.org/environment/indicators-modelling-outlooks/MFA-Guide.pdf. Accessed August 18, 2016.

Otero, I., Marull, J., Tello, E., Diana, G. L., Pons, M., Coll, F., et al. (2015). Land abandonment, landscape, and biodiversity: Questioning the restorative character of the forest transition in the Mediterranean. *Ecology and Society, 20,* 7. doi:10.5751/ES-07378-200207.

Parfitt, J., Barthel, M., & Macnaughton, S. (2010). Food waste within food supply chains: Quantification and potential for change to 2050. *Philosophical Transactions of the Royal Society B, 365,* 3065–3081. doi:10.1098/rstb.2010.0126.

Patel, R. (2012). The long green revolution. *The Journal of Peasant Studies, 40,* 1–63. doi:10.1080/03066150.2012.719224.

Pelletier, N., Audsley, E., Brodt, S., Garnett, T., Henriksson, P., Kendall, A., et al. (2011). Energy intensity of agriculture and food systems. *Annual Review of Environment and Resources, 36,* 223–246. doi:10.1146/annurev-environ-081710-161014.

Perfecto, I., & Vandermeer, J. (2010). The agroecological matrix as alternative to the land-sparing/agriculture intensification model. *Proceedings of the National Academy of Sciences of the United States of America, 107,* 5786–5791. doi:10.1073/pnas.0905455107.

Perignon, M., Masset, G., Ferrari, G., Barré, T., Vieux, F., Maillot, M., Amiot, M. J. & Darmon, N. (2016). How low can dietary greenhouse gas emissions be reduced without impairing nutritional adequacy, affordability and acceptability of the diet? A modelling study to guide sustainable food choices. *Public Health Nutrition* (published on-line first). doi:10.1017/S1368980016000653Phalan.

Phalan, B., Onial, M., Balmford, A., & Green, R. E. (2011). Reconciling food production and biodiversity conservation: Land Sharing and land sparing compared. *Science, 333,* 1289–1291. doi:10.1126/science.1208742.

Piketty, T., & Saez, E. (2013). Top incomes and the great recession: Recent evolutions and policy implications. *IMF Economic Review, 61,* 456–478. doi:10.1057/imfer.2013.14.

Pimbert, M. (2009). *Towards food sovereignty*. Workin Paper 141 of the International Institute for Environment and Development (IIED), London. Available at: http://hdl.handle.net/10535/5851 . Accessed August 11, 2016.

Pingali, P. (2012). Green revolution: Impacts, limits, and the path ahead. *Proceedings of the National Academy of Sciences of the United States of America, 109,* 12302–12308. doi:10.1073/pnas.0912953109.

Polanyi, K. (2001). *The great transformation. The political and economic origins of our time*. Boston (MA): Beacon Press.

Ponisio, L. C., M'Gonigle, L. K., Mace, K. C., Palomino, J., de Valpine, P., & Kremen, C. (2015). Diversification practices reduce organic to conventional yield gap. *Proceedings of the Royal Society of London B, 282,* 20141396. doi:10.1098/rspb.2014.1396.

Pracha, A. S., & Volk, T. A. (2011). An edible Energy Return on Investment (EEROI) analysis of wheat and rice in Pakistan. *Sustainability, 3,* 2358–2391. doi:10.3390/su3122358.

Ravera, F., Scheidel, A., Dell'Angelo, J., Gamboa, G., Serrano, T., Mingorría, S., et al. (2014). Pathways of rural change: An integrated assessment of metabolic patterns in emerging ruralities. *Environment, Development and Sustainability, 16,* 811–820. doi:10.1007/s10668-014-9534-9.

Ravetz, J., & Funtowicz, S. (2015). Post-normal science. In: S. Meisch, J. Lundershausen, L. Bossert, & M. Rockoff (Eds.), *Ethics of science in the research for sustainable development* (pp. 99–112). Baden-Baden: Nomos. doi:10.5771/9783845258430-99.

Reap, J., Roman, F., Duncan, S., & Bras, B. (2008a). A survey of unresolved problems in life cycle assessment. Part 1: goal and scope and inventory analysis. *International Journal of Life Cycle Assessment, 13,* 290–300. doi:10.1007/s11367-008-0009-9.

Reap, J., Roman, F., Duncan, S., & Bras, B. (2008b). A survey of unresolved problems in life cycle assessment. Part 2: Impact assessment and interpretation. *The International Journal of Life Cycle Assessment, 13*, 374. doi:10.1007/s11367-008-0009-9.

Reinert, E. S. (2004). *Globalization, economic development and inequality: An alternative perspective*. Cheltenham: Edward Elgar.

Renting, H., Marsden, T. K., & Banks, J. (2003). Understanding alternative food networks: Exploring the role of short food supply chains in rural development. *Environment and Planning A, 35,* 393–411. doi:10.1068/a3510.

Reyes-García, V., Martí, N., McDade, T., Tanner, S., Vadez, V. (2007). Concepts and methods in studies measuring individual ethnobotanical knowledge. *Journal of Ethnobiology*, *27*, 182–203. doi:10.2993/0278-0771(2007)27[182:CAMISM]2.0.CO;2.

Riolo, R. L., Cohen, M. D., & Axelrod, R. (2001). Evolution of cooperation without reciprocity. *Nature, 414,* 441–443. doi:10.1038/35106555.

Rosset, P. M. (2011). Preventing hunger: Change economic policy. *Nature, 479,* 472–473. doi:10.1038/479472a.

Rosset, P. M. (2013). Re-thinking agrarian reform, land and territory in La Via Campesina. *Journal of Peasant Studies, 40*(4), 721–775. doi:10.1080/03066150.2013.826654.

Rosset, P. M., & Martínez-Torres, M. E. (2012). Rural social movements and agroecology: Context, theory, and process. *Ecology and Society, 17*(3), 17. doi:10.5751/ES-05000-170317.

Roy, P., Nei, D., Orikasa, T., Xu, Q., Okadome, H., & Nakamura, N., et al. (2009). A review of life cycle assessment (LCA) on some food products. *Journal of Food Engineering, 90,* 1–10. doi:10.1016/j.jfoodeng.2008.06.016.

Sage, C. (2013). The inter-connected challenges for food security from a food regimes perspective: Energy, climate and malconsumption. *Journal of Rural Studies, 29,* 71–80. doi:10.1016/j.jrurstud.2012.02.005.

Sand, P. H. (2004). Sovereignty bounded: Public trusteeship for common pool resources? *Global Environmental Politics, 4,* 47–71. doi:10.1162/152638004773730211.

Sayer, J., Sunderland, T., Ghazoul, J., Pfund, J. L., Sheil, D., Meijaard, E., Venter, M., et al. (2013). Ten principles for a landscape approach to reconciling agriculture, conservation, and other competing land uses. *Proceedings of the National Academy of Sciences, 110*(21), 8349–8356. doi:10.1073/pnas.1210595110.

Scarborough, P., Allender, S., Clarke, D., Wickramasinghe, K., & Rayner, M. (2012). Modelling the health impact of environmentally sustainable dietary scenarios in the UK. *European Journal of Clinical Nutrition, 66,* 710–715. doi:10.1038/ejcn.2012.34.

Schlich, E., & Fleissner, U. (2003). Comparison of regional energy turnover with global food. *The International Journal of Life Cycle Assessment, 8,* 252–252. doi:10.1007/BF02978482.

Schlich, E., & Fleissner, U. (2005a). The ecology of scale: Assessment of regional energy turnover and comparison with global food. *The International Journal of Life Cycle Assessment, 10,* 171–172. doi:10.1065/lca2005.02.200.

Schlich, E., & Fleissner, U. (2005b). the ecology of scale: Assessment of regional energy turnover and comparison with global food. *The International Journal of Life Cycle Assessment, 10,* 219–223. doi:10.1065/lca2004.09.180.9.

Schneider, M., & McMichael, P. (2010). Deepening, and repairing, the metabolic rift. *Journal of Peasant Studies, 37,* 461–484. doi:10.1080/03066150.2010.494371.

Schroll, H. (1994). Energy-flow and ecological sustainability in Danish agriculture. *Agriculture, Ecosystems & Environment, 51,* 301–310. doi:10.1016/0167-8809(94)90142-2.

Schmidt, M. W. I., Torn, M. S., Abiven, S., Dittmar, T., Guggenberger, G., Janssens, I. A., et al. (2011). Persistence of soil organic matter as an ecosystem property. *Nature, 478,* 49–56. doi:10.1038/nature10386.

Schöggl, J. P., Fritz, M. M. C., & Baumgartner, R. J. (2016). Toward supply chain-wide sustainability assessment: A conceptual framework and an aggregation method to assess supply chain performance. *Journal of Cleaner Production, 131,* 822–835. doi:10.1016/j.jclepro.2016.04.035.

Sen, A. K. (1959). The choice of agricultural techniques in underdeveloped countries. *Economic Development and Cultural Change, 7,* 279–285. http://www.jstor.org/stable/1151637.

Shanin, T. (1972). *The awkward class. Political sociology of peasantry in a developing society: Russia 1910–1925.* Oxford: Clarendon Press.

Shanin, T. (Ed.). (1987). *Peasants and peasant societies.* Oxford: Basil Blackwel.

Shiva, V. (2000). *Stolen harvest: The hijacking of the global food supply.* London: Zeed Books.

Shiva, V. (2005). *Earth democracy: Justice, sustainability, and peace.* Cambridge MA: South End Press.

Shiva, V., & Gitanjali, B. (Eds.). (2002). *Sustainable agriculture and food security: The impact of globalisation.* New Delhi: Sage.

Singh, S.J., Ringhofer, L., Haas, W., Krausmann, F., Lauk, C., & Fischer-Kowalski, M. (2010). *Local studies manual: A researcher's guide for investigating the social metabolism of rural systems.* Social Ecology Working Paper 120, IFF Social Ecology, Vienna. Available at: http://www.uni-klu.ac.at/socec/downloads/WP120_Web.pdf. Accessed October 14, 2016.

Smil, V. (2000). *Feeding the world: A challenge for the twenty-first century.* Cambridge (Massachusetts): The MIT Press.

Smil, V. (2003). *China's past, China's future: Energy, food, environment.* New York: Routledge.

Smil, V. (2012). *Harvesting the biosphere. What we have taken from nature.* Cambridge (Massachusetts): The MIT Press.

Snyder, C. S., Bruulsema, T. W., Jensen, T. L., & Fixen, P. E. (2009). Review of greenhouse gas emissions from crop production systems and fertilizer management effects. *Agriculture, Ecosystems & Environment, 133,* 247–266. doi:10.1016/j.agee.2009.04.021.

Soto, D., Infante-Amate, J., Guzmán, G. I., Cid, A., Aguilera, E., García, R., et al. (2016). The social metabolism of biomass in Spain, 1900–2008: From food to feed-oriented changes in the agro-ecosystems. *Ecological Economics, 128,* 130–138. doi:10.1016/j.ecolecon.2016.04.017.

Spokas, K. A. (2010). Review of the stability of biochar in soils: Predictability of O: C molar ratios. *Carbon Management, 1,* 289–303. doi:10.4155/CMT.10.32.

Stiglitz, J. (2012). *The price of inequality: How today's divided society endangers our future.* New York: W.W. Norton & Co.

Stiglitz, J. (2015). *The great divide.* Harmondsworth: Penguin.

Suh, S. (2005). Theory of materials and energy flow analysis in ecology and economics. *Ecological Modelling, 189*(3–4), 251–269. doi:10.1016/j.ecolmodel.2005.03.011.

Swift, M. J., Izac, A. M. N., & van Noordwijk, M. (2004). Biodiversity and ecosystem services in agricultural landscapes—are we asking the right questions? *Agriculture, Ecosystems & Environment, 104,* 113–134. doi:10.1016/j.agee.2004.01.013.

Swinburn, B. A., Sacks, G., Hall, K. D., McPherson, K., Finegood, D. T., Moodie, M. L., et al. (2011). The global obesity pandemic: Shaped by global drivers and local environments. *Lancet, 378*(9793), 804–814. doi:10.1016/S0140-6736(11)60813-1.

Tadei, F. (2014). *Extractive institutions and gains from trade: Evidence from colonial Africa.* CEPR, NBER and Università Bocconi Workin Paper No. 536. Available at: ftp://ftp.igier.unibocconi.it/wp/2014/536.pdf. Accessed September 5, 2016.

Taylor, W. P. (1934). Significance of Extreme or intermittent conditions in distribution of species and management of natural resources, with a restatement of liebig's law of minimum. *Ecology, 15,* 374–379. doi:10.2307/1932352.

Tello, E., Garrabou, R., Cussó, X., Olarieta, J. R., & Galán, E. (2012). Fertilizing methods and nutrient balance at the end of traditional organic agriculture in the Mediterranean bioregion. Catalonia (Spain) in the 1860s. *Human Ecology: An Interdisciplinary Journal, 40,* 369–383. doi:10.1007/s10745-012-9485-4.

Tello, E., Galán, E., Sacristán, V., Cunfer, G., Guzmán, G. I., González de Molina, M., et al. (2016). Opening the black box of energy throughputs in farm systems: A decomposition analysis between the energy returns to external inputs, internal biomass reuses and total inputs consumed (The Vallès County, Catalonia, c.1860 and 1999). *Ecological Economics, 121,* 160–174. doi:10.1016/j.ecolecon.2015.11.012.

Tello, E., Martínez, J. L., Jover-Avellà, G., Olarieta, J. L., García-Ruiz, R., González de Molina, M. et al. (2017). The Onset of the English Agricultural Revolution: Climate Factors and Soil Nutrients. *Journal of Interdisciplinary History, 47*(4), 445–474. doi:10.1162/JINH_a_01050.

Tello, E., Valldeperas, E., Ollés, N., Marull, J., Coll, F., Warde, P., et al. (2014). Looking backwards into a Mediterranean edge environment: Landscape changes in El Congost Valley (Catalonia), 1850-2005. *Environment and History, 20,* 347–384. doi:10.3197/096734014X14031694156402.

Theurl, M. C. (2016). Local food systems and their climate impacts: A life cycle perspective. In J. Niewöhner, A. Bruns, P. Hostert, T. Krueger, J.Ø. Nielsen, H. Haberl, C. Lauk, J. Lutz, D. Müller (Eds.), *Land use competition. Ecological, economic and social perspectives* (pp. 295–309). Cham (ZG): Springer International Publishing Switzerland.

Theurl, M. C, & Schaffartzik, A. (2016). Method Précis: Life cycle assessment. In: H. Haberl, M. Fischer-Kowalski, F. Krausmann, & V. Winiwarter (Eds.), *Social ecology: Society-nature relations across time and space* (pp. 253–256). New York: Springer. doi:10.1007/978-3-319-33326-7_10.

Tilman, D. (1999). Global environmental impacts of agricultural expansion: The need for sustainable and efficient practices. *Proceedings of the National Academy of Sciences of the United States of America, 96,* 5995–6000. doi:10.1073/pnas.96.11.5995.

Tilman, D., Cassman, K. G., Matson, P. A., Naylor, R., & Polasky, S. (2002). Agricultural sustainability and intensive production practices. *Nature, 418,* 671–677. doi:10.1038/nature01014.

Tilman, D., Reich, P. B., Knops, J., Wedin, D., Mielke, T., & Lehman, C. (2001). Diversity and productivity in a long-term grassland experiment. *Science, 294,* 843–845. doi:10.1126/science.1060391.

Tilman, D., Wedin, D., & Knops, J. (1996). Productivity and sustainability influenced by biodiveristy in grassland ecosystems. *Nature, 379,* 718–720. doi:10.1038/379718a0.

Thompson, J., & Scoones, I. (2009). Addressing the dynamics of agri-food systems: An emerging agenda for social science research. *Environmental Science & Policy, 12,* 386–397. doi:10.1016/j.envsci.2009.03.001.

Timmermann, C., & Robaey, Z. (2016). Agrobiodiversity under different property regimes. *Journal of Agricultural and Environmental Ethics, 29,* 285–303. doi:10.1007/s10806-016-9602-2.

Toledo, V. M. (2002). Agroecología, sustentabilidad y reforma agraria: la superioridad de la pequeña producción familiar. *Agroecologia e Desenvolvimento Rural Sustentável, 3*(2), 27–36. Available at: http://www.pvnocampo.com.br/agroecologia/victor_toledo_escreve_sobre_agroecologia.pdf.

Toledo, V. M., Ortiz-Espejel, B., Cortés, L., Moguel, P., Ordoñez, M. D. J. (2003). The multiple use of tropical forests by indigenous peoples in Mexico: A case of adaptive management. *Conservation Ecology, 7*(3), 9. Available at: http://www.consecol.org/vol7/iss3/art9.

Toledo, V. M., Stepp, J. R., Wyndham, F. S., Zarger, R. K. (2002). Ethnoecology: A conceptual framework for the study of indigenous knowledge of nature. In: Ethnobiology and biocultural diversity. In *Proceedings of the 7th International Congress of Ethnobiology* (pp. 511–522). Athens, Georgia, USA, October 2000. Georgia: University of Georgia Press.

Toledo, V. M., & Barrera-Bassols, N. (2008). *La memoria biocultural. La importancia ecológica de las sabidurías tradicionales*. Icaria: Barcelona.

Toledo, V. M., Barrera-Bassols, N., García Frapolli, E., & Alarcón Chaires, P. (2007). Manejo y uso de la biodiversidad entre los mayas yucatecos. CONABIO. *Biodiversitas, 70,* 10–15.

Tress, B., Tress, G., Décamps, H., & d'Hauteserre, A. M. (2001). Bridging human and natural sciences in landscape research. *Landscape Urban Planning, 57,* 137–141. doi:10.1016/S0169-2046(01)00199-2.

Tripp, R. (2008). Agricultural Change and low-input technology. In S. Snapp & B. Pound (Eds.), *Agricultural systems: Agroecology and rural innovation for development* (pp. 129–160). Amsterdam: Elsevier.

Tscharntke, T., Clough, Y., Wanger, T. C., Jackson, L., Motzke, I., Perfecto, I., et al. (2012a). Global food security, biodiversity conservation and the future of agricultural intensification. *Biological Conservation, 151,* 53–59. doi:10.1016/j.biocon.2012.01.068.

Tscharntke, T., Klein, A. M., Kruess, A., Steffan-Dewenter, I., & Thies, C. (2005). Landscape perspectives on agricultural intensification and biodiversity-ecosystem service management. *Ecology Letters, 8,* 857–874. doi:10.1111/j.1461-0248.2005.00782.x.

Tscharntke, T., Tylianakis, J. M., Rand, T. A., Didham, R. K., Fahring, L., Batáry, P., et al. (2012b). Landscape moderation of biodiversity patterns and processes—eigth hypotheses. *Biological Reviews, 87,* 661–685. doi:10.1111/j.1469-185X.2011.00216.x.

Ulanowicz, R. U. (1986). *Growth and development. Ecosystems phenomenology.* Dordrecht: Springer.

Ulanowicz, R. U. (2001). Information theory in ecology. *Computers & Chemistry, 25,* 393–399. doi:10.1016/S0097-8485(01)00073-0.

Ulanowicz, R. U., Goerner, S. J., Lietaer, B., & Gomez, R. (2009). Quantifying sustainability: Resilience, efficiency and the return of information theory. *Ecological Complexity, 6,* 27–36. doi:10.1016/j.ecocom.2008.10.005UNEP.

UNEP, (2016). Global Material Flows and Resource Productivity. An Assessment Study of the UNEP International Resource Panel. Paris: United Nations Environment Programme. Available at: http://unep.org/documents/irp/16-00169_LW_GlobalMaterialFlowsUNEReport_FINAL_160701.pdf. Accessed August 20, 2016.

Van den Bergh, J. C. J. M., & Verbruggen, H. (1999). Spatial sustainability, trade and indicators: An evaluation of the 'ecological footprint'. *Ecological Economics, 29,* 61–72. doi: 10.1016/S0921-8009(99)00032-4.

Van der Ploeg, J. D. (2009). *The New Peasantries. Struggles for autonomy and sustainability in an era of empire and globalization.* London: Earthscan.

Van der Ploeg, J. D. (2010). The food crisis, industrialized farming and the imperial regime. *Journal of Agrarian Change, 10,* 98–106. doi:10.1111/j.1471-0366.2009.00251.x.

Van der Ploeg, J. D. (2013). *Peasants and the art of farming: A Chayanovian manifesto.* Halifa: Fernwood.

Vandermeer, J. (2000). Theoretical ecology meets agroecology: Towards an Ecological approach to agroecosystems. *Ecology, 81,* 1758–1759. doi:10.1890/0012-9658(2000)081[1758:TEMATA]2.0.CO;2.

Vitousek, P. M., Ehrlich, P. R., Ehrlich, A. H., & Matson, P. A. (1986). Human appropriation of the products of photosynthesis. *BioScience, 36,* 368–373. doi:10.2307/1310258.

Wang, Y., & Lobstein, T. (2006). Worldwide trends in childhood overweight and obesity. *International Journal of Pediatric Obesity, 1,* 11–25. doi:10.1080/17477160600586747.

Winqvist, C., Bengtsson, J., Aavik, T., Berendse, F., Clement, L. W., Eggers, S., et al. (2011). Mixed effects of organic farming and landscape complexity on farmland biodiversity and biological control potential across Europe. *Journal of Applied Ecology, 48,* 570–579. doi:10.1111/j.1365-2664.2010.01950.x.

World Health Organization. (2000a). *Obesity: Preventing and managing the global epidemic. WHOTechnical Report Series 894*. Geneva: World Health Organization.

World Health Organization. (2000b). Nutrition for Health and Development. A global agenda for combating malnutrition. WHO Dist. General 00.6. Geneva: World Health Organization (WHO)/ Nutrition for Health and Development (NHD)/Sustainable Development and Healthy Environments (SDE).

Wrbka, T., Erb, K. H., Schulz, N. B., Peterseil, J., Hahn, C., & Haberl, H. (2004). Linking pattern and process in cultural landscapes. An empirical study based on spatially explicit indicators. *Land Use Policy, 21,* 289–306. doi:10.1016/j.landusepol.2003.10.012.

Young, T., & Burton, M.P. (1992). *Agricultural sustainability: Definition and implications for agricultural and trade policy*. Rome: FAO Economic and Social Development Paper 110.

## Author Biographies

**Enric Tello** is an Agricultural and Environmental Historian at the University of Barcelona. He leads a multidisciplinary research team of the international project on "Sustainable Farm Systems: Long-Term Socio-Ecological Metabolism in Western Agriculture" (SFS) funded by the Social Sciences and Humanities Research Council of Canada. Along with the SFS Partnership Grant, this team has developed new methodologies like a multi-EROI approach to Energy Flow Accounting of Agroecosystems, an Energy-Landscape Integrated Analysis (ELIA), and a Sustainable Farm Reproduction Analysis (SFRA) which aim at carrying out Environmental History research as a transdisciplinary meeting point where some of the main research topics on Sustainable Science can be addressed from a long-term perspective—e.g. assessing how bio-economically circular farm systems are, what ecosystem services they offer, and what environmental impacts they cause.

**Manuel González de Molina** holds a Ph.D. in History and is a full professor in Environmental History. Currently he works as a head of the Agro-Ecosystems History Laboratory (Pablo de Olavide University, Seville, Spain). He has developed new interpretive and methodological instruments for evaluating agrarian sustainability from a metabolic and agroecological point of view. He is Director of the research-oriented Masters programme in "Agroecology, a Sustainable Approach of Organic Agriculture", and a member of the editorial board of the ISI-refereed journals *Sustainability* and *HistoriaAgraria* He has been vice-president of the Spanish Society for Organic Agriculture (SEAE) and was in charge of the Department of Organic Farming at the Regional Government of Andalusia (2004–2008).

# Chapter 3
# Biophysical Analysis of Agri-Food Systems: Scales, Energy Efficiency, Power and Metabolism of Society

**Tiziano Gomiero**

**Abstract** In this chapter, I discuss some important theoretical issues that should be addressed when attempting a sustainability assessment of the biophysical performance of agri-food systems. Assessing the sustainability of an agri-food system is a complex matter due to the multi-functional nature of agriculture and the multi-scale nature of the relations between agroecosystems and socio-economic systems. The complexification of the agri-food system makes it difficult to clearly establish what is local, and what can be described as a short food chain, and we might be confronted with cases where, according to certain environmental criteria, products imported from a long distance can perform better than locally produced ones. I argue that a fund/flow analysis is a very useful approach when assessing the pressure on biophysical systems and monitoring their health (stock/flow, and flow/flow may also be useful approaches). Energy is the cornerstone of any living system, including societies. I discuss how energy efficiency and energy flow (power) need to be understood as mutually dependent factors, and how they play a key role in interfacing with the performance of the agri-food and socio-economic systems. If we address societies as living systems, we adopt the concept of metabolism as a useful approach to study the functioning of the biophysical characteristics of agri-food systems and societies alike. In this chapter, I review the history of the concept of metabolism in this context and the approaches taken by the two main schools of thought on the topic. As many definitions are found in literature on the subject, I try to suggest two possible definitions, taking into account the specific approaches. The Vienna school, led by Marina Fischer-Kowalski, embraces a stock/flow and flow/flow approach, considering society as a black box. I would suggest naming this approach "steady-state social metabolism", or, for short, "social metabolism". The Barcelona school, led by Mario Giampietro, addresses the relation between fund/flow patterns taking place in a society (its internal organisation), and the fund/flow patterns taking place in its environment as a co-evolutionary, self-organising process. I would suggest naming this approach "co-evolutionary societal-ecological metabolism", or, for short, "societal metabolism". Finally,

---

T. Gomiero (✉)
Treviso, Veneto, Mogliano Veneto (TV), Italy
e-mail: tiziano.gomiero@libero.it

 
E. Fraňková et al. (eds.), *Socio-metabolic Perspectives on the Sustainability of Local Food Systems*, Human-Environment Interactions 7,
https://doi.org/10.1007/978-3-319-69236-4_3

I stress that a biophysical analysis has to be carried out in parallel with an analysis of the socio-economic dimension of a society, as the two dimensions are strictly correlated. I point out that in order to develop a more sustainable agri-food system, we have to intervene in the very functioning of society, and that the complex nature of society's metabolism has to be carefully addressed.

**Keywords** Agri-food system · Biophysical indicators · Multi-criteria approach · Fund/flow analysis · Social metabolism · Societal metabolism · Energy efficiency · Power

## 3.1 Introduction

This book aims at providing a sustainability assessment of specific food productionpractices concerning local food production and their impacts at a local level, while also considering their global implications, focusing in particular on the biophysical features of local food systems (Fraňková et al., Chap. 1 of this volume). In this chapter, I discuss how the assessment of local farming and agri-food systems needs to be performed considering both their biophysical characteristics (e.g., yield, environmental impact) and the functioning of society as a whole (e.g., total food produced by the agricultural system, labour productivity quantified as tons per working hour). The chapter will focus on modern societies, i.e., those societies where farmers somehow interact with the rest of society, as part of a society itself, for example, by buying goods (e.g., tools, food items) and services (medical assistance, attending school) and selling part of what they produce to the market (be it crops, livestock or other items). Isolated self-sufficient communities (i.e., hunter-gatherers, horticulturists, pastoralists) display very different characteristics. Although part of what is discussed in this chapter might also apply in those cases, nevertheless, a different approach might be best suited to properly address those social systems.

A food system can be defined as a chain of activities, from production in the field to consumption, with particular emphasis on processing and marketing and the multiple transformations of food that these stages entail (Tansey and Worsley 1995; Ericksen 2008; Lang et al. 2009). To emphasise the role of agriculture, "agriculture and food system" is often used, shortened to "agri-food system" (see FAO 2016; OECD/FAO 2016).

The agri-food system links three different aspects of life: (1) biological, i.e., the living processes used to produce and consume food and their ecological sustainability, (2) economic and political, i.e., the power and control that different groups exert over the different parts of the system, and (3) social, i.e., the personal relations, community values, and cultural traditions which affect people's use of food (Tansey and Worsley 1995).

The assessment of the sustainability of food production and agri-food systems is thus a complex task, as it requires the addressing of different but related issues at the same time: (1) the multiple identities of the food system in relation to how it is perceived by the different stakeholders, (2) the multi-scale (i.e., farm, farming system, country) and dynamic nature of the agri-food system, and (3) the interwoven relationship between the functioning of the agri-food system and the functioning of society as a whole (Tansey and Worsley 1995; Gomiero et al. 1997; Giampietro 2004; Lang et al. 2009; Giampietro et al. 2014; Gomiero 2017).

In the case of modern societies, the complexification of the agri-food system makes it difficult to clearly establish what is local, and what can be described as a short food chain (Brunori et al. 2016). Furthermore, a shorter food chain might not necessarily result in a lower environmental impact. In actual fact, we might be confronted with cases where products imported from a long distance can perform better according to certain environmental criteria than locally produced ones (El-Hage Scialabba and Müller-Lindenlauf 2010; Payen et al. 2015; Brunori 2016).

Nevertheless, shortening food chains and moving toward more locally based food systems might bring many environmental, social and economic benefits. Over the last decades, agricultural intensification has brought with it some paradoxical results: massive overproduction with subsidies provided by governments to dispose of any surplus (lately biofuels have taken that role), environmental contamination due to agrichemicals, soil degradation and loss of biodiversity (Foley et al. 2011; Gomiero et al. 2011a; Gomiero 2016). At the same time, a huge amount of food is being wasted along the food chain, and a large share of crops is fed to livestock (with an impact on GHG emissions), rural areas are depopulated and losing their cultural identity, while farmers are facing economic problems. At the society level, food-related diseases are on the increase and there is a poor awareness of our dependence on soil and agriculture. Therefore, it is urgent to rethink the functioning of our agri-food system, in order to make it more sustainable and socially fair, which could actually bring great benefits to the whole of society (Foley et al. 2011; Gomiero et al. 2011a).

We also have to be aware of how the agricultural and food system constrains the structure and functioning of society (i.e., number of individuals, quality of diet, health, distribution of labour across the economic sectors, type and quality of services) (Giampietro 2004; Giampietro et al. 2014). Therefore, changes in the agri-food system (e.g., land productivity, labour productivity, total food supply, the price of food) have to be carefully analysed in relation to how they may affect a society's metabolism (e.g., population, number of people, employment structure, demand for energy and materials). On the other hand, we have to be aware that the functioning of a society puts pressure on the agri-food system to supply the necessary amount of produce for its sustainment (e.g., by increasing taxes, providing subsidies, or, where possible/needed, by importing the necessary amount of food). In this chapter, I will also discuss the usefulness of studying the agri-food system by addressing society's metabolism.

A society can be thought of as an organism. A society interacts with its environment in order to sustain its metabolism and reproduce through the reproduction of its individuals and the pattern of time and resource use. Scholars refer to society's metabolism as the physical (both material and energy) flows between a society and its environment, and within a society itself (Martinez-Alier 1990; Georgescu-Roegen 1971; Ayres 1994; Ayres and Ayres 1996; Fischer-Kowalski 1998a, b; Giampietro et al. 2013).

The chapter is organised as follows: Sect. 3.2 discusses the fuzzy distinction between local and global agri-food systems in the present-day agri-food system. Section 3.3 deals with the multi-scale nature of agri-food systems, and discusses the importance of adopting a multi-criteria approach to the assessment of agri-food system sustainability. Section 3.4 shares some reflections on the biophysical dimension of sustainability, and discusses the usefulness of the fund/flow approach in assessing biophysical sustainability. Section 3.5 discusses the concepts of energy efficiency and energy flow (power) as key interface indicators, necessary to better understand the links between the performance of the agri-food system and the biophysical and socio-economic characteristics of societies. Section 3.6 addresses the usefulness of the concepts of social and societal metabolism in the assessment of agri-food system sustainability. Section 3.7 draws some conclusions on the above.

## 3.2 Local Versus Global Is a Fuzzy Distinction in Present-Day Farming: Implications for the Assessment of the Sustainability of Local, Short Food Supply Chains

In this section, I argue that there is no established definition of what constitutes "local food", and a "local agri-food system" may be hard to find due to the highly interdependent world we are living in.

### *3.2.1 On the Difficulties of Distinguishing Local from Global*

Local agri-food systems, short food supply chains, alternative food networks, and direct sales, have been indicated as strategies able to increase the "biophysical" efficiency of the food chain,[1] i.e., the amount of resources used and their impact per unit of output (i.e., energy required per unit of produce, GHGs emissions per unit of

[1]Of course, efficiency also concerns the socio-economic side of productive activity, for example labour, capital.

produce), and to improve the income of farmers, reducing the environmental impact of food production, guaranteeing better food quality, while at the same time enhancing the social fabric of rural communities and of society at large (Goodman et al. 2001; Martinez et al. 2010; Kneafsey et al. 2013; Johnson et al. 2013; CSM 2016). Johnson et al. (2013) point out that focus on locally sourced foods and efforts to convince consumers to "buy local" are not new concepts, and that, in the USA, "state grown" or "locally grown" programmes were introduced as early as the 1930s, and such programmes now exist in most U.S. states. In Europe, in recent decades, several EU member states have developed legal frameworks and incentives to support such types of food chain (Kneafsey et al. 2013). A review carried out by the USDA (Martinez et al. 2010) found that farmers' markets have positive impacts on local economies, with a significant multiplier effect for jobs (amounting to 1.4–1.8 jobs, that is, each full-time equivalent job created at farmers' markets supported around half of a full-time equivalent job in other sectors of the local economy). Studies also show that consumers consider locally produced food to be of better quality and are willing to pay a higher price for them (for a review see Ekanem et al. 2016). It thus seems that local food systems should also be encouraged through adequate policies, given that shortening the food chain may result in a more sustainable agriculture, able to enhance important social and cultural values. It has to be pointed out that, often, local food systems are characterised by the adoption of organic agricultural practices, which greatly reduce the use of potentially harmful agrochemicals in crops and of antibiotics in livestock, while enhancing soil fertility and providing other environmental benefits (Gomiero et al. 2011b; Gomiero 2013; Reganold and Wachter 2016).

Nevertheless, there is no established definition of what constitutes a "local food" (Martinez et al. 2010; Johnson et al. 2013; Kneafsey et al. 2013). Kneafsey et al. (2013), summarising the main literature on the topic, define Local Food Systems as those systems where the production, processing, trade and consumption of food occur in a defined geographical area of limited size (corresponding to a about a 20–100 km radius, depending on the sources and opinions), and Short Supply Chains as those supply chains where the number of intermediaries is minimised, the ideal scenario being a direct contact between the producer and the consumer. It is also not simple to distinguish local from global in a globalised world (Robinson and Schmallegger 2015; Brunori et al. 2016). A farm in Italy selling its product on the local market may still use fuels extracted in the Middle East, have machineries produced in the USA, use pest control compounds produced in Germany, plant seeds from The Netherlands, adopt agricultural practices developed elsewhere, use technological equipment designed in the USA or Japan and produced in China, and employ manpower from Africa or Eastern Europe. It might also benefit from subsidies coming from the European Common Agricultural Policy. Brunori et al. (2016), in their sustainability analysis of local versus global food chains, point out that "local" can be quite a fuzzy concept, and conclude that "*… differences between "local" and "global" are dispersed along a local–global continuum, and that in real life local and global do not always belong to separate settings or domains.*" (Brunori et al. 2016, p. 17) (see also the special issue edited by Brunori 2016).

### 3.2.2 Local Does not Automatically Mean More Sustainable

Intuitively, locally produced food may seem more sustainable than food coming from far away. For example, thinking of distance may induce us to think that locally produced food should require less energy, and therefore produce less GHGs, than produce imported from far away, making it more energetically and environmentally sustainable. While in many cases this may indeed hold true, nevertheless, we should not make this assumption or generalise too much. Due to the effect of the economy of scale, or environmental comparative advantage, there are cases where agricultural products imported from far away may have a lower environmental impact than the same products produced locally. Vegetables produced in greenhouses, intensive livestock, and products that are cold-stored in a controlled environment for several months (for apples the typical storage time is 6 to 12 months—The Guardian 2003; Gile 2013)[2] may be much more costly in energy terms when produced in Northern Europe than when imported from Southern Europe or from other continents (Wells 2001; Saunders et al. 2006; Williams 2007; El-Hage Scialabba and Müller-Lindenlauf 2010; Payen et al. 2015; Foley 2016).

Saunders et al. (2006), for instance, report that in the case of dairy and sheep meat production, New Zealand is far more energy efficient than the UK, even including transport costs: twice as energy efficient in the case of dairy, and four times as energy efficient in the case of sheep meat. Wells (2001) found that New Zealand dairy production was on average less energy intensive than in North America or Europe, even though on-farm primary energy input had doubled in 20 years and energy ratio (outputs/inputs) had increased by 10%. Williams (2007) reported that Dutch $CO_2$ emissions for rose cultivation were about 6 times larger than those caused when producing them in Kenya and delivering the product to Europe. Payen et al. (2015), carried out a detailed LCA comparing off-season tomato production for the French market, considering the production of tomatoes in Morocco under non-heated greenhouses, and the French cropping systems, characterised by heated greenhouses with high levels of inputs. The authors considered as many inputs and impacts as possible, and found that the French heated greenhouses required about 8 times more energy (fossil fuels) per unit of produce, and were about 4 times more damaging to human health and to ecosystems than Moroccan tomatoes. Concerning the recent trendy narrative of "hyper-local" food, where crops are grown inside a building, whether a warehouse or an office building etc., some authors argue that such production systems are very energy intensive and very expensive to run (Hamm 2015; Foley 2016).

Of course, the choices made by the consumers play an important part in shaping the characteristics of the agri-food system of a region. For example, consumers may choose to eat according to the seasons and therefore rely on local produce when they are available. On the other hand, climatic and social characteristics may force a

[2]Post-harvest cold storage in a controlled atmosphere is widely used for many kinds of fruits and vegetables (The Guardian 2003; Gile 2013).

society to import some produce because some crops may not be grown locally, or, even if they can be, the supply may not meet the demand due to the limited agricultural land available.[3] Cost of the produce is another key factor that plays a part in determining what consumers buy. Again, due to specific local conditions, some crops may be too expensive to produce because of low yields, high amounts of inputs to be used, high risk of pests or climate extremes.

It is also not obvious that farmers are better off when selling their products locally. Soper (2016) makes the case that while it might be true for many farmers operating in the USA or EU (the Global North), for poor farmers operating in the Global South selling their products on the international market (in the Global North) may guarantee a better income.

## 3.3 The Multi-Scale Nature of Agri-Food Systems: Adopting a Multi-Criteria Approach to the Sustainability Assessment of Agri-Food Systems

In this section, I provide some definitions that may help better understand the nested hierarchical structure of agriculture and its elements, and briefly address the complex relation linking farming activities with society at large. I also address the fact that when discussing agri-food systems we should be aware that many different stakeholders are involved. Stakeholders may have different perceptions concerning what should be sustained and why.

### *3.3.1 The Multi-Scale Nature of Agri-Food Systems*

A **farm** can be defined as the physical support where agriculture is carried out (the land area, the buildings etc., used for growing crops and rearing animals). A **farm system** can refer to the household, its resources, resource flows and their interactions within the individual farm. The biophysical, socio-economic and human elements of a farm are interdependent, and thus sustainability issues should address the farm as operating on multiple domains, criteria and scales. A **farming system** is defined as a population of individual farm systems that have broadly similar resource bases, enterprise patterns, household livelihoods and constraints, and for which similar development strategies and interventions would be appropriate (e.g., small-scale low-input agriculture, large-scale cereal cultivation). Thus, a farming system falls within a typology of farm systems, or a typology of households (e.g., small 5–10 ha farms, mainly producing cereals using organic practices, or large

[3]Germany may not be food self-sufficient even if all the agricultural land were to grow wheat and pea in rotation and no animals were reared (Gomiero 2017).

50–100 ha farms, producing cereals using conventional intensive agricultural practices). A farming system can comprise anything from a small number of farms (e.g., large-scale landholdings farming commodity crops), up to millions of households (e.g., millions of Asian small rice growers).

Following a hierarchical approach, a farm is characterised by a specific farm system, with features corresponding to a generalised farming system (a farm system typology), operating within a given **agricultural system**, i.e., the technical and socio-economic characteristics of the sector (e.g., level of mechanisation, economic pressure). Mazoyer and Roudart (2006), distinguish between different agricultural systems in relation to the power capacity applied, and their relative performances in term of areas farmed per worker and tons produced per worker. The authors distinguish, for example: manual cultivation, animal-drawn cultivation using the plough with fallowing, animal-drawn cultivation using the plough without fallowing, mechanised animal-drawn cultivation using the plough without fallowing, as well as four different typologies (stages) of mechanisation/industrialisation. The performances of such agricultural systems range from 1 ha per farmer producing 1 ton of cereals per worker per year, in the case of manual cultivation, to the hundreds of ha per farmer, with a productivity of more than 500 tons per worker per year in the case of high mechanisation. For a discussion on these issues and a list of definitions see for example Beets (1990), McConnell and Dillon (1997), Gomiero et al. (1997), FAO (2001, 2017), Gomiero and Giampietro (2001), Giampietro (2004), Mazoyer and Roudart (2006). Table 3.1 provides a visual summary of the definitions.

Within a society, a given farm or farming system can be characterised by its performance in relation to a series of domains and indicators. Such performance can be assessed against certain benchmarks according to the scope of the assessment. We have to be aware that the performance of a farm or farming system is linked to the biophysical and socio-economic systems in which agricultural activities take place. Benchmarks are also determined or constrained by those characteristics.

In Fig. 3.1, a simple representation is provided showing a multi-criteria, multi-scale representation of the characteristics of a farm or farming system. The two domains on the upper section concern the specific characteristics of the farm or farming system (a few indicators relating to socio-economic and biophysical performance). The two domains below concern the characteristics of the context/society in which the farm, or farming system, operates. The usefulness of this approach is that it forces the analysis to keep track, in parallel, of both the characteristics of the farm, or farming system, and the biophysical and socio-economic characteristics of the region and society.

We can also use this approach to compare farms, or farming systems, operating in different contexts (for further theoretical discussion on this topic see Giampietro and Pastore 1999; Gomiero and Giampietro 2001; Giampietro 2004; Gomiero et al. 2006; Madrid et al. 2013; Serrano-Tovara and Giampietro 2014).

Eventually, the mix of farming systems operating in a region and the characteristics of the agricultural system determine what the agriculture sector supplies to society and its performance level.

**Table 3.1** Different scales and approaches by which we can address agricultural activities

| Term | Definition | Example | |
|---|---|---|---|
| Farm | The physical support where agriculture is carried out (the actual land area of the farm) |  (a) | |
| Farm system | The systemic characteristics that define the activity of the farm (relations within the farm and between the farm and the external world) | 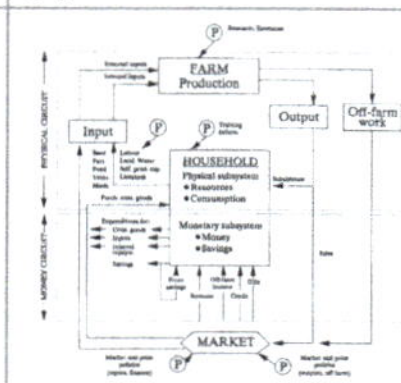  (b) | |
| Farming system | A farm system typology |  Large scale cereal monoculture (c) |  Small scale vegetable polyculture (d) |
| Agricultural system | The technical and socio-economic characteristics of the agricultural sector of a country or region |  Intensive, high input, high power (c) |  Low input, low power (e) |

(a) *Source* https://commons.wikimedia.org/wiki/File:Ontario_farm.jpg (wiki commons)
(b) *Source* http://www.fao.org/Wairdocs/ILRI/x5547E/x5547e0c.gif
(c) *Source* https://en.wikipedia.org/wiki/Combine_harvester (wiki commons)
(d) *Source* https://en.wikipedia.org/wiki/Women_and_agriculture_in_Sub-Saharan_Africa (wiki commons)
(e) *Source* https://en.wikipedia.org/wiki/Working_animal (wiki commons)

Total production (in tons or kcal), land productivity (tons or kcal per ha), and labour productivity (tons or kcal per hr of labour), are the main biophysical parameters that characterise the performance of the agricultural sector. These parameters are also linked to the socio-economic characteristics of the society in determining production costs and the final price of food for consumers. Therefore, they are interface parameters between the biophysical and socio-economic dimensions of the agri-food system. Farming, in order to be a viable activity (with the farmers able to stay in the market), has to achieve a level of performance that meets some minimum thresholds in terms of, for example, land productivity, labour productivity, and the economic return of the investment. The values of such thresholds respond to the pressure exerted by society, that is to say, the constraints posed by the characteristics of the socio-economic system (Grigg 1992; Gomiero et al. 1997; Giampietro 1997, 2004; Pryor 2005; Mazoyer and Roudart 2006; Giampietro et al. 2012, 2013).

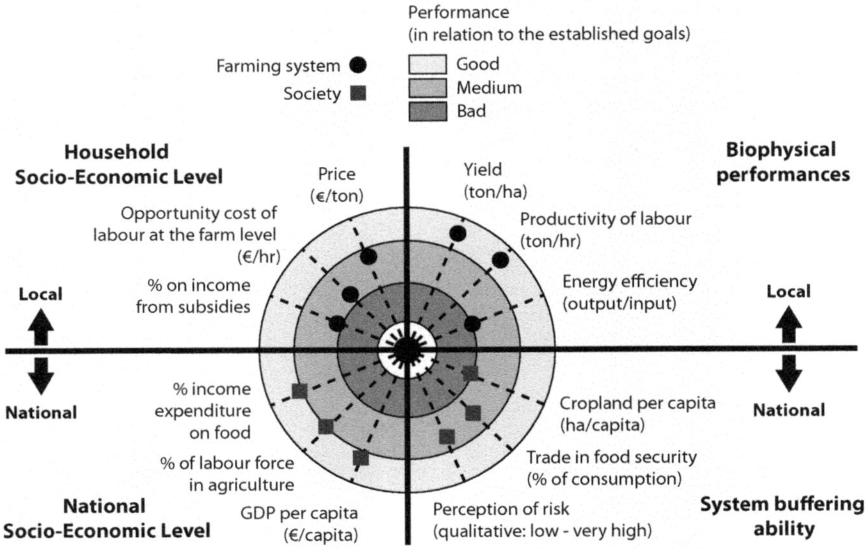


**Fig. 3.1** An example of a possible multi-criteria, multi-scale representation linking the characteristics of a farm (e.g., a specific small-scale horticultural farm) or a farming system typology (e.g., average characteristics of the small-scale horticulture farms) with the characteristics of the society where the activity is performed. The assessment of performance is of course in relation to the established goals: for example, for a farmer achieving high yields and high labour productivity may be a positive thing, even if that comes at the cost of reducing energy efficiency. For a society, to have a low percentage of the working force in agriculture may be a positive thing, as more working time is available to be invested in other productive sectors (figure by the author)

The relations between the farming/agri-food system and society are very complex, concern many different domains, scales and actors and are characterised by many complex feedbacks (Fig. 3.2).

The agri-food system is based on the integration of the different farming systems contributing to the supply of produce required by society, at some specific level of performance (i.e., demand for natural resources and technical capital, land and labour productivity). Performance levels are constrained on one hand by the overall performance of society (that means, for example, that in order to be viable,[4] farming activity should provide farmers with an income that is not too low compared to the rest of society), and, on the other, by the agroecological characteristics of the land (i.e., climate, soil fertility, water supply). We should also be aware that food has a strong cultural component. Therefore, what society demands (and the agri-food system supplies) may also be constrained by the local food culture, which in turn may have been shaped by a co-evolutionary process between society and its

[4]See Sect. 3.5.1 for definitions of the terms viability, feasibility and desirability.

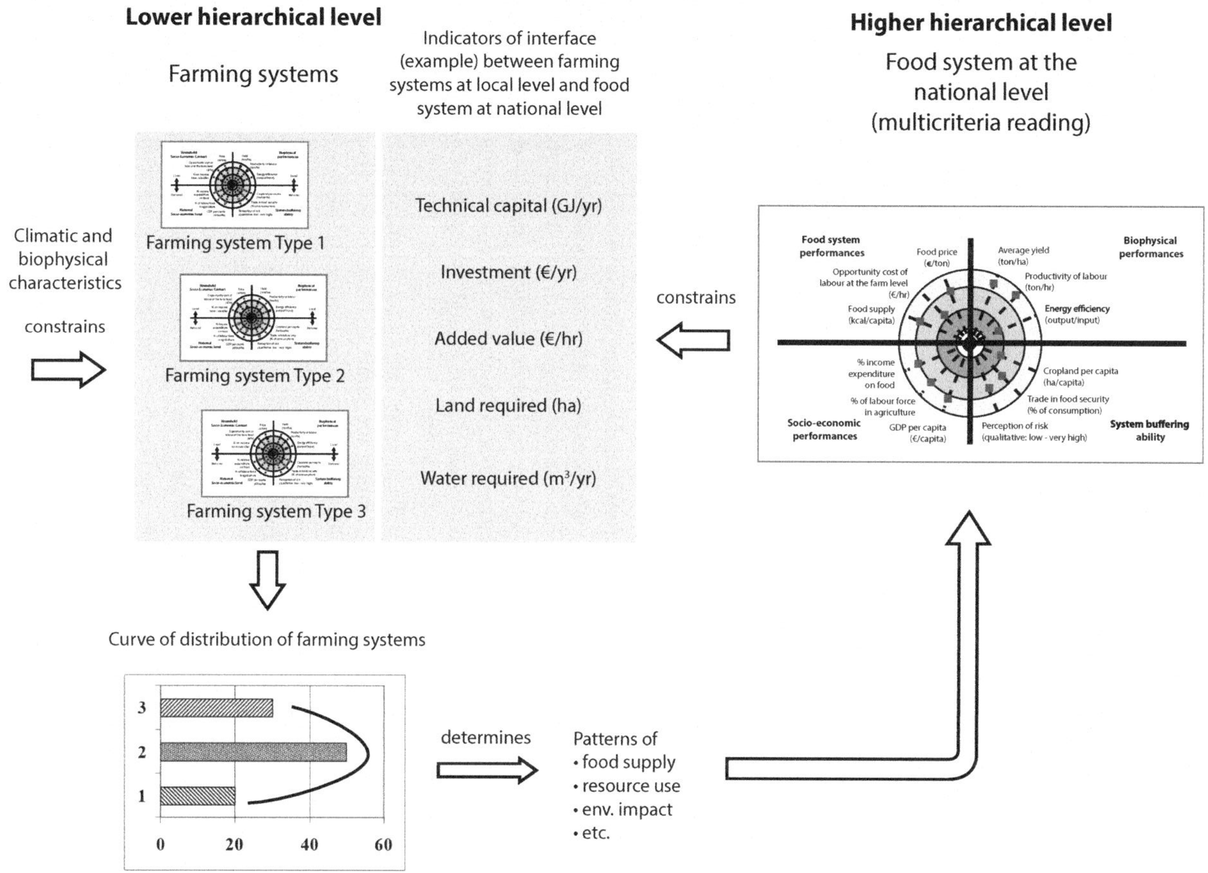


**Fig. 3.2** The relations between the farming/agri-food system and society are complex, involving many different domains, scales, and generating complex feedbacks. It has to be noted that the production systems are also embedded in historical and evolutionary dimensions that characterise a society and its behaviours (i.e., identity, culture, goals) that may play an important role in determining the structure and functioning of the agri-food systems. See text for further explanations (figure by the author)

ecological environment (i.e., food taboos) (Harris 1998; Mintz and Du Bois 2002; Pilcher 2012). Furthermore, what can be grown in a given place, and the techniques and technology required, can determine the characteristics of the local culture and the actual organisation of society (Bray 1994; Diamond 1998).

### 3.3.2 The Multiple Identities of Agri-Food Systems

Any modern agri-food system[5] involves many different stakeholders (i.e., producers, consumers, industries, distributors, researchers, policy makers), who at the same time influence the practice and define the meaning of what is understood as an agri-food system and its sustainability, according to different and non-equivalent criteria and narratives. Some studies, based on literature reviews and interviews with representative stakeholders (Gamboa et al. 2016, see also Chap. 10 in this volume; Brunori 2016), report that stakeholders engaged with different sets of attributes when assessing the sustainability of food production, according to the narrative chosen. For example, Gamboa et al. (2016) report that when dealing with the commodity narrative (food as commodity) the most relevant attributes were added value, productivity, job creation and cost of production. When dealing with the environmental narrative (the relation of food production with the environment) the most relevant attributes addressed were energy efficiency, GHG emissions and biodiversity. Kelemen et al. (2013) report that the perception of biodiversity and its role in sustainable agriculture may vary between different typologies of farmers (i.e., organic vs. conventional), and the countries they are from. The authors conclude that farmers' perceptions of biodiversity seem strongly embedded in their everyday lives, and linked to farming practices. Unfortunately, it seems that studies concerning the real multi-stakeholders assessment of a food system have yet to be carried out.

A further challenge lies in deciding the "proper" set of performance attributes to be addressed by the analysis, and in defining the goals to be achieved (Giampietro 2004; Lang et al. 2009; Brunori et al. 2016; Scharber and Dancs 2016). For example, farmers may wish to increase their share of the end price of a product (usually quite low), competing with the other (usually much stronger) actors in the food chain (i.e., industry, retail), while consumers may wish to purchase products at a lower price. Reducing the cost of food may be of interest to other economic sectors of society, as these would benefit from the savings made by consumers. It should be the role of policy makers to manage such conflicting interests, guaranteeing society its food supply. They should also ensure that food quality meets established safety standards and that food production has a reduced environmental impact.

[5]There are still small societies (i.e., isolated communities of hunter and gatherers and horticulturists) where the complexity of the communities is very limited compared to industrialised countries.

## 3.4 The Biophysical Dimension of Sustainability

In this section, I will deal with two issues related to the biophysical dimension of agri-food systems, concerning specifically the production side, i.e., the agroecosystem.[6] The first issue concerns the use of indicators. The second concerns the fund/flow approach and its usefulness in the analysis of agroecosystem sustainability.

### *3.4.1 Addressing Key Criteria*

A large number of indicators can be used to assess the performance and sustainability of an agricultural and agri-food system. The FAO report "*Sustainability Assessment of Food and Agriculture systems*" (FAO 2013) provides a long list of possible indicators concerning the environment, economy, social issues and governance. Many more can be thought of as well. Thus, the criteria and sets of indicators that we may choose for an assessment depend on the scope of the assessment. We should also be aware that measuring indicators has a cost, which limits what can be done. Indicators can often represent a good proxy for the assessment of the overall health of an agroecosystem. Soil Organic Matter (SOM), for example, is rather simple to measure and can provide important information on soil health. It can also be used as a proxy for the health of soil biodiversity (higher SOM means more activity by soil organisms, from bacteria to earthworms) (Bot and Benites 2005; Wall et al. 2012).

In a farm, the diversity of wild plant species, the diversity of crops (agro-biodiversity) and the percentage of non-farmed land covered by vegetation are also good proxies for overall biodiversity (these characteristics are linked to the presence of invertebrate and vertebrate species). The ecological context in which a farm operates, is, of course, highly important too (Tscharntke et al. 2012). A small, well-managed farm may provide little support for biodiversity if it is found in a degraded agricultural context, such a region characterised by large-scale intensive monocultures and a species-poor landscape.

The use of agrochemicals is another key indicator that can be used as a proxy for the environmental impact of agricultural activities. The use of pesticides (including herbicides) is of particular importance. It has to be pointed out that in life cycle inventories pesticides and herbicides are accounted for in terms of energy or GHGs, and in this way they appear negligible compared to other production costs, such as fertilisers and fuel. The assessment of such compounds has to be done firstly considering their poisonous characteristics.

[6]According to Altieri (2002, p. 8) "*Agroecosystems are communities of plants and animals interacting with their physical and chemical environments that have been modified by people to produce food, fibre, fuel and other products for human consumption and processing*".

Energy efficiency may also be a useful indicator (see Sect. 3.5 for details on this issue). Properly assessing energy efficiency can be a difficult exercise, due, for example, to the difficulty associated with defining system boundaries and tracking energy use (output/input) in complex cropping patterns over many years (Gomiero et al. 2011b). Such an exercise is much easier to carry out for simply managed conventional farming systems than for often complex organic systems. It has to be pointed out that, even when a conventional systems is found to be more energy efficient than a comparable organic or low input system, other criteria should be addressed, such as soil health, environmental contamination, and biodiversity.

### 3.4.2 The Fund/Flow Approach: The Importance of Addressing Funds in Sustainability Analyses

The fund/flow model was proposed by economist Georgescu-Roegen (1971), in order to overcome the limits of the stock and flow models used in economics. Georgescu-Roegen (1971) provides a detailed discussion on the meaning of the terms stock, fund and flow. Here we report the synthesis provided by Giampietro et al. (2014):

***Stock***: "*Referring to reservoirs or buffer of flows, which change their identity (they are depleted or filled) during the time duration of the analysis.*" (p. 37). Stock elements are, for example: fossil fuels, minerals, total agricultural land.

***Flow***: "*Those elements which are either produced (appear) or consumed during the analytical representation. They reflect the choice made by the analyst when deciding what the system does and how it interacts with its context.*" (p. 223). Flow elements are, for example: food, energy, water and money. Energy is the key flow that keeps all biological processes going. Solar energy flows into the ecosystems, or agri-ecosystems, and is absorbed by plants to fix $CO_2$. Biomass is then transformed by herbivores, and so on, while in the process energy is degraded from high to low quality.

***Fund***: "*Structural elements whose "identity" remain the same during the analytical representation. They reflect the choice made by the analyst when deciding what the system is and what the system is made of.*" (p. 223). Fund elements are, for example: population, work force, technological capital, managed land. Considering agricultural activities, a fund element is soil, or soil organic matter. A reduction in the content of soil organic matter means that the soil is under stress and is losing its "identity", that is to say, its structure and function (Fig. 3.3).

The fund/flow model approach addresses processes happening at different scales. At the macro-scale, it analyses aggregate (average) flows per year at the society level (i.e., total food demand, total food produced in terms of biomass, total energy supply, total working time). At the meso-scale, one finds the analysis of flows per year at the level of single activities (i.e., agriculture, industry, services, household sector). At the micro-scale, it measures the performance of the actual activities (in the case of agriculture: yield per ha, kcal per hour of work etc.). The fund/flow approach helps (or, rather, forces to) raise awareness concerning the relation between the

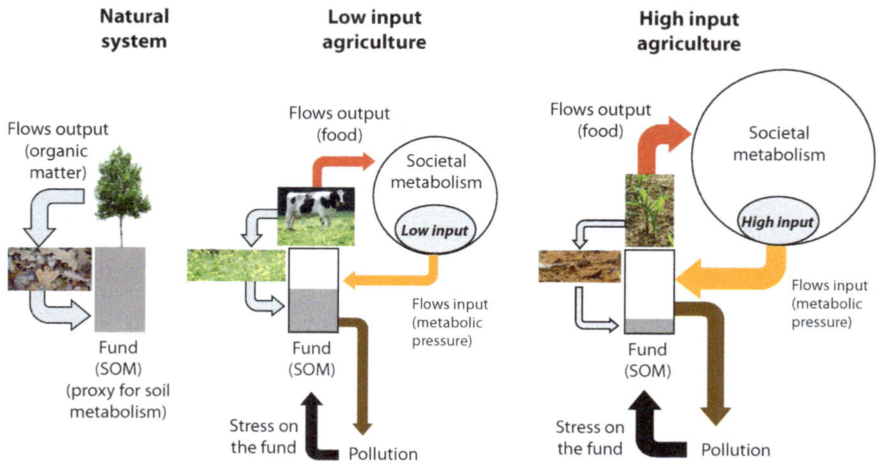


**Fig. 3.3** The fund/flow model allows addressing of the impact of flows (metabolic pressure) on funds, in this case Soil Organic Matter (SOM). In order to preserve ecological system functionality (metabolism), a minimum level of funds have to be preserved. A too low level of SOM ignites a self-reinforcing process of soil erosion that can lead to soil loss and desertification (figure by the author)

metabolic processes taking place in agroecosystems and those taking place at the society level. For example, organic matter (SOM) is a key fund, and its formation rate depends on the metabolism of the ecosystem (or agroecosystem). In order to preserve such a fund, it is of key importance to monitor the rate of nutrient uptake by crops (a flow) and the rate of SOM depletion (C loss, another flow), as that will impact on soil metabolism and on the ability to regenerate the fund (SOM). Reducing the SOM percentage in soil compromises the soil's ability to retain water and sustain biological activity, causing it to become more prone to erosion (Fig. 3.3).

Using SOM content in pristine environments as a reference point (as the fund), by adopting a fund/flow approach we may monitor how agricultural practices impact on soil health. SOM can be used as a proxy for soil metabolism and its health. How SOM varies in relation to the agricultural practices adopted can inform us about how such practices are affecting its health, and in turn, the reproducibility of the fund and related environmental services. For example, a SOM percentage below 2% makes soil highly prone to erosion and crops heavily affected by drought, due to its limited water holding capacity (Rusco et al. 2001; Bot and Benites 2005; Gomiero 2013, 2016). High pressure on this fund, as in the case of high-input agricultural practices, leads to loss of SOM. At present, in the Mediterranean region, it is common to find SOM between 0 and 1% (JRC 2010). Fund degradation also requires an increasing of the amount of input flows (e.g., irrigation, due to the lower water holding capacity, and artificial fertilisers to address lower soil fertility).

By losing SOM, soil also reduces its capacity to regenerate SOM itself, as it needs organic matter to sustain its own metabolism and the living organisms inhabiting it.

Soil metabolism has to be preserved in order for soil to be able to provide the services necessary to sustain agricultural activities in the long run (i.e., fertility, biodiversity, water holding ability, carbon sequestration, resistance to erosion).

Other approaches are also used to study the biophysical dimension of agri-food systems: the stock/flow and the flow/flow approaches. A stock/flow relation is employed when the flow of a resource relies on a stock (i.e., on a non-renewable resource), such as fossil fuels used to produce fuels for the tractors (and other machinery), fertilisers and other agro-chemicals[7] (considering the agri-food system we should also include the fossil fuels used in transportation, storage and in the food industry), or phosphate rocks from which phosphate fertilisers are extracted,[8] and fossil water (or paleowater) used for irrigation.[9] A stock/flow approach can help assess the energy efficiency of human activities in relation to the depletion of fossil fuels (i.e., energy intensity, how much we can produce with a given amount of fossil energy). A flow/flow approach can be applied to the assessment of the efficiency in the use of solar energy, or of other types of renewable energy carriers (i.e., wind, water).

## 3.5 Energy Efficiency and Power in the Assessment of Agri-Food System Performance: Interfacing Society's Biophysical and Socio-Economic Dimensions

In this section, I discuss how energy efficiency and power need to be understood as mutually dependent factors and play a key role in interfacing agri-food system performance and the performance of the socio-economic system.

### 3.5.1 *Energy Efficiency and Productivity in Agriculture*

By ***energy efficiency***, we mean the ratio between the energy output and energy input (joule per joule, or kcal per kcal).[10] Concerning the energy efficiency of the

[7]Actually, fossil fuels are renewable but over geological times, that is to say tens/hundreds of millions of years, a timespan that is not compatible to the life time of human societies or the human species.

[8]Phosphorus has no substitute in food production. Phosphorus security is emerging as one of the twenty-first century's greatest global sustainability challenges (Cordell and White 2014).

[9]Fossil water, also known as paleowater, is water that has been contained in some undisturbed space, usually groundwater in an aquifer, for thousands of years. Fossil water aquifers are either poorly replenished or not replenished at all, making groundwater in those aquifers a non-renewable resource (Steward et al. 2013).

[10]An important indicator of energy efficiency is the Energy Return On Investment indicator (EROI, or EROEI, the Energy Return on Energy Invested). EROI refers to how much energy is returned from one unit of energy invested in an energy-producing activity (Hall et al. 1992, 2011;

**Table 3.2** Energy efficiency comparison of maize production in Mexican subsistence agriculture versus industrial agriculture in the USA (data refer to the 1970s) [Based on data from Pimentel and Pimentel (1979), Pimentel and Hall (1984), Pimentel and Giampietro (1994)]

| Indicator | Mexico subsistence agriculture | USA industrial agriculture | Ratio USA/ Mexico |
|---|---|---|---|
| Productivity of the land Yield (t/ha) | 1.9 | 7.4 | 4 |
| Energy Efficiency Output/ input (kcal/kcal) | 10:1 | 2:1 | 0.2 |
| Productivity of labour Power (kg/hr) | 1.7 | 740 | 435 |

agriculture sector, studies carried out in the 1970s pointed out that industrial agriculture had become largely dependent on fossil energy, and, in turn, was much more energy inefficient than traditional agriculture (Pimentel and Pimentel 1979, 2008; Smil et al. 1982; Stout 1984; Pimentel and Hall 1984; Hall et al. 1992, 2014; Wilson 1992; Smil 2000; Pelletier et al. 2011). As an example, Table 3.2 provides a comparison of the energy performance of traditional maize production in Mexico and intensive maize production in the USA during the 1970s.

While Mexican traditional farming achieved an output/input ratio of 10:1, the output/input of industrial agriculture is only 2:1 (Pimentel and Pimentel 1979; Pimentel and Hall 1984). We may further argue that if the externalities of industrial farming were accounted for (e.g., pollution, soil erosion), its energy efficiency may appear even lower. Following these findings, environmentalists called for a return to more traditional and energetically efficient agricultural practices, while agricultural experts claimed that technology advancements had the potential to reduce agriculture sector energy demands.[11]

Nevertheless, another key issue here is that maize farming in the USA, although 5 times less energy efficient, is 4 times more productive on a per ha basis (land productivity), and, what's more striking (and important as it affects society's overall organization), it is 435 times more productive on a per hour of work basis (labour productivity). The higher productivity of the land means that more food can be produced per unit of surface, generating a high amount of food supply for society,

Giampietro and Mayumi 2009; Pelletier et al. 2011). See Hall et al. (2011, 2014), for a review of the different approaches to the calculation and Tello et al. (2016) for a further development of the concept in the case of agriculture.

[11]Nevertheless, while the use of more and better technology may help increase labour productivity, it requires a lot of energy too. Computerised tractors, satellites, and other technologies do not come cheap in energy terms, Genetic Modified Organisms are a risky enterprise, and seem to lead to more problems (e.g., pest resistance, increased use of agrochemicals, soil degradation, a worrisome monopoly on seeds, health issues). Technology is also costly in economic terms, and its adoption tends to follow a process of fast-decreasing marginal returns (increasingly large investments compared to prices, leading to reduced net earnings). Large estates can cope with the problem because of their economies of scale. Smaller farmers face bankruptcy instead, and have to quit their jobs, their farms bought by large estates, which get even larger (Mazoyer and Roudart 2006).

which can sustain a larger population, providing enough surplus to also feed large livestock herds. In the USA, agriculture generates a supply able to feed all its 350 million people, 90 million cattle and calves, 200 million hogs (older swine) and pigs, 8.5 billion broilers and 350 million egg-laying hens (USDA 2014). The surplus allows for (or requires) the burning of 50% of their maize via bioethanol (Gomiero 2015), and selling (even dumping) part of its production abroad. High labour productivity allows all this production to be obtained with less than 1% of the USA's labour force (CIA 2016). Increasing labour productivity in the agricultural sector allows society to shift its working time toward other activities, for example towards a process of industrialisation, increasing services, schooling, and possibly enjoying more free time.

We therefore face a paradox: the higher the socio-economic development of a society, the lower the energy efficiency of its agriculture (Pimentel and Hall 1984; Martinez-Alier 1990; Smil 1991; Hall et al. 1992; Giampietro 2004; Pelletier et al. 2011). Wilson (1992, p. 218) argued that "*…while agriculture has become efficient technologically, it has become less efficient in its use of energy.*" The author pointed out that although the use of a tractor to produce food reduces the energy efficiency of the process, it still boosts the national economy. Such a paradox can be explained by the different characteristics of the energy carriers on which different societies rely upon. Subsistence societies rely on solar energy, whose flow is a constant (and where plant biomass is the only energy carrier both for humans and animals), while industrial societies rely mainly on fossil fuels and can accelerate the speed of energy consumption (Georgescu-Roegen 1971; Fischer-Kowalski and Haberl 2007; Giampietro et al. 2012). In order to study this process of social change we should focus on the concept of power.

### 3.5.2 The Role of Fossil Fuels in Boosting the Power of Agri-Food Systems: The Effect on the Structure of Society

By ***power*** I mean the speed at which energy is delivered and provides work (unit: joule per second, or watt). The level of power that a society can develop constrains its performances.[12] As for agriculture, the power level achievable poses fundamental constraints to farming activities. Some critical events, such as ploughing, sowing and harvesting are required to be performed within a very narrow temporal window. Harvest time is possibly the most critical time in the agricultural year. Once the crop matures, harvesting has to be performed as fast as possible, that is to say at maximum power, to prevent crops from being spoiled by bad weather or

[12]For a technical discussion concerning power and its relation to the development of a society, I refer the reader to Odum (1971), Smil (1994, 2003, 2008), Giampietro and Mayumi (2009), Giampietro et al. (2013); Giampietro and Diaz-Maurin (2014).

pests. That is why during harvest time, in traditional agricultural societies, all the people who are able to work join in the fieldwork from the very early morning to very late evening. This example should make us aware that some activities are so important that they represent true bottlenecks for both individual households and for the whole of society. Such bottlenecks greatly influence how a society is organised in terms of culture, technology, reproductive pattern, risk perception etc. (Ellen 1982; Beets 1990; Chambers 1997; Pryor 2005; Mazoyer and Roudart 2006; Giampietro et al. 2013). Addressing such issues is fundamental in order to grasp the complex functioning of agricultural activities and, in turn, to plan sound agricultural policies.

Energy efficiency does not inform us of the speed with which energy and matter are supplied, as efficiency does not concern time. Nevertheless, in the case of living systems, it does not matter that the agri-food system has an energy efficiency of 100:1 if it cannot supply the energy at, say, 2000 kcal/day per capita to people, to cover their basic metabolic needs; it would not be a viable agri-food system.

With regard to agriculture, we may reason that energy efficiency is a fundamental indicator to assessing the ***feasibility*** of a food production system, that is to say, the compatibility of the effort with the external constraints imposed by the environment. If energy efficiency is greater than 1, we gain from our investment. Nevertheless, energy efficiency alone does not suffice to properly inform us about the ***viability*** of a food production system. i.e., the compatibility with internal constraints, in this case society's demand for food (Giampietro 2004; Giampietro et al. 2013, 2014; Sorman and Giampietro 2013). Giampietro (2004) and Giampietro et al. (2013, 2014) argue that the concept of ***desirability*** is also important when addressing sustainability—there can be solutions that people might not like or accept even if they are physically feasible and viable, for example, because of food taboos for people who follow a religion, or specific cultural habits.

In order to be able to sustain the structure and functioning of a household or a society, food production has to be compatible with the internal constraints posed by the system's energetic and nutritional requirements. Table 3.3 provides a

**Table 3.3** Energy supply, agriculture productivity and percentage of agricultural labour in agriculture- and industry-based societies [figures from Smil (2000), Giampietro (2004), Giampietro and Mayumi (2009)]

| Society | % Agricultural labour | Agricultural sector productivity$^{(En)}$ (kg cereal/hr labour) | Productivity of the energy supply system$^{(Ex)}$ (GJ/hr labour) |
|---|---|---|---|
| Agriculture-based society | 60–80 | ~2 | 0.5 |
| Industry-based society | 2–6 | 200–300 | 50 |

(En): "Endosomatic energy" is the energy that sustains human metabolism (consumed by humans inside their bodies), (Ex): "Exosomatic energy" is the energy consumed by the whole society (outside human bodies)

comparison between the characteristics of the agricultural sector of industrialised countries (i.e., Europe or North America, the Global North), characterised by a very high level of agricultural mechanisation and labour productivity (high power level applied per worker), and the pattern of metabolism of countries characterised by a very low level of agricultural mechanisation and labour productivity, the case for many countries in the Global South. The comparison concerns the rate of energy flow (power level) of the energy and of the agriculture sector.

In countries of the Global North, the low percentage of working population actively involved in agriculture allows the allocation of working time to industry, the services sector, and to have a large number of people attending school for many years, a large number of pensioners, and much more time for leisure. In the case of many countries of the Global South, most of the people (of all ages) spend most of their time in food production. Increasing the land and labour productivity of the agricultural sector (the primary sector in economic terms), allows for the creation of new kinds of jobs in the industrial (the secondary sector) and in the services sectors (the tertiary sector). Such reorganisation is generally coupled with a reduction in the population's reproductive rate, which helps sustain the accumulation of capital and the transition to industrialisation. If that does not happen, an increase in the productivity of the agricultural sector may just boost population growth, outweighing the benefits of increased productivity.

Higher labour productivity, when not coupled with the development of other productive sectors where people can be employed, may cause massive unemployment among the rural population, leading to dramatic social and economic problems.[13] That is the case in many present-day countries of the Global South, where the fast process of land concentration and industrialisation of agriculture, not coupled with the fast creation of jobs in other productive sectors, are creating a mass of poor and marginalised people pouring into fast-growing city slums. (For a discussion of these topics see for example Ellen 1982; Grigg 1992; Cochrane 1993; Giampietro 2004; Pyor 2005; Mazoyer and Roudart 2006; Timmer 2009; Giampietro et al. 2013; González de Molina and Toledo 2014).

We should also note that, due to the large energy consumption that characterises industrialised countries, the agricultural sector's demand for energy, even if high when compared to traditional systems, is relatively quite limited (Smil 1991, 1994; Wilson 1992; Beckman et al. 2013). In the USA, for example, in the 1980s, agriculture accounted for about 2% of the total U.S. energy consumption (Wilson 1992), and in recent years for even less than 2% (Beckman et al. 2013), while the energy intensity of U.S. farm output slightly decreased (Beckman et al. 2013); since the 1990s, the use of synthetic nitrogen has been more or less stable (USDA 2016). That means that the USA increased energy use in other sectors of society. Camargo

[13]The impact of technology on highly populated countries of the Global South has been a concern since the 1960s. The issue was addressed by renowned scholar Ernst Schumacher, who, in the 1970s, popularised the concept of "intermediate technology", a technology that is appropriate in order to increase productivity in countries of the Global South, but only to a certain extent—to prevent the spread of massive unemployment (Schumacher 1973).

et al. (2013), report an on-farm energy use accounting for 0.8% of the total US energy use, an incredibly small percentage (considering that the USA population is about 320 million people). Nevertheless, Canning et al. (2010) report that the whole agri-food system (which includes agricultural activities, transformation, packaging, transportation, catering sector, etc.) accounts for about 15–16% of the country's total energy consumption.

## 3.6 The Usefulness of Concepts of Social and Societal Metabolism in the Sustainability Assessment of Agri-Food Systems

In this section, I briefly review the history of the concept of metabolism in the study of human societies. I also try to review the terminology, as different terms have been, and are, used (by the same author) to address the study of human society metabolism. I then analyse the importance of the concept of power in the study of human metabolism.

### *3.6.1 Historical Overview*

Early studies on the relation between the characteristics of the agricultural system and society can be traced back to the second part of the 18th century in the works of Podolinskij and Sacher, who worked contemporarily without knowing each other's work (Altieri 1987). Their work concerned the energy efficiency of agriculture and the energy flow (energy content in food) supplied by agriculture to sustain societies, as well as the role of human agricultural work in society (an early notion of power in the agricultural system). Podolinskij based his work on the model of human physiology; the functioning of a society resembled the functioning of human physiology and could be studied as such. Sacher worked on both the idea of energy efficiency and energy flow in agriculture and society. Sacher was also aware that the use of coal-fuelled machinery greatly increased the productivity of labour, reducing, in turn, the number of farmers in the society, freeing time that could become available to perform other tasks (Altieri 1987). Furthermore, Sacher's attempt to carry out a detailed analysis tracing all agricultural system inputs and outputs also represents the precursor of present-day Life Cycle Analysis. In view of their work, Podolinskij and Sacher may be considered the precursors of the study of society's metabolism (I had an exchange of opinions about this with prof. Juan Martinez-Alier, who agrees with such a statement). It was Marx who used the term *societal metabolism* (Fischer-Kowalski 1998a; González de Molina and Toledo 2014), as an analogy to describe the exchange of commodities and the relation of production within society (Altieri 1987; Fischer-Kowalski 1998a). Nevertheless,

Marx never referred to energy (Altieri 1987, Chap. 14), and the final theory developed by Marx and Engels was flawed, "… *incapable of adequately conceptualising the ecological condition and limits of human need-meeting interactions with nature*" (Benton 1989, p. 63, in Fischer-Kowalski 1998a). Altieri (1987, Chap. 14) argues that although Engels understood the difference between the energy stock in coal and the energy flow from the sun, he did not realise the importance of the biophysical (energetic) approach taken by Podolinskij and did not incorporate it in his work.

An important impulse to the study of the relation between energy and society was provided, among others, by the work of the Ukrainian mathematician Alfred J. Lotka (1880–1949), British chemist Frederick Soddy (1877–1956), Spanish ecologist Ramon Margalef (1919–2004), Rumanian economist Nicholas Georgescu-Roegen (1906–1994) and US ecologist H.T. Odum (1924–2002) (Altieri 1987; Fischer-Kowalski and Haberl 2007; González de Molina and Toledo 2014; Giampietro et al. 2013).

Energy studies grew along with the process of industrialisation and the development of anthropological studies of human societies (ecological anthropology, human ecology, ecological economics) (Ellen 1982; Altieri 1987). Energy studies were then further boosted by the 1970s energy crisis (Smil 1991; Giampietro et al. 2014), although, from the 1980s onwards, interest in the topic faded away, to regain attention only recently (Giampietro et al. 2014).

The term ***societal metabolism*** was introduced in scientific literature by Austrian sociologist Fischer-Kowalski (1997, 1998a, b),to indicate a discipline studying the development of societies within a biophysical narrative, through the study of energy and material flows, an alternative to the neo-classical solely monetary-based approach. Fischer-Kowalski derived the concept of societal metabolism from the concept of ***industrial metabolism***"*which focuses on the flow of materials and energy in modern industrial society through the chain of extraction, production, consumption, and disposal.*" (Fischer-Kowalski 1998a, p. 62). The term *industrial metabolism* was first used by the physicist Robert Ayres for a joint conference of the United Nations University and UNESCO in Tokyo in 1988 (Fischer-Kowalski and Haberl 2015).

### *3.6.2 Definitions and Schools of Thought*

Since its appearance, different terms: *societal, social, socioeconomic, society's metabolism*, have been used to refer to the study of the metabolism of societies, by different scholars and by the same scholars in different publications. Some authors noted that this fact, coupled with an improper or unclear use of the term metabolism, might generate some confusion (Giampietro 2014; Pauliuk and Hertwich 2015).

*Societal, social, socioeconomic, society's metabolism* have been used by Fischer-Kowalski and colleagues (Fischer-Kowalski and Haberl 2007; Krausmann et al. 2008). While in the early works *societal metabolism* is used, later on *social metabolism* has been preferred, and the authors refer to them as synonymous.

Fischer-Kowalski also used *socioeconomic metabolism* as synonymous with social/societal metabolism; "*society's metabolism, or (interchangeably) socioeconomic metabolism*" Fischer-Kowalski (1998b, p. 108; see also Fischer-Kowalski and Haberl 2007; Krausmann et al. 2008).

On the other hand, Giampietro and colleagues use the term *societal metabolism* exclusively. Giampietro (personal communication) defines societal metabolism as society's specific structural characteristics, i.e., the metabolic patterns associated with a pattern of funds and flows (operating on different hierarchical levels), which is related to the characteristics of the pattern of ecosystem funds and flows (external environment). This is the reason why he named its accounting approach ("*a diagnostic-simulation tool*" Giampietro, 2014 p. 12) "Multi-Scale Integrated Analysis of Societal and Ecosystem Metabolism" (MuSIASEM) (for details, see Giampietro et al. 2012, 2014, 2013).

It falls beyond the scope of this work to provide an in-depth analysis of these different schools of thought. Nevertheless, I would like to suggest how different approaches to the study of a society's metabolism, namely the Vienna School (lead by prof. Marina Fischer-Kowalski) and the Barcelona School (lead by prof. Mario Giampietro) could be defined. Broadly speaking, both authors started by envisaging society's metabolism as the flow of matter and energy exchanged between society and nature (e.g., Fischer-Kowalski, 1998a, b; Giampietro et al. 2000, 2014). Nevertheless, there are differences in their theoretical development and subsequent accounting methodologies.

### 3.6.3 The Fischer-Kowalski Approach—"Steady-State Social Metabolism"

Fischer-Kowalski's approach focuses on the analysis of energy and material stocks and flows, trying to connect such processes with society's economic dimension. Fischer-Kowalski and Haberl (2015, p. 100) point out that "*Addressing a social system's metabolism means looking upon its economy in terms of biophysical stocks and flows*". Such an approach, which categorises the elements comprising a society in terms of stocks and flows (a stock-flow model), is based on the narrative proposed by steady-state economy scholars such as Kenneth Boulding and Herman Daly, who integrated the two previous models in use in economics, the "flow model" and the "stock model" (Giampietro and Lomas 2014; Georgescu-Roegen 1971). The reason might lay in their focus on the fossil fuel crisis that characterised the 1970s.

The approach considers society as (1) a black box, within which flows of matter and energy pass through and are disposed of, and (2) a steady-state entity, without addressing its dynamics; it takes a picture of what is happening at a given moment in time.

By taking a black box approach, the model provides an immediate energy and material flow accounting, and can monitor the impact of social metabolism on stocks. On the other hand, it may not be able to address the evolutionary process of society's self-organisation and its impact on the metabolism of the supporting system (the environment). This is because it fails to address the fact that ecosystems also have a metabolism. The latter metabolisms do not rely on stocks (as is the case with our industrial society) but on flows and funds. Flows correspond to the solar energy entering the ecosystem, and the cycle of water and organic matter. Funds are the structural elements that constitute the functional compartments of the ecosystem and that have to be preserved in order for the ecosystem to function. For example, we can consider living organic matter in the soil (ton/ha) as a fund, as it keeps the soil alive and properly functioning in relation to the health of the ecosystem (Gomiero 2015). By introducing the concept of fund, we have to face the fact that different funds operate on different time scales. This is also true of societies, when such a concept is adopted (Georgescu-Roegen 1971; Giampietro et al. 2013; Giampietro 2014).

I would suggest naming the approach taken by Fischer-Kowalski as "steady-state social metabolism", or, in short, "social metabolism", referring to "social" as it tends to focus on society as a black box.

### *3.6.4 The Giampietro Approach—"Co-Evolutionary Societal-Ecological Metabolism"*

In the approach taken by Giampietro, stocks are still important, but flow/fund analysis is considered of key importance as well, because of the dynamic, co-evolutionary process that links the development of societies and the dynamics of the environment-supporting system.

According to Giampietro, a society is to be considered as a *dissipative self-organising hierarchical system*. Societies and ecosystems can be described as nested hierarchical systems, where each level (i.e., the person, the household, the society, or the soil, the field, the landscape) is characterised by a metabolism that operates on a different time scale. Society interacts with, and modifies, the environment in order to guarantee the necessary flows of matter and energy to sustain its metabolism (the characteristics of its funds, i.e., population, labour distribution and productivity, etc.). On the other hand, changes in the characteristics of the environment (patterns of funds and flows) affect a society's metabolism.

Using the organism metaphor, Giampietro's approach tries to describe society by its (1) "anatomy", that is, by defining its organs (i.e., agriculture, industry, services, leisure, household activity, education), considering their size in terms of the total human time allocated, and by its (2) "physiology", addressing the specific metabolism of each organ (by adopting a fund/flow model). The metabolism of a society is thus coupled with the metabolism of the environment, and the mutual interactions between the two systems are studied (see Giampietro et al. 2013, p. 192).

This is certainly an in-depth approach, which may help study of the dynamic and self-organising nature of the interaction between social and environmental systems. Some problems with this approach might lay in the fact that different observers may have different (reasonable and legitimate) opinions about how to define the typologies of time use; time spent in education can be accounted as an investment to supply future services by a researcher and as a cost for society by another. Again, it is possible to distinguish between services provided by private or public sectors, as they might have different characteristics. What can be considered as a fund or a stock may also possibly vary according to the issue we are addressing and the timeframe we use (an issue already discussed at length by Georgescu-Roegen 1971). Total agricultural land might be regarded as a stock when the quantity of land is addressed. For example, desertification may cause a reduction in land stocks, while forest clearing may increase the stock, to an existing maximum that will be eventually impossible to overcome (maximum stock of land available). Nevertheless, it might be regarded as a fund when addressed in terms of quality (for example by considering the total organic matter in relation to fertility or as carbon sink). To give an example, organic matter can be reduced when soil is overexploited and increased when proper agroecological practices are adopted. It is therefore crucial to define clearly how the model is constructed and what is being addressed.

I would suggest naming the approach taken by Giampietro "co-evolutionary societal-ecological metabolism", or, for short, "societal metabolism", referring to "societal" as it focuses on the processes taking place within the inner structure and the functioning of a society (its anatomy and physiology). Co-evolutionary, as it addresses the relation between fund/flow patterns taking place in a society, and the fund/flow patterns taking place in the environment as a self-organising process.

### 3.6.5 A Final Remark

I wish to stress that the actual theoretical foundations, models and tools developed by the cited authors are more complex than it has been possible to summarise in this work, therefore I refer the reader to the original works to better grasp the theoretical and applicative aspects of those approaches. Here I wish to point out that there are some important tools available that may help better framing of the sustainability assessment of agricultural practices and agri-food systems. Also, as apparent in this book (Chaps. 5–11), there are many variations and possible crosscuts of the two basic approaches as practised by different scholars in the field, who do not necessarily follow the particular approaches of the founders, and develop them further.

## 3.7 Conclusion

The sustainability assessment of agri-food systems is a complex and challenging task, due to the multi-functional nature of agri-food systems and the complex relations between the agri-food system and the biophysical and socio-economic domains that characterise a society. The scope of this paper is to address some key issues that are highly relevant when carrying out a sustainability assessment of agri-food systems.

In today's world, local and global are interconnected, albeit at varying degrees, and it is impossible to find entirely local agri-food systems (and even if there were such a case, this strictly local system might be affected by the deposition of nitrogen and other chemicals transported by the wind). We should thus be aware that "local" does not automatically mean more sustainable. Sound assessments should be carried out to determine the real biophysical efficiency and impact of "local" agri-food systems. The choices made by consumers may play an important role in leading the agro-food system toward a more sustainable path. Nevertheless, the biophysical characteristics of a region (e.g., climate, land available) may force a society to rely on the global market in order to properly feed itself. Assessing the sustainability of agri-food systems is complex also because many stakeholders, who may have different and legitimate points of view and goals, are involved (e.g., consumers may aim to spend less on food, while farmers may prefer to sell their produce at higher prices).

The characteristics of a farm or farming system have to be understood by addressing the constraints posed by local pedoclimatic and agroecological characteristics, as well as within the structure and functioning of agricultural and socio-economic systems (including historical and cultural aspects). A farmer may content himself with achieving food self-sufficiency, without producing any surplus for the market. If land fertility can be preserved, that model of production is sustainable for the farmer. Could we conclude that such a farm system represents a sustainable model for an entire society? In a society where part of the population is urbanised and other productive sectors exist, this model of production is not sustainable at all, as it fails to supply food for those people not involved in farming. A society (or better its urban population) may actively prevent this behaviour from spreading, by, for example, putting pressure on farmers to produce more by increasing the level of taxation (as has been the case since the formation of early complex societies). On the other hand, a society that put too much pressure on its farmers, and in turn on its soil, might end up losing it all, as was the case for some ancient civilizations (Hillel 1991; Montgomery 2007).

There are many biophysical indicators that can used to assess the sustainability of local agri-food systems, some of them might be highly cost-effective, as, aside from providing key information on system performance they can serve as proxy indicators for other issues. Soil organic matter is a key indicator with which to assess the health status of soil, and can be a proxy for soil biodiversity and other soil characteristics. Fund/flow analysis is a very useful approach when assessing the pressure on biophysical systems and monitoring their health (stock/flow and flow/flow may also be useful approaches when addressing non-renewable resources such as fossils fuels, or flows such as the flow of energy and material in the agri-food system). With regard to biophysical performance indicators, energy efficiency and energy flow are two key indicators. Aside from assessing the performance of an agri-food system, such indicators allow the linking of agri-food systems with the structure and functioning of society, i.e., with its metabolism.

A society can be thought of as an organism, interacting with its environment in order to sustain its metabolism. Scholars refer to society's metabolism as the physical (both material and energy) flows between a society and its environment, and within society itself. With regard to the study of the metabolism of societies, two basic approaches have been developed. The early approach was developed by Fischer-Kowalski, and uses a stock/flow and flow/flow methodology, considering society as a black box. I would suggest naming it "steady-state social metabolism", or, for short, "social metabolism". A more recent approach was envisaged by Giampietro, who addresses the relation between the fund/flow patterns taking place in a society (its internal organisation), and the fund/flow patterns taking place in the environment as a co-evolutionary, self-organising process. I would suggest naming such an approach "co-evolutionary societal-ecological metabolism", or, for short, "societal metabolism".

Present day agri-food systems, especially in the Global North, seem locked in a perverse loop. The agricultural sector is overproducing while farmers see their income dramatically eroded. Farmers are subsidised to intensively grow crops that are then burned (along with taxpayers' money). Soil is being degraded at a fast pace, agrochemicals are contaminating the environment and posing a threat to human health. Even in the Global North, more and more people have problems affording to buy food or having a healthy diet, while a large percentage suffer from obesity and related conditions.

Obviously, we need a careful, complete and systemic review of the dysfunctions that affect the functioning of the agri-food system in order to make it more sustainable. In this chapter, I discussed some issues concerning the biophysical side of the agri-food system that may be relevant in order to carry out such an analysis. Nevertheless, as I stressed in the chapter, a biophysical analysis has to be carried out in parallel with the analysis of a society's socio-economic dimension, as the two dimensions are strictly correlated. In order to develop a more sustainable agri-food system, we have to intervene in the very functioning of a society, and the complex nature of the metabolic processes that keep societies functioning has to be carefully addressed.

## References

Altieri, M. (1987). *Agroecology: The science of sustainable agriculture*. Boulder, CO, USA: Westview Press.

Altieri, M. A. (2002). Agroecology: The science of natural resource management for poor farmers in marginal environments. *Agriculture, Ecosystems & Environment, 93*(1–3), 1–24.

Ayres, R. U. (1994). Industrial metabolism: Theory & policy. In R. U. Ayres & E. S. Udo (Eds.), *Industrial Metabolism: Restructuring for Sustainable Development*, Chapter 1: 3–20, Tokyo, Japan: United Nations University Press.

Ayres, R. U., & Ayres, L. W. (1996). *Industrial ecology: Closing the materials cycle*. Aldershott, UK: Edward Elgar.

Beckman, J., Borchers, A., & Jones, C. A. (2013). Agriculture's supply and demand for energy and energy products, EIB-112, U.S. Department of Agriculture, Economic Research Service, May 2013. http://www.ers.usda.gov/media/1104145/eib112.pdf.

Beets, W. C. (1990). *Raising and sustaining productivity of smallholder farming system in the tropics*. Alkmaar, The Netherlands: AgBe Publishing.

Benton, T. (1989). Marxism and natural limits: An ecological critique and reconstruction. *New Left Review, 178*, 51–86.

Bot, A., & Benites, J. (2005). The importance of soil organic matter. Key to drought-resistant soil and sustained food production. FAO, Rome, http://www.fao.org/docrep/009/a0100e/a0100e00.htm#Contents.

Bray, F. (1994). *The rice economies: Technology and development in Asian societies*. Berkeley, USA: University of California Press.

Brunori, G. (Ed.). (2016). Special issue: Sustainability performance of conventional and alternative food chains. *Sustainability*, http://www.mdpi.com/journal/sustainability/special_issues/conventional-and-alternative-food-chains.

Brunori, G., Galli, F., Barjolle, D., Van Broekhuizen, R., Colombo, L., Giampietro, M., et al. (2016). Are local food chains more sustainable than global food chains? Considerations for assessment. *Sustainability, 8*(5), 27. doi:10.3390/su8050449.

Camargo, G. G. T., Ryan, M. R., & Richard, T. (2013). Energy use and greenhouse gas emission from crop production using the farm energy analysis toll. *BioScience, 63,* 263–273.

Canning, P., Charles, C. A., Huang, S., Polenske, K. R., Waters, A. (2010). Energy use in the U.S. food system; Economic Research Report No. 94; United States Department of Agriculture: Washington DC, USA. http://www.ers.usda.gov/publications/err-economic-research-report/err94.aspx.

Chambers, R. (1997). *Whose reality counts?: Putting the first last*. London, UK: Intermediate Technology Publications.

CIA (Central Intelligence Agency, USA). (2016). Labor force—by occupation. https://www.cia.gov/library/publications/the-world-factbook/fields/2048.html.

Cochrane, W. W. (1993). *The development of American agriculture: A historical analysis* (2nd ed.). Minneapolis, MN, USA: University of Minnesota Press.

Cordell, D., & White, S. (2014). Life's Bottleneck: Sustaining the world's phosphorus for a food secure future. *Annual Review of Environment and Resources, 39,* 161–188.

CSM (Civil Society Mechanism). (2016). Connecting smallholders to markets: An analytical guide. Committee on World Food Security. http://www.fao.org/fileadmin/templates/cfs/Docs1516/cfs43/CSM_Connecting_Smallholder_to_Markets_EN.pdf.

Diamond, J. (1998). *Guns, germs and steel: A short history of everybody for the last 13,000 years*. New York, USA: Vintage.

Ekanem, E., Mafuyai, M., & Clardy, A. (2016). Economic importance of local food markets: Evidence from the literature. *Journal of Food Distribution Research, 47,* 57–64.

El-Hage Scialabba, N., & Müller-Lindenlauf, M. (2010). Organic agriculture and climate change. *Renewable Agriculture and Food Systems, 25,* 158–169.

Ellen, R. (1982). *Environment, subsistence and system*. New York, USA: Cambridge University Press.

Ericksen, P. J. (2008). Conceptualizing food systems for global environmental change research. *Global Environmental Change, 18,* 234–245.

FAO (The Food and Agriculture Organization of the United Nations). (2001). Farming systems and poverty: Improving farmers' livelihoods in a changing world. FAO, Rome, Italy, Available online ftp://ftp.fao.org/docrep/fao/003/y1860e/y1860e.pdf. Accessed June 10, 2016.

FAO (The Food and Agriculture Organization of the United Nations). (2013). Sustainability Assessment of Food and Agriculture systems (SAFA). FAO, Rome, Italy. http://www.fao.org/fileadmin/templates/nr/sustainability_pathways/docs/SAFA_Indicators_final_19122013.pdf.

FAO (The Food and Agriculture Organization of the United Nations). (2016). Energy, agriculture and climate change: Towards energy-smart agriculture. FAO, Rome, Italy. http://www.fao.org/3/a-i6382e.pdf.

FAO (The Food and Agriculture Organization of the United Nations). (2017). Analysis of farming systems. http://www.fao.org/farmingsystems/description_en.htm.

FAO and ITPS. (2015). Status of the World's Soil Resources (SWSR). Main Report, Food and Agriculture Organization of the United Nations and Intergovernmental Technical Panel on Soils, Rome, Italy. http://www.fao.org/documents/card/en/c/c6814873-efc3-41db-b7d3-2081a10ede50/.

Fischer-Kowalski, M. (1997). Society's metabolism: On the childhood and adolescence of a rising conceptual star. In M. Redclift & G. Woodgate (Eds.), *The International Handbook of Environmental Sociology* (pp. 119–37). Edward Elgar, Cheltenham.

Fischer-Kowalski, M. (1998a). Society's metabolism. The intellectual history of materials flow analysis, Part I, 1860–1970. *Journal of Industrial Ecology, 2,* 61–78.

Fischer-Kowalski, M. (1998b). Society's Metabolism. The intellectual history of materials flow analysis, Part II, 1970–1998. *Journal of Industrial Ecology, 2,* 107–136.

Fischer-Kowalski, M., & Haberl, H. (2007). Conceptualizing, observing and comparing socioecological transitions. In M. Fischer-Kowalski & H. Haberl (Eds.), *Socioecological transitions and global change: Trajectories of social metabolism and land use* (pp. 1–30). Edward Elgar, Cheltenham.

Fischer-Kowalski, M., & Haberl, H. (2015). Social metabolism: A metrics for biophysical growth and degrowth. In J.Martinez-Alier & R.Muradian (Eds.), *Handbook of Ecological Economics* (pp. 100–138). Edward Elgar, Cheltenham.

Foley, J. (2016). Local food is great, but can it go too far? https://the-macroscope.org/local-food-is-great-but-can-it-go-too-far-ba686abe2ab7#.lasuswvds.

Foley, J. A., Ramankutty, N., Brauman, K. A., Cassidy, E. S., Gerber, J. S., Johnston, M., et al. (2011). Solutions for a cultivated planet. *Nature, 478,* 337–342.

Gamboa, G., Kovacic, Z., Di Masso, M., Mingorría, S., Gomiero, T., Rivera-Ferré, M., et al. (2016). The complexity of food systems: Defining relevant attributes and indicators for the evaluation of food supply chains in Spain. *Sustainability, 8,* 515.

Georgescu-Roegen, N. (1971). *The entropy law and the economic process*. Cambridge, USA: Harvard University Press.

Giampietro, M. (1997). Socioeconomic constraints to farming with biodiversity. *Agriculture, Ecosystems and Environment, 62*(2,3), 145–167.

Giampietro, M. (2004). *Multi-scale integrated analysis of agroecosystems*. Boca Raton, FL, USA: CRC Press.

Giampietro, M. (2014). The scientific basis of the narrative of societal and ecosystem metabolism. In M. Giampietro, R. J. Aspinall, J. Ramos-Martin, & S. G. F. Bukkens (Eds.), *Resource accounting for sustainability assessment: The nexus between energy, food, water and land use*. New York, USA: Routledge.

Giampietro, M., Aspinall, R. J., Ramos-Martin, J., & Bukkens, S. G. F. (Eds.). (2014). *Resource accounting for sustainability assessment: The nexus between energy, food, water and land use*. New York, USA: Routledge.

Giampierto, M., & Diaz-Maurin, F. (2014). Energy grammar. In M. Giampietro, R. J. Aspinall, J. Ramos-Martin, & S. G. F. Bukkens (Eds.), *Resource accounting for sustainability assessment: The nexus between energy, food, water and land use* (pp. 90–115). Routledge: New York, USA.

Giampietro. M., & Lomas, P. (2014). Interface of societal and ecosystem metabolism. In M. Giampietro, R. J. Aspinall, J. Ramos-Martin, & S. G. F. Bukkens (Eds.), *Resource accounting for sustainability assessment: The nexus between energy, food, water and land use* (pp. 33–48). Routledge: New York, USA.

Giampietro, M., Mayumi, K., & Martinez-Alier, J. (2000). Introduction to the special issues on societal metabolism: Blending new insights from complex system thinking with old insights from biophysical analyses of the economic process. *Population and Environment, 22*, 97–10.

Giampietro, M., & Mayumi, K. (2009). *The biofuel delusion: The fallacy of large scale agro-biofuels production* (p. 318). Earthscan: London, UK.

Giampietro, M., Mayumi, K., & Sorman, A. H. (2012). *The metabolic pattern of societies: Where economists fall short*. Taylor & Francis Group: Routledge.

Giampietro, M., Mayumi, K., & Sorman, A. H. (2013). *Energy analysis for a sustainable future*. Routledge, London, UK: Taylor & Francis Group.

Giampietro, M., & Pastore, G. (1999). Multidimensional reading of the dynamics of rural intensification in China: The Amoeba approach. *Critical Review in Plant Sciences, 18*(3), 299–329.

Gile, A. (2013). The science of cold apple storage. Modern farmer. http://modernfarmer.com/2013/08/the-science-of-cold-apple-storage/.

Gomiero, T. (2013). Alternative land management strategies and their impact on soil conservation. *Agriculture, 3*, 464–483.

Gomiero, T. (2015). Are biofuels an effective and viable energy strategy for industrialized societies? A reasoned overview of potentials and limits. *Sustainability, 7*, 8491–8521.

Gomiero, T. (2016). Soil degradation, land scarcity and food security: Reviewing a complex challenge. *Sustainability, 8*, 281; doi:10.3390/su8030281.

Gomiero, T. (2017). Agriculture and degrowth: State of the art and assessment of organic and biotech-based agriculture from a degrowth perspective. *Journal of Cleaner Production*. doi:10.1016/j.jclepro.2017.03.237. In press.

Gomiero, T., & Giampietro, M. (2001). Multiple-scale integrated analysis of farming systems: The Thuong Lo commune (Vietnamese uplands) case study. *Population and Environment, 22*(3), 315–352.

Gomiero, T., Giampietro, M., Bukkens, S. M., & Paoletti, G. M. (1997). Biodiversity use and technical performance of freshwater fish culture in different socio-economic context: China and Italy. *Agriculture, Ecosystems and Environment, 62*(2,3), 169–185.

Gomiero, T., Giampietro, M., & Mayumi, K. (2006). Facing complexity on agroecosystems: A new approach to farming system analysis. *International Journal Agricultural Resources, Governance and Ecology, 5*, 116–144.

Gomiero, T., Pimentel, D., & Paoletti, M. G. (2011a). Is there a need for a more sustainable agriculture? *Critical Reviews in Plant Science, 30*, 6–23.

Gomiero, T., Pimentel, D., & Paoletti, M. G. (2011b). Environmental impact of different agricultural management practices: Conventional versus organic agriculture. *Critical Reviews in Plant Science, 30*, 95–124.

González de Molina, M., & Toledo, V. M. (2014). *The social metabolism. A socio-ecological theory of historical change*. Germany: Springer, Dordrecht.

Grigg, D. (1992). *The transformation of the agriculture in the west*. Oxford, UK: Basil Blackwell.

Hall, C. A. S., Cleveland, C. J., & Kaufmann, R. (1992). *Energy and resource quality*. Niwot, Colorado, USA: University of Colorado.

Hall, C. A. S., Dale, B. E., & Pimentel, D. (2011). Seeking to understand the reasons for different energy Return on investment (EROI) estimates for biofuels. *Sustainability, 3*, 2413–2432.

Hall, C. A. S., Lambert, J. G., & Balogh, S. B. (2014). EROI of different fuels and the implications for society. *Energy Policy, 64*, 141–152.

Hamm, M. W. (2015). Feeding cities—with Indoor vertical farms? http://www.fcrn.org.uk/fcrn-blogs/michaelwhamm/feeding-cities-indoor-vertical-farms.

Harris, M. (1998). *Good to eat: Riddles of food and culture*. Long Grove, Illinois, USA: Waveland Press.

Hillel, D. (1991). *Out of the Earth: Civilization and the life of the soil*. Berkeley, CA, USA: University of California Press.

Johnson, R., Alison, R., & Cowan, T. (2013). The role of local food systems in U.S. Farm Policy. Congressional Research Service, Prepared for Members and Committees of Congress, R42155. https://www.fas.org/sgp/crs/misc/R42155.pdf.

JRC (Joint Research Centre). (2010). The State of soil in Europe. A contribution of the JRC to the EEA Environment State and Outlook Report—SOER 2010. Joint Research Centre, EC, Ispra (VA), Italy. http://publications.jrc.ec.europa.eu/repository/bitstream/JRC68418/lbna25186enn.pdf.

Kelemen, E., Nguyen, G., Gomiero, T., Kovács, E., Choisis, J.-P., Choisis, N., et al. (2013). Farmers' perceptions on biodiversity: Lessons learnt from a discourse-based qualitative valuation study. *Land Use Policy, 35,* 318–328.

Kneafsey, M., Venn, L., Schmutz, U., Balázs, B., Trenchard, L., Eyden-Wood, T., et al. (2013). Short food supply chains and local food systems in the EU. A state of play of their socio-economic characteristics. European Commission, EUR 25911—Joint Research Centre—Institute for Prospective Technological StudiesLuxembourg: Publications Office of the European Union. http://ipts.jrc.ec.europa.eu/publications/pub.cfm?id=6279.

Krausmann, F., Erb, K. -H., Gingrich, S., Lauk, C., & Haberl, H. (2008). Global patterns of socioeconomic biomass flows in the year 2000: A comprehensive assessment of supply, consumption and constraints. *Ecological Economics, 65,* 471–487.

Lang, T., Barling, D., & Caraher, M. (2009). *Food policy: Integrating health, environment and society*. London, UK: Earthscan.

Madrid, C., Cabello, V., & Giampietro, M. (2013). Water-use sustainability in socioecological systems: A multiscale integrated approach. *BioScience, 63,* 14–24.

Martinez, S., Hand, M., Da Pra, M., Pollack, S., Ralston, K., Smith, T., et al. (2010). Local food systems: Concepts, impacts, and Issues. ERR 97, U.S. Department of Agriculture, Economic Research Service, May 2010. http://www.ers.usda.gov/media/122868/err97_1_.pdf.

Martinez-Alier, J. (1990). *Ecological economics, energy, environment and society*. Oxford, UK: Blackwell.

Mazoyer, M., & Roudart, L. (2006). *The history of world agriculture: From the Neolithic age to the present crisis* (p. 2006). Earthscan: London, UK.

McConnell, D. J., & Dillon, J. L. (1997). *Farm management for Asia: A system approach*. FAO, Rome, p. 355. http://www.fao.org/documents/card/en/c/0358a77f-37f0-5297-8b84-503398b45f21/.

Mintz, S. W., & Du Bois, C. M. (2002). The anthropology of food and eating. *Annual Review of Anthropology, 31,* 99–119.

Montgomery, D. R. (2007). *Dirt: The erosion of civilization*. Berkeley, CA, USA: University of California Press.

Odum, H. T. (1971). *Environment, power, and society*. New York, USA: Wiley.

OECD/FAO. (2016). OECD-FAO Agricultural Outlook 2016–2025, OECD Publishing, Paris. http://dx.doi.org/10.1787/agr_outlook-2016-en.

Payen, S., Basset-Mens, C., & Perret, S. (2015). LCA of local and imported tomato: An energy and water trade-off. *Journal of Cleaner Production, 87,* 39–148.

Pauliuk, S., & Hertwich, E.G. (2015). Socioeconomic metabolism as paradigm for studying the biophysical basis of human societies. *Ecological Economics, 119,* 83–93.

Pelletier, N., Audsley, E., Brodt, S., Garnett, T., Henriksson, P., Kendall, A., et al. (2011). Energy intensity of agriculture and food systems. *Annual Review of Environment and Resources, 36,* 223–246.

Pilcher, J. (Ed.). (2012). *The oxford handbook of food history*. Oxford, UK: Oxford University Press.

Pimentel, G., & Giampietro, M. (1994). Food, land, population and the U.S. economy. Carrying Capacity Network 2000 P Street, N.W., Suite 240 Washington DC. http://www.dieoff.com/page55.htm#TABLE_6.

Pimentel, D., & Hall, C. W. (Eds.). (1984). *Food and natural resources*. San Diego, USA: Academic Press.

Pimentel, D., & Pimentel, M. (1979). *Food, energy, and society*. New York, USA: Wiley.

Pimentel, D., & Pimentel, M. (2008). *Food, Energy, and Society* (3rd ed.). Boca Raton, FL, USA: CRC Press.

Pryor, F. L. (2005). *Economic system of foraging, agricultural and industrial societies*. New York, USA: Cambridge University Press.

Reganold, J. P., & Wachter, J. M. (2016). Organic agriculture in the twenty-first century. *Nature Plants, 1,* 15221.

Robinson, G., & Schmallegger, D. (Eds). (2015). *The globalisation of agriculture*. Edward Elgar Handbook.

Rusco, E., Jones, R., & Bidoglio, G. (2001). Organic matter in the soils of Europe: Present status and future trends. European Soil Bureau Soil and Waste Unit Institute for Environment and Sustainability JRC Ispra, Italy. http://publications.jrc.ec.europa.eu/repository/bitstream/JRC24739/EUR%2020556%20EN.pdf.

Saunders, C., Barber, A., & Taylor, G. (2006). *Food Miles—Comparative Energy/Emissions Performance of New Zealand's Agriculture Industry.* Research Report No. 285 Agribusiness & Economics Research Unit, Lincoln University, New Zealand. http://www.jborganics.co.nz/saundersreport.pdf. Accessed October 20, 2016.

Scharber, H., & Dancs, A. (2016). Do locavores have a dilemma? Economic discourse and the local food critique. *Agriculture and Human Values, 33,* 121–133.

Schumacher, E. F. (1973). *Small is beautiful*. New York, USA: Harper & Row.

Serrano-Tovar, T., & Giampietro, M. (2014). Multi-scale integrated analysis of rural Laos: Studying metabolic patterns of land uses across different levels and scales. *Land Use Policy, 36,* 155–170.

Smil, V. (1991). *General energetics*. New York, USA: Wiley.

Smil, V. (1994). *Energy in the world history*. USA: Westview Press, Boulder, Co.

Smil, V. (2000). *Feeding the World: A challenge for the twenty-first century*. Cambridge, MA, USA: The MIT Press.

Smil, V. (2003). *Energy at the crossroads*. Cambridge, MA, USA: The MIT Press.

Smil, V. (2008). *Energy in nature and society: General energetics of complex systems*. Cambridge, USA: The MIT Press.

Smil, V. Power Density Primer. (2010). Available online http://www.vaclavsmil.com/wpcontent/uploads/docs/smil-article-power-density-primer.pdf. Accessed January 10, 2015.

Smil, V., Nachman, P., & Long II, T. V. (1982). *Energy analysis and agriculture: An application to U.S. Corn Production*. Westview Press, Boulder, CO., USA.

Soper, R. (2016). Local is not fair: Indigenous peasant farmer preference for export markets. *Agriculture and Human Values, 33,* 537. doi:10.1007/s10460-015-9620-0.

Sorman, A. H., & Giampietro, M. (2013). The energetic metabolism of societies and the degrowth paradigm: Analyzing biophysical constraints and realities. *Journal of Cleaner Production, 38,* 80–93.

Steward, D. R., Bruss, P. J., Yang, X., Staggenborg, S. A., Welch, S. M., & Apley, M. D. (2013). Tapping unsustainable groundwater stores for agricultural production in the High Plains Aquifer of Kansas, projections to 2110. *PNAS, 110,* E3477–E3486.

Stout, B. A. (1984). *Energy use and management in agriculture*. MA, USA: Breton Publishing.

Timmer, C.P. (2009). *A World without Agriculture: The Structural Transformation in Historical Perspective.* Washington, DC, USA: The AEI Press.

Tansey, G., & Rajotte, T. (2008). *The future control of food*. London, UK: Earthscan.

Tansey, G., & Worsley, T. (1995). *The food system: A guide*. London, UK: Earthscan.

Tello, E., Galán, E., Sacristán, V., Cunfer, G., Guzmán, G. I., González de Molina, M., et al. (2016). Opening the black box of energy throughputs in farm systems: A decomposition analysis between the energy returns to external inputs, internal biomass reuses and total inputs consumed (the Vallès County, Catalonia, c.1860 and 1999). *Ecological Economics, 121,* 160–174.

The Guardian. (2003). Just how old are the 'fresh' fruit & vegetables we eat? https://www.theguardian.com/lifeandstyle/2003/jul/13/foodanddrink.features18.

Tscharntke, T., Tylianakis, J. M., Rand, T. A., Didham, R. K., Fahrig, L., Batáry, P., et al. (2012). Landscape moderation of biodiversity patterns and processes—Eight hypotheses. *Biological Reviews, 87,* 661–685.

Turner, B. L. I. I., & Brush, S. B. (1987). *Comparing farming systems.* New York, USA: The Guilford press.

USDA. (2016). Fertilizer Use & markets. https://www.ers.usda.gov/topics/farm-practices-management/chemical-inputs/fertilizer-use-markets/.

USDA (United States Department of Agriculture). (2014). Census of agriculture—2012 Census Publications. https://www.agcensus.usda.gov/Publications/2012/Full_Report/Volume_1,_Chapter_1_US/usv1.pdf.

Wall, D. H., Bardgett, R. D., Behan-Pelletier, V., Herrick, J. E., Jones, T. H., Ritz, K., et al. (Eds.). (2012). *Soil ecology and ecosystem services.* Oxford, UK: Oxford University Press.

Wells, C. (2001). Total energy indicators of agricultural sustainability: Dairy farming case study. Final report. Prepared for Ministry of Agriculture and Forestry by University of Otago, Dunedin. http://www.maf.govt.nz/mafnet/publications/techpapers/techpaper0103-dairy-farming-case-study.pdf. Accessed October 20, 2016.

Williams, A. (2007). Comparative study of cut roses for the British market produced in Kenya and the Netherlands. http://www.fcrn.org.uk/sites/default/files/Cut_roses_for_the_British_market.pdf. Accessed October 20, 2016.

Wilson, P. (1992). The inputs to agriculture. In C. R. W. Spedding (Ed.), *Fream's principles of food and agriculture* (pp. 204–227). Cambridge, MA, USA: Blackwell.

Young, A. (1998). *Land resources: Now and for the future.* Cambridge, UK: Cambridge University Press.

## Author Biography

**Tiziano Gomiero** holds an M.Sc. in Nature Science from Padua University, Italy (1993). He continued his studies at Florence University, Italy, where he earned a Master's degree in Wildlife Management, and at the Autonomous University of Barcelona, Spain, where he earned a Master's degree in Ecological Economics and a Ph.D. in Environmental Science (2005), working on modelling for integrated analysis of farming system. He has worked on several international projects in Europe, Asia and Latin America, concerning rural development, farming system analysis, organic agriculture, biodiversity, and genetic engineering in agriculture. He is also interested in the sustainability of biofuels, soil and biodiversity conservation, and ecological economics. On these topics, he has published about sixty papers and chapters in books, of which thirty papers have been published in international scientific journals. From 2006 to 2011, he served as adjunct professor at Padua University (Italy), teaching Ecology and Agroecology and human ecology. At present, he works as a consultant for research projects and for public institutions.

# Chapter 4
# The Energy–Landscape Integrated Analysis (ELIA) of Agroecosystems

**Joan Marull and Carme Font**

**Abstract** Over the last century, we have seen an unprecedented growth in both global food production and associated socio-environmental conflicts, connected to increasingly industrialized farm systems and a decline in biodiversity. The objective of this chapter is to bring together an integrated methodology, applicable to different spatial scales (from regional to local), to deal with the long-term socio-metabolic balances and changes in the ecological functionality of farm systems. We propose an Intermediate Disturbance-Complexity model of agroecosystems to assess how different levels of human appropriation of photosynthetic production affect the functional landscape structure that hosts biodiversity on a regional scale. We have developed an Energy-Landscape Integrated Analysis that allows us to measure both the energy storage represented by the complexity of internal energy loops, and the energy information held in the whole network of socio-metabolic energy flows, in order to correlate both with the energy imprint in the landscape patterns and processes that sustain biodiversity on a local scale. Further research could help to reveal how and why different management strategies of agroecosystems lead to key turning-points in the relationship between energy flows, landscape functioning and biodiversity. There is no doubt that this research will be very useful in the future to help design more worldwide sustainable food systems.

**Keywords** Food-biodiversity dilemma · Land-sharing debate · Intermediate disturbance hypothesis · Human appropriation of net primary production · Energy return of investment · Sustainable farm systems

J. Marull (✉)
Barcelona Institute of Regional and Metropolitan Studies, Autonomous University of Barcelona, 08193 Bellaterra, Spain
e-mail: joan.marull@uab.cat

C. Font
Department of Mathematics, Autonomous University of Barcelona, 08193 Bellaterra, Spain
e-mail: carme.font.moragon@gmail.com



E. Fraňková et al. (eds.), *Socio-metabolic Perspectives on the Sustainability of Local Food Systems*, Human-Environment Interactions 7,
https://doi.org/10.1007/978-3-319-69236-4_4

## 4.1 Sustainable Agroecosystems: The Global Food-Biodiversity Dilemma

### *4.1.1 Introduction*

In the last century, we have seen unprecedented growth in both global food production and associated socio-environmental conflicts, connected to increasingly industrialised and globalised farm systems (Mayer et al. 2015). Farm systems are facing a global challenge amidst a socio-metabolic transition (Schaffartzik et al. 2014) that places them in a dilemma, between increasing land use intensity to meet the growing demand of food, feed, fibres and fuels (Godfray et al. 2010), and attempting to avoid a loss in biodiversity (Cardinale et al. 2012). The industrialisation of agriculture through the 'green revolution', which spread from the 1960s onwards, has been a major cause of this loss (Matson et al. 1997; Tilman et al. 2002).

However, it is increasingly acknowledged that well-managed agroecosystems can play a key role in biodiversity maintenance (Tscharntke et al. 2005). From a land-sharing perspective on biological conservation (Perfecto and Vandermeer 2010, see also Chaps. 1 and 2 in this volume), there is a claim for wildlife-friendly farming, able to provide complex agroecological matrices. A heterogeneous and well-connected land matrix could maintain a high richness of species in cultural landscapes (Jackson et al. 2012). Depending on land use intensity and the type of farming, agroecosystems may either enhance or decrease biodiversity (Swift et al. 2004). In turn, the adaptive capacities to farming disturbances and agroforestry vary across species and biomes (Gabriel et al. 2013).

In order to solve the global food-biodiversity dilemma, the relationship between species richness and land use patterns must be scrutinized, according to the quantity and quality of anthropogenic disturbances applied by farmers across the landscape (Fischer et al. 2008; Phalan et al. 2011). If human society wants to maintain all types of ecosystem services in the future, we need appropriate methods, operative criteria and meaningful indicators to assess when, where and why the energy throughput driven by farmers increases or decreases the mosaic pattern of cultural landscapes and their capacity to hold biodiversity (Gliessman 1990; Pierce 2014).

### *4.1.2 Objective*

The main objective of this chapter is to present an integrated methodology, applicable on different spatial scales (from regional to local), to deal with the long-term socio-metabolic balances and changes in the ecological functionality of the land matrix. We introduce an Intermediate Disturbance–Complexity (IDC) model of agroecosystems, aimed at assessing how different levels of human appropriation of net primary production affect the landscape's capacity to host biodiversity on a regional scale. Finally, we propose an Energy–Landscape Integrated Analysis (ELIA) of agroecosystems that assesses both the complexity of internal energy loops, and the information held in the whole network of socio-metabolic energy flows, so as to correlate this energy-information interplay with the functional landscape structure on a local scale.

## 4.2 Intermediate Disturbance–Complexity (IDC)

### *4.2.1 The Intermediate Disturbance Hypothesis in Agroecosystems*

The intermediate disturbance hypothesis is a non-equilibrium explanation which describes the state of biodiversity in ecosystems (Wilson 1990). There are different definitions of disturbance (Van der Maarel 1993), but a common one is the destruction (or harvest) of biomass (Calow 1987), leading to the opening up of space and resources for recolonising species—an approach that foregrounds the variation of its spatial extent in ecosystem communities (Wilson 1994). Species coexistence would require spatially patchy disturbance that leads to a trade-off between species that are able to perform best at different stages of post-disturbance succession (Chesson and Huntly 1997). At intermediate disturbance frequencies, both competitive and dispersal species may coexist (Barnes et al. 2006).

Agroecosystems may provide important habitats for a variety of species, especially by offering a considerable number of ecotones (Benton et al. 2003), as well as a permeable land matrix, allowing dispersion among local populations (Shreeve et al. 2004). Due to the edge effect and high connectivity, a complex landscape may host greater biodiversity than landscapes with a uniform character (Harper et al. 2005). In order to manage agroecosystems, farmers invest both energy and their knowledge, which shapes the spatial patterns of a landscape embodied with a bio-cultural heritage (Marull et al. 2016b). The impact of anthropogenic disturbance, resulting from farming activities on biodiversity, may be either positive or negative, depending on the quantity and quality of these socio-metabolic flows and the vertical/horizontal complexity of land uses (Swift et al. 2004). Understanding and managing these patchy mosaics in favour of sustainability requires an interdisciplinary approach to bio-cultural diversity (Parrotta and Trosper 2012) embedded in agroecological landscapes (Matthews and Selman 2006).

### *4.2.2 The Landscape Agroecology Approach*

We present a model of how landscape functional structure is affected by different levels of anthropogenic disturbances on ecosystems when farmers alter *Net Primary*

*Production* (*NPP*) through land use change. The model considers landscape as a complex evolving system. Therefore, we will employ both the conceptual and methodological approaches that allow us to consider the whole land matrix (*X*) as a dynamic (agroeco)system (Marull et al. 2010):

$$T = F(X) = \{V \text{ is open of } T, V \subset X\}$$

where *X* is the land matrix, that is to say the total land area under study, *T* is taken to be a discrete topology: every subset (*V*) of *X* is open to socio-metabolic flows (i.e., energy, materials) in the system of *T*. The area defined in this way is continuous and quantifiable. Therefore, the formal expression of *X* starts off from the gathering of all the points (land-units; $r_i$) in the scope of a given study area:

$$X = \cup\, r_i, i \in I$$

Thus, we have adopted the proposed landscape continuum model[1] as a starting point (Fischer and Lindenmayer 2006). The method relies on topological analysis of land cover maps, it has been entirely formalised using mathematical language, and it has been developed and implemented using geographic information systems (GIS).

The conservation of heterogeneous and well-connected land matrix, with a synergic interplay between human energy disturbances and landscape complexity, would be able to hold high species richness (Marull et al. 2016a). In order to test the intermediate disturbance hypothesis in agroecosystems (the relationship between social-metabolism and landscape complexity), we can analyse a set of land cover metrics (landscape patterns and processes from land cover maps) as a function of unharvested *NPP*. Then we obtain a new variable *Le* (*Landscape Ecology Metric*) using Principal Components Analysis (PCA). Once we have *Le*, we perform a regression analysis with the *Human Appropriation of Net Primary Production—HANPP*, a measure of the extent to which humans modify the amount of *NPP*, that is, the energy available for the agroecosystem's trophic chains after biomass harvest (Haberl et al. 2007).

### 4.2.3 Assessing Landscape Patterns and Processes

We use the *Shannon-Wiener Index* (*H'*), commonly used in ecology as a landscape pattern attribute (Vranken et al. 2015), to account for land matrix heterogeneity

[1]Fischer and Lindenmayer (2006) argued that land matrix and landscape heterogeneity are fundamentally important, and deserve equal attention as habitat or protection patches, especially in human modified landscapes.

(i.e., horizontal structure), based on two components: the number and proportion of different patch types.

$$H' = -\sum_{i=1}^{k} p_i \log_k p_i$$

where $k$ is the total number of land-units in the study area, and $p_i$ is the proportion of the land-unit $i$ in a specific sample cell. $H'$ reaches its highest value when: $p_i = \frac{1}{k}$ for $i = 1, \ldots, k$ (i.e., all land-units are equally probable).

The assessment of landscape processes (Lindenmayer and Fischer 2007) is based on the ecological connectivity model proposed by Marull and Mallarach (2005). This assessment relies on defining a set of *Ecological Functional Areas* (*EFA*) and a computational model of cost distance (through an impedance land matrix), which includes the effect of anthropogenic barriers, considering the type of barrier, the range of distances and the kind of land cover involved. The model defines a basic *Ecological Connectivity Index* ($ECI_b$) in a normalized scale from 0 to 10. This $ECI_b$ emphasizes the role played by the land matrix:

$$\mathrm{ECI_b} = 10 - 9[\ln(1 + \mathrm{x_i})/\ln(1 + \mathrm{x_t})^3]$$

where $x_i$ is the value of the sum of the cost distance by pixel and $x_t$ the maximum theoretical cost distance, then, $ECI_a$ is the absolute *Ecological Connectivity Index*:

$$ECI_a = \sum_{m=1}^{m=n} ECI_b/m$$

where $m$ is the number of *EFA* considered; this index helps to emphasize the role played by all sorts of *EFA* in keeping up ecological connectivity (Pino and Marull 2012).

### 4.2.4 The Intermediate Disturbance Complexity Model

A focus on landscape ecological functionality stresses the spatial dimension of biodiversity, the interplay between human disturbances and land matrix heterogeneity-connectivity, and the role of agroecological land management in ecosystem service provision (Tscharntke et al. 2005). This perspective relies on the interplay between different disturbances and land use diversity as the key mechanism that actually matters in biodiversity maintenance (Loreau et al. 2010).

However, much of this biodiversity is only perceived at scales larger than plot or farm level, and depends on the landscape-wide heterogeneity of land covers.

We work with sample cells from land cover maps, so that:

$$\sum_{i=1}^{k} p_i = 1$$

where $p_i$ is the proportion (%) of the land unit $i$ in a specific cell, and $k$ is the number of land units. We will refer to $p$ as vector: $p = (p_1, \ldots, p_k)$.

We calculate a modification of the Shannon-Wiener Index, the *Landscape Heterogeneity* (*L*), to capture the equi-diversity (richness and evenness) of habitats in sample cells. *L* is used for looking at agroecosystems as the spatial 'imprint' of their social metabolism (energy flows).

$$L = \left( - \sum_{i=1}^{k} p_i \log_k p_i \right) (1 - p_u)$$

where $k$ is the number of different land covers (potential habitats). The existence of built-up landcover $p_u$ results in a loss of potential habitats. Thus, $p_i$ is the proportion of non-urban land covers $i$ in every cell. *L* can be improved, including landscape processes (i.e., ecological connectivity), when data is available, using the following algorithm:

$$Le = \left( aL + b\frac{ECI}{10} \right) 1/(a + \mathrm{b})$$

In this way, we obtain a new indicator, the *Landscape Ecology Metric* (*Le*) (Marull et al. 2015), capturing landscape patterns (*L*, heterogeneity) and processes (*ECI*, connectivity), using PCA (where $a$ and $b$ are the empirical coefficients of both components).

We use *HANPP* as a measure of human disturbance (i.e., affecting vertical structure), where *NPP* is the net amount of biomass produced annually by autotrophic organisms that constitutes the nutritional basis for food chains. *HANPP* is calculated using the following identities (Haberl et al. 2014):

$$\mathrm{HANPP} = \mathrm{HANPP_{luc}} + \mathrm{HANPP_{harv}}$$
$$\mathrm{HANPP_{luc}} = \mathrm{NPP_0} - \mathrm{NPP_{act}}$$

where $HANPP_{harv}$ is the *NPP* appropriation through harvest, and $HANPP_{luc}$ is the change of *NPP* through human-induced land conversions. $HANPP_{luc}$ is defined as the difference between the *NPP* of the potential ($NPP_0$) and actual ($NPP_{act}$) vegetation. *HANPP* is associated with each land unit of the study area, so that *HANPP* is calculated as the weighted sum of some fixed land unit coefficients ($w_i$) by the proportion of surface occupied by each land unit within the sample cell:

$$HANPP = \sum_{i=1}^{k} w_i p_i$$

where $w_i$ denotes the weight of land unit $i$ and $p_i$ the proportion of land unit $i$ in the study area. Variations in *HANPP* not only depend on the variations of $p_i$, but on the variations of $w_i$ as well.

The result is that we have one $H'$ and *HANPP* value for each cell (Marull et al. 2016b).

Looking at the figure *HANPP-H'*, any sample data on these variables (obtained from the same land unit cartographic data) must bear some relationship (Fig. 4.1b). The issue is how to interpret the sample data according to the density of pair values of *HANPP-H'*. We assume that Fig. 4.1b draws the shape of all possible values adopted by the relationship between farming disturbance and land cover diversity, where the actual values of disturbance-complexity interplays of a given landscape can be represented. *HANPP* expresses an average number for each sample cell, but can be obtained with different land cover combinations. Hence, it is important to include land cover metrics in *IDC* calculation (Marull et al. 2017).

Finally, the IDC model combines the landscape structure (*L*) with the biomass available to other species *(1 − HANPP/100)*:

$$\mathrm{IDC} = \mathrm{L}(1 - \mathrm{HANPP}/100)$$

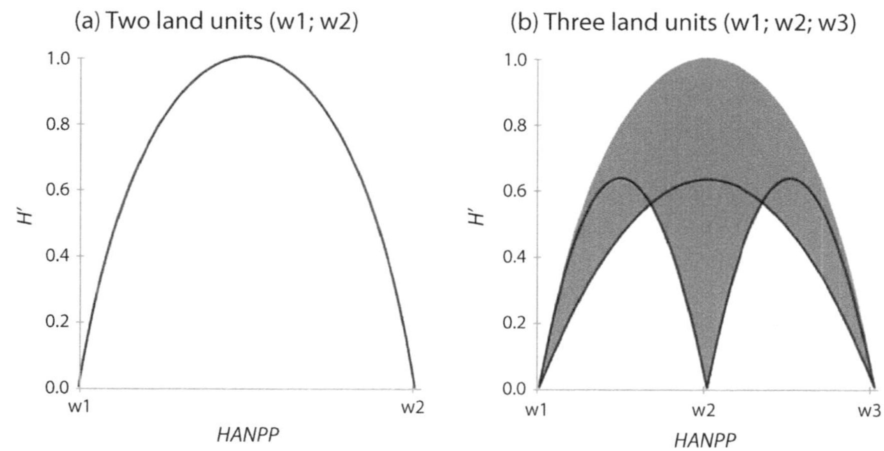


**Fig. 4.1** Shannon-Wiener index ($H'$)—Human Appropriation of Net Primary Production (*HANPP*) ($HANPP = \sum_{i=1}^{k} w_i p_i$) theoretical dispersion graphics for two (**a**) (*H'-HANPP* dispersion graphic with $w_1 = 0$ and $w_2 = 1$) and three (**b**) [*H'-HANPP* dispersion graphic with $w_1 = 0$, $w_2 = 1/2$ and $w_3 = 1$; black lines when $p_i = 0$ (Sect. 4.2.4)] land units

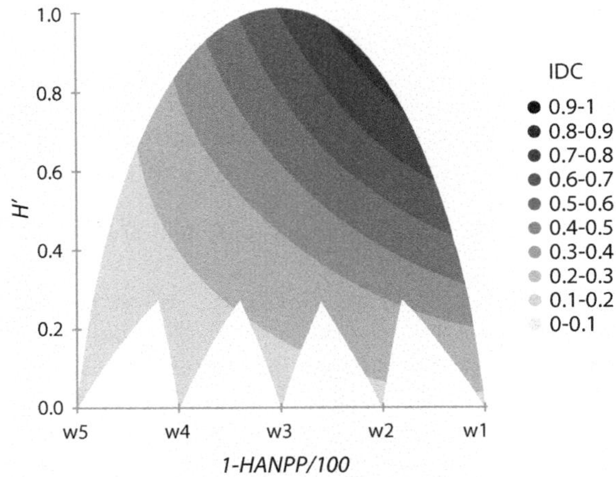


**Fig. 4.2** Theoretical values of the Intermediate Disturbance Complexity (*IDC*) model [IDC = H′(1 − HANPP/100) (Sect. 4.2.4)]. Relationship between Shannon-Wiener Index (*H′*) and Human Appropriation of Net Primary Production (*HANPP*)

where *L* is the energy 'imprint' in the landscape structure (*L* can be substituted by *Le*, including functional attributes of the landscape). *IDC* ranges from 0 to 1, even though its maximum value depends on where the weights ($w_i$) of *HANPP* are displayed.

IDC tries to improve our understanding of the functioning of agroecosystems and the subsequent effects on biodiversity on a regional scale (Marull et al. 2016a), revealing how and why different management leads to turning points in the relationship of the energy profile (*HANPP*) with the landscape structure (Fig. 4.2).

## 4.3 Energy–Landscape Integrated Analysis (ELIA)

### *4.3.1 The Energy-Information Hypothesis in Agroecosystems*

Living systems are capable of using their metabolism in order to maintain or even increase their organisation (Schrödinger 1944). They attain a far-from-thermodynamic equilibrium, set up with the organised information that allows the transferring of energy while maintaining their complexity, reproducing themselves, and evolving (Gladyshev 1999; Ulanowicz 2003). Applying this approach to agroecosystems requires analysis of: (1) the energy throughput and the level of closeness of socio-metabolic cycles; (2) the information carried by the spatially differentiated patterns of this energy flowing across the land matrix; and (3) the landscape heterogeneity to which the species are adapted (Ho and Ulanowicz 2005).

As in any other ecosystem, in agroecosystems landscape complexity arises in nature because the dissipation of energy in space leads to the formation of self-organised structures, and to a historical succession, ruled by adaptive selection (Morowitz 2002). Thanks to the internal biophysical cycles that link organisms, these agroecosystems can increase their own complexity, increase temporal energy storage and decrease entropy. This set of emergent properties translates into the integrated spatial heterogeneity and biodiversity of landscapes (Ho 2013). Their sustainability is directly related to the information-complexity interplay, and inversely related to energy dissipation (Prigogine 1996).

Consequently, agroecosystems can be seen as the historically changing outcome of the interplay between socio-metabolic flows (Haberl 2001), land use patterns set up by farmers, and their ecological functioning (Wrbka et al. 2004). Despite the long-lasting work carried out on energy analysis of farm systems, which has revealed a substantial decline in the energy returns of agro-industrial management brought about by the massive consumption of cheap fossil fuels (Giampietro et al. 2013), the role played by socio-metabolic energy throughput, as one of the main driving forces of contemporary Land Cover/Land Use Change (LCLUC), is not yet well-understood (Peterseil et al. 2004). ELIA intends to link the agroecological accounting of energy flows (Tello et al. 2016) and the study of LCLUC from a landscape ecology standpoint (Marull et al. 2016b). This requires specifying and measuring the pattern of energy flows and the information held in agroecosystems.

### 4.3.2 Energy Flows of an Agroecosystem as a Graph

ELIA represents the energy flows in an agroecosystem as a graph, where specific energy flows are 'nodes' whose 'edges' (i.e., links between them) represent their interaction. Figure 4.3 shows how the phytomass obtained from solar radiation through the autotrophic production by plants, that accounts for the *actual Net Primary Production* ($NPP_{act}$), is the energy source for heterotrophs living there (Vitousek et al. 1986). From this starting point, we analyse the pattern adopted by the energy processes subsequently carried out, the internal loops they generate, the final product extracted, and the external inputs introduced from outside the agroecosystem. The graph shows the three subsystems—internal energy loops ("forestry"—green; "farmland"—red; and "livestock"—purple) included in a mixed farming agroecosystem.

The whole biomass included in $NPP_{act}$, that becomes available for all species, is split into *Unharvested Biomass* (*UB*) and the share of *Net Primary Production harvested* ($NPP_h$) (Fig. 4.3). The *UB* remains in the same place where it has been primary produced to feed the farm-associated biodiversity. It becomes a source of the whole *Agroecosystem Total Turnover* (*ATT*) that closes the first cyclical sub-system called forestry, because it allows for the production of $NPP_{act}$, again through the trophic net of non-domesticated species, either in the edaphic processes of the soil or aboveground.

This does not mean, however, that the entire *NPP* which has been appropriated by farmers leaves the agroecosystem. In turn, $NPP_h$ is subdivided into *Biomass*

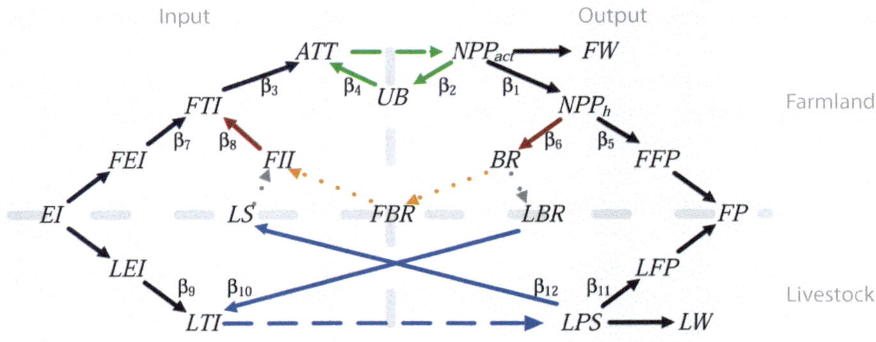


**Fig. 4.3** Graph model of interlinked energy carriers flowing in a mixed-farming agroecosystem [The colours of the arrows represent the 'forestry' (green), 'farmland' (red) and 'livestock' (purple) subsystems (Sects. 4.3.2 and 4.3.3)]. Variables: Actual Net Primary Production ($NPP_{act}$); Unharvested Biomass (*UB*); Harvested Net Primary Production ($NPP_h$); Biomass Reused (*BR*); Farmland Biomass Reused (*FBR*); Livestock Biomass Reused (*LBR*); Farmland Final Produce (*FFP*); External Input (*EI*); Farmland External Input (*FEI*); Livestock External Input (*LEI*); Livestock Total Input (*LTI*); Livestock Produce and Services (*LPS*); Livestock Final Produce (*LFP*); Livestock Services (*LS*); Final Produce (*FP*); Agro-ecosystem Total Turnover (*ATT*); Farmland Total Input (*FTI*); Farmland Internal Input (*FII*). $\beta_i$'s are the incoming-outgoing coefficients. Relationships between variables: $NPP_{act} = UB + LP$; $NPP_h = BR + FFP$; $BR = FBR + LBR$; $EI = FEI + LEI$; $LTI = LEI + LBR$; $LPS = LP + LS$; $FP = FFP + LFP$; $ATT = FTI + UB$; $FTI = FII + FEI$; $FII = FBR + LS$

*Reused* (*BR*) inside the agroecosystem and *Farmland Final Produce* (*FFP*) that goes outside, to be consumed by humans. The *BR* share is an important flow that remains within the agroecosystem as a farmer's investment, addressed to maintain two basic renewable funds: livestock and soil fertility. Hence, *BR* closes the second basic loop called farmland subsystem.

Then *BR* is split into the share that goes to feed the domesticated animals as *Livestock Biomass Reused* (*LBR*), which is added to the whole amount of *Livestock Total Inputs* (*LTI*), and *Farmland Biomass Reused* (*FBR*) which adds up to *Farmland Total Inputs* (*FTI*) as seeds, green manure and other vegetal fertilizers (Fig. 4.3). In this way, the farmland subsystem, which comes from the $NPP_{act}$ in the forestry subsystem, becomes linked to the third livestock subsystem. These energy linkages in the graph enable us to see how they relate to an integrated land use management.

Afterwards, *LBR* flows to domestic animal bioconversion and then splits into *Livestock Final Produce* (*LFP*) and internal *Livestock Services* (*LS*), obtained by farmers as draft power and manure (both make up *Livestock Produce and Services LPS*). The two subsequent loops, called farmland and livestock subsystems, are partially closed within the agroecosystem, while offering a *Final Produce* (*FP*) to be consumed outside—as well as receiving a lower or higher amount of *External Inputs* (*EI*). Therefore, *UB*, *BR* and *LS* provide the internal flows that lead to a stronger or weaker 'loopiness' in the pattern of energy networks of agroecosystems (Fig. 4.3).

Figure 4.3 shows the three subsystems included in one agroecosystem that becomes the outline of a mixed farming that integrates cropping and forestry with livestock breeding. The complexity reached and the information needed to run an integrated mixed farming like this is much higher than with the three subsystems carried out separately. We use this graph model to calculate the level of energy storage within the agroecosystem provided by its 'loopiness', as well as the information embedded in this network of flows.

### 4.3.3 Energy Carriers Stored Within Agroecosystems

Agroecosystems have cyclical aspects because the outputs of one subsystem can serve as inputs for the next subsystem (Fig. 4.3). This, in turn, provides the base for its 'loopiness,' that allows the storage of energy carriers and information within the dissipative structure (Ho and Ulanowicz 2005). There is an exception to this rule though, when some energy flows circulating inside the agroecosystem are turned into what Odum (1993) named a 'resource out of place'. As seen in Fig. 4.3, sometimes a fraction of $NPP_{act}$ can be wasted. The same may happen with a fraction of the *LPS*, such as dung slurry coming from agro-industrial feedlots, that is spread out in excess on cropland and finally contaminates the water table. If they exist, *Farmland Waste* (*FW*) and *Livestock Waste* (*LW*) do not contribute to the renewal of the agroecosystem's funds as these neither enhance its internal complexity, nor meet human needs.

In the integrated graph (Fig. 4.3) we can identify six subprocesses. In all of these, the energy flows can be differentiated between the portion that remains within the agroecosystem and the other portion that goes to other subsystems or out of the whole agroecosystem. Accordingly, there is always a pair of incoming-outgoing energy flows for each sub-process of the agroecosystem. Hence, we propose twelve coefficients ($\beta_i$) along the edges of the graph.

$$\beta_1 = \frac{NPP_{\mathrm{h}}}{NPP_{act}}, \beta_2 = \frac{UB}{NPP_{act}}, \beta_3 = \frac{FTI}{ATT}, \beta_4 = \frac{UB}{ATT}, \beta_5 = \frac{FFP}{NPP_h}, \beta_6 = \frac{BR}{NPP_h},$$
$$\beta_7 = \frac{FEI}{FTI}, \beta_8 = \frac{FII}{FTI}, \beta_9 = \frac{LEI}{LTI}, \beta_{10} = \frac{LBR}{LTI}, \beta_{11} = \frac{LFP}{LPS}, \beta_{12} = \frac{LS}{LPS}.$$

These $\beta_i$'s account for the proportion in which every flow is split into two at each crossroads within the network. Then, we can differentiate between even and odd $\beta_i$'s, where the even $\beta_i$'s account for the energy flows looping inside the agroecosystem, and the odd ones are leaving it. An advantage of using $\beta_i$'s is that they are bounded (between 0 and 1), which allows us to compare different case studies or long-term examples. Here, in order to ensure that $\beta_i + \beta_{2i} = 1$, $NPP_{act}$ has been taken as the sum of *UB* and $NPP_h$, this means, without waste. The same has been done for *LPS*, that is, the sum of *LS* and *LFP*.

In Fig. 4.3 we differentiate between three types of arrows. Solid arrows show the energy flows that represent the internal and external exchange of energy carriers. Dashed arrows indicate energy flows that require biological conversion (i.e., photosynthesis). Finally, point-line arrows show energy carriers that are not diverted inside or outside but remain as 'resources out of place' (i.e., waste).

### 4.3.4 Making Agroecosystems' Energy Graphs Spatially-Explicit

Once we have the agroecosystem's energy network graph (Fig. 4.3), we are interested in the relationships of the evolving complexity of the internal energy loops, with the information they contain, and the diachronic LCLUC. The next step is converting the incoming-outgoing coefficients ($\beta_i$'s) to their land matrix expressions, by calculating the mean territorialised values of energy flowing across each land use (i.e., in $\text{MJ} \cdot \text{ha}^{-1}$). In the following, we explain how.

In most energy flows, there are no difficulties when assigning a value for each land use (data come from EROIs; Tello et al. 2016) if they form part of the first two subsystems (forestry and farmland; Fig. 4.3). In the livestock subsystem, the key point is to set the weight of the whole internal energy loop which corresponds to each land use by taking into account the part of the animal bioconversion that goes to each type of farmland. In order to allocate the full energy costs of livestock to different land uses, we not only take into account the values of *LS* (manure and traction), but also *LW* (dung wasted).

Then, in order to link this network of energy flows with the land matrix, we have to relate ingoing and outgoing flows, measured in the same spatial unit of analysis. Remember that our aim is to analyse the agroecosystem's energy pattern of flows, as a dissipative structure (Prigogine 1996). Hence, what is relevant here is not only the magnitude of each energy flow but also two other things highlighted by our model-graph: (1) the specific part of the network that provides the enhancement of its complexity; and (2) the increasing information embedded in this network. According to Ho and Ulanowicz (2005), the most relevant flows are the loop producers that have to be separated from the entropy producing flows. For this reason, we use $\beta_i^j$ as a first variable defined as the quotient of the energy flow relation *i* associated with the land use *j* (see Sect. 4.3.3). All variables of the energy flow graph (Fig. 4.3) are expressed for each land use *j*. Thus, for each sample cell we have $\beta_i$.

$$\beta_i = \sum_{j=1}^{k} \beta_i^j p_j,$$

where $p_j$ is the proportion of the land use $j$ in the corresponding sample cell of a particular case study, and $k$ is the number of different land uses it contains. Starting from this spatially-explicit $\beta_i$'s, we can calculate the complexity and information carried with energy flows in order to analyse its relationship with landscape patterns.

### 4.3.5 From Energy Flows to Landscape Patterns Through Information

Once we have defined how to account for spatially-explicit energy flows, we can introduce the three indicators that we are going to use in ELIA. They are ordered hierarchically, according to the logical string that goes from the interplay between energy and information to landscape patterns. Energy storage can be seen as the harnessing of dissipation thanks to the farmers' efforts to generate and improve energy loops (Ulanowicz 2003) in traditional and organic farming. The intervention of those farmers' labour also means that the looping of these biomass reuses is not produced randomly through space, because it is driven by information. Depending on the information delivered by farmers' labour the energy flows are directed in one or another way across the land matrix with different intensities. It is because energy carriers flow across different land covers following a deliberate pattern that they imprint a specific mosaic that we recognize as a cultural landscape.

Therefore, energy reinvestment and storage driven by farmers' knowledge produces an effect on landscape patterns and processes (Sect. 4.2.3). ELIA combines the following three indicators: the complexity attained through the energy storage of internal loops in an agroecosystem (*E*); the information embedded in the energy network of flows (*I*); and the landscape functional structure (*L*). The 'loopiness' of energy carriers driven by farmers through *UB*, *BR* and *LS* flows (Fig. 4.3) is a measure of *E*, that expresses the energy potentially available for all food chains existing in the agroecosystem. We are going to start measuring *E* as the quantity of energy remaining in the system (Sect. 4.3.6), and then we will measure *I*, which allows the farmers to reproduce the agricultural metabolism thanks to the information embedded in the system (Sect. 4.3.7). *I* is a measure of how evenly distributed the set of pairwise incoming-outgoing flows of the graph are. Both indicators, *E* and *I*, assess characteristics of human-made structures that allow us to analyse the energy flows of agroecosystems and bring to light the energy-information interplay (Sect. 4.3.8). These variables are then related with *L*, considering them to be the landscape 'imprint' of energy flows produced by social metabolism (Sect. 4.3.9).

### 4.3.6 Measuring Energy Storage as the Complexity of Internal Energy Loops

We understand agroecosystem complexity as the differentiation of dissipative structures that allows for diverse potential ranges in their behaviour (Tainter 1990). At the same time, the more complex the space-time differentiation is, the more energy is stored within a system (Ho and Ulanowicz 2005). Hence, higher mean values of even $\beta_i$'s entail that agroecosystems are increasing in complexity because the different cycles are coupled together and the residence time of the stored energy is increased thanks to a greater number of interlinked transformations looping inside. Accordingly, our way of calculating complexity is as follows:

$$E = \frac{\beta_2 + \beta_4}{2} k_1 + \frac{\beta_6 + \beta_8}{2} k_2 + \frac{\beta_{10} + \beta_{12}}{2} k_3.$$

$$k_1 = \frac{UB}{UB + BR + LS}, k_2 = \frac{BR}{UB + BR + LS}, k_3 = \frac{LS}{UB + BR + LS},$$

where the coefficients $k_1, k_2, k_3$ account for the share of reusing energy flows that are looping through each of the three subsystems (Fig. 4.3).

The formula used implies that $E$ remains within the range $[0, 1]$. $E$ close to 0 implies low reuse of energy flows—a behaviour that usually corresponds to an agro-industrial management which is highly dependent on external inputs and with maximum levels of *HANPP*. $E$ close to 1 implies more internal energy loops, meaning that a high share of energy flows harvested are reused within the agroecosystem—a behaviour usually associated with organic farming with lower dependence on external inputs, lower biomass extraction as *FP*, and also moderate levels of *HANPP*.

$E$ assesses the amount of all the energy flows that go inside the agroecosystem, relative to the total amount of energy flowing across each one of the three subsystems of the network structure. Hence $E$ measures the proportion of energy stored on the land coming from each loop considered sequentially. That is, taking into account that a share of the flow from the first loop can still be redirected inside again when flowing across the two subsequent loops. When we account for the three loops, we are adopting a landscape standpoint that is focused on what happens with the energy flowing across different land units driven by farmers, and we name this proportion of energy stored on the land *Energy Storage* ($E$).

Let us compare *HANPP* and $E$ to analyse two important variables of both models, IDC and ELIA (used at regional and landscape scales, respectively). Expanding the formula of $E$ and assuming the territorialised expression of each beta, we have,

$$
\begin{aligned}
E &= (\beta_2+\beta_4)\frac{k_1}{2}+(\beta_6+\beta_8)\frac{k_2}{2}+(\beta_{10}+\beta_{12})\frac{k_3}{2} \\
&= \frac{1}{2}(k_1\beta_2+k_1\beta_4+k_2\beta_6+k_2\beta_8+k_3\beta_{10}+k_3\beta_{12}) \\
&= \frac{1}{2}\left(k_1\sum_{i=1}^{n}p_i\beta_2^i+k_1\sum_{i=1}^{n}p_i\beta_4^i+k_2\sum_{i=1}^{n}p_i\beta_6^i+k_2\sum_{i=1}^{n}p_i\beta_8^i+k_3\sum_{i=1}^{n}p_i\beta_{10}^i+k_3\sum_{i=1}^{n}p_i\beta_{12}^i\right) \\
&= \sum_{i=1}^{n}p_i\frac{1}{2}\left(k_1\beta_2^i+k_1\beta_4^i+k_2\beta_6^i+k_2\beta_8^i+k_3\beta_{10}^i+k_3\beta_{12}^i\right).
\end{aligned}
$$

Now, we can call $\alpha_i=\frac{1}{2}\left(k_1\beta_2^i+k_1\beta_4^i+k_2\beta_6^i+k_2\beta_8^i+k_3\beta_{10}^i+k_3\beta_{12}^i\right)$, so we get

$$E=\sum_{i=1}^{n}p_i\alpha_i.$$

This expression is similar to the *HANPP* formula; in fact, the two indicators have the same behaviour. Given that $\sum_{i=1}^{n}p_i=1$, in both cases we have a weighted sum of $w_i$ and $\alpha_i$, respectively.

According to the new expression of $E$, the difference between *HANPP* and $E$ devolves on the values $w_i$ and $\alpha_i$. This is because $\alpha_i$ plays the same role as $w_i$ in *HANPP*. However, we should remember that $E$ (applicable at the farm system level; between local and landscape scales) and *HANPP* (only applicable between landscape and regional–global scales in our model) have opposite meaning: high values of *HANPP* in general indicate more human appropriation and so less energy available for other species (in local-specific cases *HANPP* values can become negative; e.g., when $NPP_{act} > NPP_0$ due to fertiliser input), while high $E$ denotes just the contrary, more internal energy processes (and this means more energy available to sustain biodiversity). For this reason, it is better to compare $E$ and *1-HANPP*, once we have adjusted *HANPP* to between 0 and 1.

### 4.3.7 Measuring Information as Shown in the Energy Flow Pattern

In ELIA, Information Theory is applied to the graph model (Fig. 4.3). The equi-distribution of the energy flowing across the edges that link the nodes of this graph assumes that the information they carry cannot be known beforehand. Information can be seen as a measure of uncertainty, or the degree of freedom for the system to evolve (Prigogine 1996). This kind of information is often called structuring information that only registers the likelihood of the occurrence of a pair of events (Ulanowicz 2001). It differs from the meaningful content of the information farmers use to direct the energy flows according to a defined purpose, and also from the spatially organised information that can be measured in the land cover diversity of a farmland mosaic—or even from the information loop of considering the latter as an imprint of the former.

The *Energy Information* (*I*) is always site-specific for the unit of analysis observed, which is an important trait from a bio-cultural standpoint (Barthel et al. 2013). When ELIA registers a decrease of *I*, the information running the system has been lost or transferred from the traditional agroecological knowledge of farmers located at landscape level towards higher hierarchical scales, where other people have taken control of some important parts of the agroecosystem functioning after being linked to increasingly globalised food chains (McMichael 2011).

Accordingly, we use a Shannon-Wiener Index, adapted to be applied over each pair of $\beta_i$'s, so that this indicator shows whether the $\beta_i$'s pairs are evenly distributed or not. This measure of *I* accounts for the equi-proportionality of pairwise energy flows that exit from each node in every sub-process:

$$I = -\frac{1}{6}\left(\sum_{i=1}^{12} \beta_i \log_2 \beta_i\right)(\gamma_F + \gamma_L),$$

$$\gamma_F = \frac{UB + NPP_h}{2(UB + NPP_h + FW)}$$

$$\gamma_L = \frac{LS + LP}{2(LS + LP + LW)}$$

Base 2 logarithms are applied as the probability is dichotomous. The introduction of the information-loss coefficients $\gamma_F, \gamma_L$ ensures that *I* remains lower than 1 when the agroecosystem presents farm and/or livestock waste.

*I* values close to 1 are those with an equi-distribution of incoming and outgoing flows of the agroecosystem's network structure where the structuring information-message is high, whereas values close to 0 mean patterns of probability far from equi-distribution. *I* values close to 0 correspond to a low site-specific information content in agroecosystem functioning, which may be related to an industrialised farm system with high *HANPP* and low relevance of traditional peasant knowledge; or, by contrast, to an almost 'natural' turnover with slight *HANPP* that may also correspond at present to rural abandoned forest or pastoral areas. Conversely, agroecosystems with *I* equal to 1 are the ones with equi-distributed incoming and outgoing energy flows in each sub-process, as well as with intermediate levels of *HANPP* (Marull et al. 2015), that probably correspond to an organic mixed farming deeply embedded in local knowledge.

### 4.3.8 Interplay of Energy Storage with Information

Which configuration is adopted by the whole set of possible values that the interaction between *E* and *I* can take? As a first option, we compute some possible combinations of $\beta_i$'s, and then perform the values of $E$ and $I$ for them, supposing $\gamma_L = \gamma_F = \frac{1}{2}, k_1 = k_2 = k_3 = \frac{1}{3}$. But $E$ differentiates between the different distribution of $\beta_i$'s values into the system, while $I$ does not.

$$I(\beta_1, \beta_2, \ldots, \beta_{12}) = I\left(\beta_{\sigma(1)}, \beta_{\sigma(2)}, \ldots, \beta_{\sigma(12)}\right),$$

where $\sigma$ is a permutation of $\beta_i$'s. $I$ provides seven types of zeros. To study these zeros, we must look at each pair $\beta_i$ and $\beta_{2i}$ (see Fig. 4.4a), for $i = 1, 3, 5, 7, 9, 11$, as $I(\beta_i, \beta_{2i}) = 0$ both when $\beta_i = 1$ or $\beta_{2i} = 1$. So we find seven possible combinations that imply $I = 0$, these are $(\beta_2, \beta_4, \beta_6, \beta_8 \beta_{10} \beta_{12})$: $(0,0,0,0,0,0)$, $(1,0,0,0,0,0)$, $(1,1,0,0,0,0)$, $(1,1,1,0,0,0)$, $(1,1,1,1,0,0)$, $(1,1,1,1,1,0)$, $(1,1,1,1,1,1)$, and any permutation of them. Furthermore, some of these $\beta_i$'s combinations are unlikely, due to the fact that they do not maintain any equilibrium during loopiness.

Following Tello et al. (2016), we assume that if the energy amount of *BR* in an agroecosystem is greater than the energy content of its *EI* (*BR* > *EI*), then the ratio of *FP* over the *Total Inputs Consumed* (*TIC*) grows more for any improvement of *FP/BR* than for *FP/EI*. Hence, we can argue, from the above example, that any increase in *EI* will imply a corresponding increase in 'non loop-producers' $\beta_i$'s relations. Accordingly, we suppose that some coherence can be established between the loop-producing $\beta_i$'s (i.e., not all possible beta-combinations are equally likely).

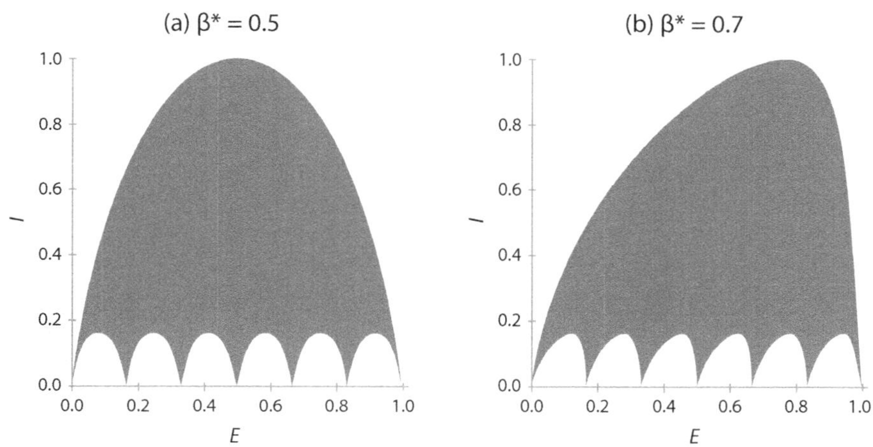


**Fig. 4.4** Theoretical relationship between complexity of internal energy loops (*E*) and information held in the network of energy flows (*I*) of an agroecosystem—measuring equi-diversity ($\beta^* = 0.5$; **a**) [We have maximum information (*I*) for an intermediate level of complexity (*E*) provided by the storage of energy carriers looping inside (Sect. 4.3.8)]. We can also see this relationship when there is an optimisation of energy flows ($I^*$) in order to maintain the agroecosystem funds over time—measuring the information farmers use to direct the energy flows to increase landscape efficiency ($\beta^* = 0.7$; **b**) [We have maximum optimisation $I(\beta^*, \beta)$ for a $max\{EI\}e = 0.6169$ (Sect. 4.3.10)]

Figure 4.4a shows the theoretical representation of interactions between *E* and *I* components. $c_i = (i - 1)/6$, represent the *E* values corresponding to $(\beta_2, \beta_4, \beta_6, \beta_8, \beta_{10}, \beta_{12})$ configurations that make $I = 0$ and all its permutations. We can see an arc that reaches its maximum value on the vertical axis (*I*) for intermediate values of *E*, in the horizontal axis. We have maximum information for an intermediate level of complexity provided by the storage of energy flows looping inside—which for the sake of simplicity we will henceforth call a 'equilibrated' agroecosystem.

At the peak point of *I* (Fig. 4.4a) we found an equi-proportionality of incoming and outgoing energy flows, a property that is not only coherent with our way of capturing the information embedded in agroecosystems, but also fits with the vector directions of optimal paths found by Tello et al. (2016) for improving their joint energy efficiency (*FP*/*TIC*), depending on whether *BR* > *EI* or the opposite. Low levels of site-specific information are found in the landscape when the agroecosystem tends either towards an agro-industrial management by increasingly relying on *EI*, or towards rural abandonment when farmers' labour and knowledge are withdrawn from it (i.e., either in highly 'intensive-industrialised' farm systems, or in former agroecosystems that presumably are being 'renaturalised'). More information embedded in cultural landscapes becomes a key resource for the future of sustainable farming that seeks to balance agricultural production with biodiversity conservation.

### 4.3.9 Energy Imprint and Landscape Pattern Modelling

In order to measure the energy imprint in the landscape functional structure, we need to introduce a land metric (see Sect. 4.2.4). We use *L* to account for landscape heterogeneity and, when data is available, *ECI* to account for landscape ecological connectivity. In this way, we obtain *Le* (Marull et al. 2015), capturing landscape patterns (*L*) and processes (*ECI*). This reveals the capacity of landscape mosaics to offer a range of habitats that sustain biodiversity (Harper et al. 2005). Much of this species richness is apparent at scales larger than plot or farm level, and depends on a landscape-wide heterogeneity of land covers.

After having defined all the ELIA indicators (*E*, *I* and *L*), we are going to analyse their relationship. We surmise that the interplay between *E* and *I* jointly leads to complexity, understood as a balanced level of intermediate self-organisation (Gershenson and Fernández 2012). We also assume that the complexity of energy flows and *L* are related to landscape ecological processes and biodiversity.

The relationship between *E*, *I* and *L* is shown in Fig. 4.5. The values have been obtained from theoretical coefficients for two extreme agroecosystems' typologies (from 'natural' to 'intensive-industrialised' scenarios). We propose $\beta_i$'s for three model types of agroecosystems (Fig. 4.5a): 'natural' ($T_1$), 'balanced' ($T_3$) and

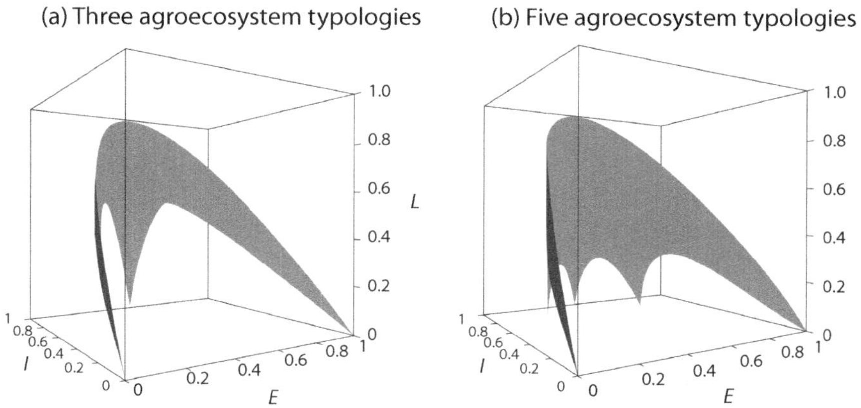


**Fig. 4.5** Theoretical relationship between complexity of internal energy loops ($E$), information held in the network of energy flows ($I$) and landscape functional structure ($L$), taking three (**a**) and five (**b**) agroecosystem typologies [The lengthening over each weight of the $E$ formula corresponds to each agroecosystem typology ($T_i$). $T_1$ corresponds to the most 'natural' agroecosystem, $T_3$ refers to a 'balanced' agroecosystem, and $T_5$ refers to an 'industrial-intensive' agroecosystem. Then, $T_2$ and $T_4$ (**b**) correspond to intermediate values (Sect. 4.3.9)]

'intensive-industrialised' ($T_5$) ones. The 'natural' agroecosystem ($T_1$) is similar to an ecosystem without human influence (i.e., low or null *HANPP*; even $\beta_i$'s are equal to one, while odd $\beta_i$'s are equal to zero); the 'balanced' agroecosystem ($T_3$) has been defined as one with an equal proportion of incoming or outgoing energy flows (i.e., intermediate *HANPP*; all $\beta_i$'s are ½); and the 'intensive-industrialised' agroecosystem ($T_5$) is defined as having no internal biomass reuses (i.e., high *HANPP*; odd $\beta_i$'s are equal to one and even $\beta_i$'s to zero). In addition, two other agroecosystem types ($T_2$ and $T_4$) have been introduced to show the results for intermediate values between the three basic types described before (see Fig. 4.5b).

Lastly, the points shown in Fig. 4.5 come from a probabilistic approximation by considering all possible land use combinations in a cell. The first form is obtained using the values of the 'natural', 'balanced' and 'intensive-industrialised' agroecosystems ($T_1$, $T_3$ and $T_5$), while in the second the intermediate agroecosystems ($T_1$, $T_2$, $T_3$, $T_4$ and $T_5$) have also been considered. As a result, Fig. 4.5 reveals the relationship between the complexity of energy flows ($E$), the information carried in them ($I$), and their joint spatial imprint in agroecosystems ($L$).

This ELIA modelling enables us to test the relationship we deem to exist between the simultaneous loss in energy throughput and landscape ecological efficiency (Marull et al. 2010). We go a step forward from previous explorations of the links between intermediate levels of socio-metabolic human disturbance, as assessed with *HANPP*, and the ecological functioning of cultural landscapes on a regional scale (Marull et al. 2015).

*ELIA* is the energy–landscape integrated analysis resulting from the model (as shown in Fig. 4.5). In order to improve its application, we propose *ELIA* as a simplified indicator that combines the landscape functional structure with the complexity of the interlinking pattern of energy flows ("loopiness") and the information carried by them, as a proxy of biodiversity in agroecosystems:

$$ELIA = \frac{(E \cdot I)L}{max\{EI\}a}$$

where $E$ is the energy storage, $I$ is the information carried by the network structure of energy flows and $L$ is the energy imprint in the landscape structure ($L$ can be substituted by $Le$; i.e., including functional attributes of the landscape). According to the ELIA model, the equilibrated $max\{EI\}e = 0.6169$ ($k_i = \frac{1}{3}$). When there is no equilibrium, the absolute $max\{EI\}a = 0.7420$ ($k_i = 1$). This product is a way of summarising the information provided by these three indicators.

Let us take a closer look at the ELIA model. First, we start with $E \cdot I$. It is easy to see that the maximum of $E \cdot I$ is equivalent to the maximum of $(-\beta_{2i} \log_2 \beta_{2i} - (1 - \beta_{2i}) \log_2(1 - \beta_{2i}))\beta_{2i}$, because each pair $(1 - \beta_{2i}, \beta_{2i})$ is mathematically independent of the other betas. So, we assume $x = \beta_{2i}$, for easier notation, and derive the previous expression and equalise to zero to obtain the $x$ that gives us the maximum of $E \cdot I$.

$$\frac{\partial}{\partial x}(-x \log_2 \mathrm{x} - (1 - x) \log_2(1 - x))x = 2x \log_2\left(\frac{1 - x}{x}\right) - \log_2(1 - x).$$

Now, $2x \log_2\left(\frac{1-x}{x}\right) - \log_2(1 - x) = 0$, implies $x = 0.7036$ and $max\{EI\}e = 0.6169$. Then we have the equilibrated maximum $E \cdot I$ for $k_i = \frac{1}{3}$, and $\gamma_F = \gamma_L = 0.5$ (i.e., subsystems equilibrium and no waste). Waste implies lower values of $I$ and consequently of $E \cdot I$. Finally, the absolute maximum of $E \cdot I$ is given when $k_i = 1$ (and the other k's are 0): $max\{EI\}a = 0.7420$. Even though this last combination is unlikely in an agroecosystem, it is possible in a theoretical mathematic case.

Once we have the maximum $E \cdot I$ to structure the landscape (energy storage and distribution) we can add the landscape energy imprint ($L$). However, there is no one unique model of landscape patterns for given values of $E$ and $I$. Hence, *ELIA* theoretically ranges from 0 to 1, for any value of the parameters considered ($k$ and $\gamma$).

### 4.3.10 Cultural Landscapes as Socio-metabolic Imprint

In order to understand the relationship between the stored energy ($E$), the information it contains ($I$) and its impression on the landscape ($L$), we have to consider a three-dimensional model (Fig. 4.5). The results can be interpreted in the sense that it is culture (the knowledge passed down from generation to generation), which allows farmers to manage the energy entering the system in the most efficient way to probably maintain a sustainable exploitation of the territory. This calls for an integrated research of coupled human-natural systems aimed at revealing complex structures and processes (Liu et al. 2007).

Our aim is to change the theoretical approach where the maximum value of $I$ is obtained according to the Information Theory (structuring information that registers the likelihood of the occurrence of a pair of events, that is the equi-distribution of $\beta' s$ in the ELIA graph: $\beta = (0.5, 0.5, \ldots, 0.5, 0.5)$), to a human-modified $I^*$ to manage a sustainable agroecosystem: so we look for a $\beta^*$ such that $I^* = 1$. For this, the following linear transformation is introduced:

$$T(\beta^*, \beta) = \begin{cases} \frac{0.5}{\beta^*} \beta, & \beta < \beta^* \\ 0.5 + \frac{0.5}{(1-\beta^*)} (\beta - \beta^*), & \beta \geq \beta^* \end{cases}$$

where $\beta^*$ are the theoretical $\beta' s$ values at which $I^*$ would be maximum. This $\beta^*$ could be modified taking into account other variables (i.e., environmental or agrological constrains).

The above piecewise function transports the interval (0, $\beta^*$) to the interval (0, 0.5), and the interval ($\beta^*$, 1) to (0.5, 1). That means that the maximum value of $I$, after this transformation, will be in $\beta^*$ instead of in 0.5. So, now we can choose any $\beta^* \in (0, 1)$ to change where the maximum value of $I$ is obtained.

Now, we can define the *Agroecosystem Information* $I^*$ index as follows:

$$I^* = I(\beta^*, \beta) = \left( -\frac{1}{6} \sum_{i=1}^{12} T(\beta_i^*, \beta_i) \log_2 T(\beta_i^*, \beta_i) \right) (\gamma_W)$$

where $\beta_i$ is a particular value according to the agroecosystem empirical data; and $\beta^*$ are the chosen value at which maximum information is archived, and $\gamma_W = \gamma_F + \gamma_L$. Furthermore, we can add the weights for each of the three subsystems of the ELIA graph (Fig. 4.3) ($\sum k_i = 1$).

$$I^* = \left( \sum_{j=1}^{3} \left( - \sum_{i=(j-1)4+1}^{4j} T(\beta_i^*, \beta_i) \log_2 T(\beta_i^*, \beta_i) \right) k_j \right) (\gamma_W).$$

In this vein, we have improved the original indicator of energy information ($I$) in an interesting way: not to measure equi-diversity according to the Information Theory (as a measure of uncertainty, or the degree of freedom for the system to evolve), but to measure the optimisation of energy flows ($\beta^*$) in order to maintain the agroecosystem funds (*UB*, *BR*, *LS*) over time. $I^*$ measures the information farmers use to direct the flows of energy carriers in the landscape according to a defined purpose (the sustainability of the farm system), losing degrees of freedom in a subtle human-nature far-from-thermodynamic equilibrium agroecosystem, with organised information that allows the transferring of energy while maintaining their complexity over time (Ulanowicz 2003).

Sustainability in agroecosystems is achieved, then, by keeping the complexity of the socio-metabolic cycles, increasing internal information while decreasing the entropy in the agroecosystem. This strategy is based on the dynamic land use heterogeneity characteristic of bio-cultural landscapes.

## 4.4 Bio-cultural Landscapes: The Land-Sharing Debate

### *4.4.1 Discussion: The Socio-metabolic Perspective*

Traditional organic farm systems with a solar-based metabolism, tended to organise their land usages according to different gradients of intensity, keeping an integrated management of the landscape because their whole subsistence depended on the land matrix functional structure. In order to offset the energy lost in the inefficient human exploitation of animal bioconversion—on which they had to depend to obtain the internal farm services of traction and manure (Guzmán and González de Molina 2009)—, traditional organic farming kept livestock breeding carefully integrated with cropland, pasture and forest spaces (Krausmann 2004).

While the organic farm management scheme of closing energy cycles within an agroecosystem has led to landscape mosaics which allow the land-sharing strategy for biological conservation (Tscharntke et al. 2012); the socio-ecological transition to agro-industrial farm systems that rely on external flows of inputs coming from underground fossil fuels has enabled society to overcome the age-old energy dependency on bioconverters (Schaffartzik et al. 2014). As a result, integrated land use management was no longer necessary—and overcoming this former necessity also led to the loss of its agroecological virtue.

The environmental damage caused worldwide by this lack of integrated management between energy flows and land uses urges societies to recover the former 'landscape efficiency' (the socio-economic satisfaction of human needs while maintaining the landscape ecological functionality) now (Marull et al. 2010). Depending on the energy storage-distribution, and how these energy flows are imprinted in the landscape, agroecosystems may either enhance or decrease biodiversity (Marull et al. 2016b). Since the lack of an integrated management of energy flows and land uses at different spatial scales is part of the current global ecological crisis, its recovery becomes crucial for more sustainable farm systems.

This line of research involves a wider and more complex approach to agroecosystems' energy efficiency. It requires not only accounting for a single input-output ratio between the final product and the external energy applied, but also looking at the harnessing of energy flows that loop within the system. The cyclical nature of these flows is important in order to grasp the emergent complexity and the information held within the agroecosystem, given that they involve an internal maximisation of less-dissipative social metabolism. The temporal energy storage that these loops allow becomes a foundation for all sustainable systems (Ho 2013).

Hence, the usual methodology of energy flow analysis of social metabolism needs to be adapted and enlarged in order to give an account of the cyclical character of agroecosystems' processes (Guzmán and González de Molina 2015) and their imprint in bio-cultural landscapes.

### 4.4.2 Conclusion

The main aim of this chapter is to explain the hypothesis that a decrease in landscape efficiency, related to a misplacing of information held by energy flows (local farmers' knowledge) and its mutual interplay with energy-loop complexity lies behind the deterioration of the energy yield of agroecosystems. This is as a result of the current crisis of the rural world, that is losing its age-old capacity to keep an integrated land use management at different space-time scales.

We have proposed an Intermediate Disturbance-Complexity (IDC) model of agroecosystems to assess how different levels of human appropriation of photosynthetic capacity affect the landscape functional structure that hosts biodiversity on a regional scale (Marull et al. 2016a).

We have also developed an Energy-Landscape Integrated Analysis (ELIA) of agroecosystems that allows us to measure the energy storage in terms of the complexity of internal energy loops, and the energy information held in the whole network of socio-metabolic energy flows. We can correlate both with the energy imprint in the landscape ecological patterns and processes that sustain biodiversity on a local scale (Marull et al. 2016b).

The combination of a spatially uneven disturbance and increased land cover heterogeneity should be able to offer more habitats for different species and ecological communities. As a result, beta-diversity (biodiversity at landscape scale) increased, and overrode the inevitable fall in alpha-diversity (at plot level) in cropland, which is the typical local impact of farm system functioning on biodiversity (Gliessman 1990). As long as this newly introduced farm-associated biodiversity (Altieri 1999) did not preclude the survival of former species richness, which would be sheltered in more undisturbed land-units, the whole process could even entail an increase in gamma-diversity (on a regional scale).

Confirming or rejecting this 'landscape efficiency' hypothesis requires further research applying IDC and ELIA to different spatial scales and time periods, and using large biodiversity datasets in order to find out where the critical thresholds in energy throughputs and the information-complexity interplay are located. This research agenda would help to reveal how and why different agroecosystem managements lead to key turning points in the relationship of the pattern of energy flows with landscape ecological functioning and biodiversity (Agnoletti 2014). There is no doubt that landscape agroecology research would be very useful for designing more sustainable food systems worldwide in the future.

## References

Agnoletti, M. (2014). Rural landscape, nature conservation and culture: Some notes on research trends and management approaches from a (Southern) European perspective. *Landscape Urban Plan, 126,* 66–73.

Altieri, M. (1999). The ecological role of biodiversity in agroecosystems. *Agriculture, Ecosystems & Environment, 74,* 19–31.

Barnes, B., Sidhu, H. S., & Roxburgh, S. H. (2006). A model integrating patch dynamics, competing species and the intermediate disturbance hypothesis. *Ecological Modelling, 194,* 414–420.

Barthel, S., Crumley, C., & Svedin, U. (2013). Bio-cultural refugia—Safeguarding diversity of practices for food security and biodiversity. *Global Environmental Change, 23*(5), 1142–1152.

Benton, T. G., Vickery, J. A., & Wilson, J. D. (2003). Farmland biodiversity: Is habitat heterogeneity the key? *Trends in Ecology & Evolution, 18,* 182–188.

Calow, P. (1987). *Evolutionary physiological ecology*. Cambridge: Cambridge University Press.

Cardinale, B. J., Duffy, J. E., Gonzalez, A., et al. (2012). Biodiversity loss and its impact on humanity. *Nature, 486,* 59–67.

Chesson, P., & Huntly, N. (1997). The roles of disturbance, mortality, and stress in the dynamics of ecological communities. *American Naturalist, 150,* 519–553.

Fischer, J., Brosi, B., Daily, G. C., et al. (2008). Should agricultural policies encourage land sparing or wildlife-friendly farming? *Frontiers in Ecology and the Environment, 6*(7), 380–385.

Fischer, J., & Lindenmayer, D. B. (2006). Beyond fragmentation: The continuum model for fauna research and conservation in human-modified landscapes. *Oikos, 112*(2), 473–480.

Gabriel, D., Sait, S. M., Kunin, W. E., et al. (2013). Food production versus biodiversity: Comparing organic and conventional agriculture. *Journal of Applied Ecology, 50,* 355–364.

Gershenson, C., & Fernández, N. (2012). Complexity and information: Measuring emergence, self-organization, and homeostasis on multiple scales. *Complexity, 18*(2), 29–44.

Giampietro, M., Mayumi, K., & Sorman, A. H. (2013). *Energy analysis for sustainable future: Multi-scale integrated analysis of societal and ecosystem metabolism*. Oxon: Routledge.

Gladyshev, G. P. (1999). On thermodynamics, entropy and evolution of biological systems: What is life from a physical chemist's viewpoint. *Entropy, 1,* 9–20.

Gliessman, S. R. (Ed.). (1990). *Agroecology: Researching the ecological basis for sustainable agriculture*. New York: Springer.

Godfray, H. C. J., Beddington, J. R., Crute, I. R., et al. (2010). Food security: The challenge of feeding 9 Billion people. *Science, 327,* 812–818.

Guzmán, G. I., & González de Molina, M. (2009). Preindustrial agriculture versus organic agriculture: The land cost of sustainability. *Land Use Policy, 26*(2), 502–510.

Guzmán, G. I., & González de Molina, M. (2015). Energy efficiency in agrarian systems from an agro-ecological perspective. *Agroecology and Sustainable Food Systems, 39,* 924–952.

Haberl, H. (2001). The energetic metabolism of societies. Part I: Accounting concepts. *Journal of Industrial Ecology, 5,* 107–136.

Haberl, H., Erb, K. H., Krausmann, F., et al. (2007). Quantifying and mapping the human appropriation of net primary production in earth's terrestrial ecosystems. *Proceedings of the National Academy of Sciences of the United States of America, 104*(34), 12942–12947.

Haberl, H., Erb, K.-H., & Krausmann, F. (2014). Human appropriation of net primary production: Patterns, trends, and planetary boundaries. *Annual Review of Environment and Resources, 39,* 363–391.

Harper, K. A., MacDonald, S. E., Burton, P. J., et al. (2005). Edge influence on forest structure and composition in fragmented landscapes. *Conservation Biology, 19,* 768–782.

Ho, M.-W. (2013). Circular thermodynamics of organisms and sustainable systems. *Systems, 1*(3), 30–49.

Ho, M.-W., & Ulanowicz, R. (2005). Sustainable systems as organisms? *BioSystems, 82*(1), 39–51.

Jackson, L. E., Pulleman, M. M., Brussaard, L., et al. (2012). Social-ecological and regional adaptation of agrobiodiversity management across a global set of research regions. *Global Environmental Change, 22*(3), 623–639.

Krausmann, F. (2004). Milk, manure, and muscle power. Livestock and the transformation of preindustrial agriculture in Central Europe. *Hum Ecol, 32*(6), 735–772.

Lindenmayer, D. B., & Fischer, J. (2007). Tackling the habitat fragmentation panchreston. *TREE, 22,* 127–132.

Liu, L., Dietz, T., Carpenter, S. R., et al. (2007). Complexity of coupled human and natural systems. *Science, 317*(5844), 1513–1516.

Loreau, M., Mouquet, N., & Gonzalez, A. (2010). Biodiversity as spatial insurance in heterogeneous landscapes. *Proceedings of the National Academy of Sciences, 100*(22), 12765–12770.

Marull, J., & Mallarach, J. M. (2005). A GIS methodology for assessing ecological connectivity: Application to the Barcelona Metropolitan Area. *Landscape Urban Plan, 71,* 243–262.

Marull, J., Pino, J., Tello, E., et al. (2010). Social metabolism, landscape change and land use planning in the Barcelona Metropolitan region. *Land Use Policy, 27*(2), 497–510.

Marull, J., Tello, E., Fullana, N., et al. (2015). Long-term bio-cultural heritage: Exploring the intermediate disturbance hypothesis in agro-ecological landscapes (Mallorca, C. 1850–2012). *Biodiversity and Conservation, 24*(13), 3217–3251.

Marull, J., Font, C., Tello, E., et al. (2016a). Towards an energy-landscape integrated analysis? Exploring the links between socio-metabolic disturbance and landscape ecology performance (Mallorca, Spain, 1956–2011). *Landscape Ecology, 31,* 317–336.

Marull, J., Font, C., Padró, R. et al. (2016b). Energy-landscape integrated analysis: A proposal for measuring complexity in internal agroecosystem processes (Barcelona Metropolitan Region, 1860–2000). *Ecological Indicators, 66,* 30–46.

Marull, J., Delgadillo, O., La Rota, M. J., et al. (2017). Socioecological transition in the Cauca river valley, Colombia (1943–2010): Towards an energy–landscape integrated analysis. Regional Environmental Change (in press).

Matthews, R., Selman, P. (2006). Landscape as a focus for integrating human and environmental processes. *Journal of Agricultural Economics, 57,* 199–212.

Matson, P. A., Parton, W. J., Power, A. G., et al. (1997). Agricultural intensification and ecosystem properties. *Science, 277,* 504–509.

Mayer, A., Schaffartzik, A., Haas, W. et al. (2015). Patterns of global biomass trade and the implications for food sovereignty and socio-environmental conflict. EJOLT Report No. 20, p. 106.

McMichael, Ph. (2011). Food system sustainability: Questions of environmental governance in the new world (dis)order. *Global Environmental Change, 21*(3), 804–812.

Morowitz, H. J. (2002). *The emergence of everything: How the world became complex.* Oxford: Oxford University Press.

Odum, E. P. (1993). *Ecology and our endangered life-support systems.* Massachusetts: Sinauer Associates.

Parrotta, J. A., & Trosper, R. L. (2012). Traditional forest-related knowledge: Sustaining communities, ecosystems and biocultural diversity. *World Forests, 12,* 1–621.

Perfecto, I., & Vandermeer, J. (2010). The agroecological matrix as alternative to the land-sparing/agriculture intensification model. *Proceedings of the National Academy of Sciences, 107*(13), 5786–5791.

Peterseil, J., Wrbka, T., Plutzar, C., et al. (2004). Evaluating the ecological sustainability of Austrian agricultural landscapes—The SINUS approach. *Land Use Policy, 21*(3), 307–320.

Phalan, B., Onial, M., Balmford, A., et al. (2011). Reconciling food production and biodiversity conservation: Land sharing and land sparing compared. *Science, 333,* 1289–1291.

Pierce, S. (2014). Implications for biodiversity conservation of the lack of consensus regarding the humped-back model of species richness and biomass production. *Functional Ecology, 28,* 253–257.

Pino, J., & Marull, J. (2012). Ecological networks: Are they enough for connectivity conservation? A case study in the Barcelona Metropolitan Region (NE Spain). *Land Use Policy, 29,* 684–690.

Prigogine, I. (1996). The end of certainty. Time, chaos and the new laws of nature. New York: The Free Press.

Schaffartzik, A., Mayer, A., Gingrich, S., et al. (2014). The global metabolic transition: Regional patterns and trends of global material flows, 1950–2010. *Global Environmental Change, 26,* 87–97.

Schrödinger, E. (1944). *What is life?.* Cambridge: Cambridge University Press.

Shreeve, T. G., Dennis, R. L. H., & Van Dick, H. (2004). Resources, habitats and metapopulations —Whither reality? *Oikos, 106,* 404–408.

Swift, M. J., Izac, A. M. N., & van Noordwijk, M. (2004). Biodiversity and ecosystem services in agricultural landscapes—Are we asking the right questions? *Agriculture, Ecosystems & Environment, 104*(1), 113–134.

Tainter, J. (1990). *The collapse of complex societies.* Cambridge: Cambridge University Press.

Tello, E., Galán, E., Sacristán, V., et al. (2016). Opening the black box of energy throughputs in agroecosystems: A decomposition analysis of final EROI into its internal and external returns (The Vallès County, Catalonia, c. 1860 and 1999). *Ecological Economics, 121,* 160–174.

Tilman, D., Cassman, K. G., Matson, P. A., et al. (2002). Agricultural sustainability and intensive production practices. *Nature, 418,* 671–677.

Tscharntke, T., Klein, A. M., Kruess, A., et al. (2005). Landscape perspectives on agricultural intensification and biodiversity-ecosystem service management. *Ecology Letters, 8,* 857–874.

Tscharntke, T., Clough, Y., Wanger, T. C., et al. (2012). Global food security, biodiversity conservation and the future of agricultural intensification. *Biological Conservation, 151,* 53–59.

Ulanowicz, R. E. (2001). Information theory in ecology. *Computers & Chemistry, 25,* 393–399.

Ulanowicz, R. E. (2003). Some steps toward a central theory of ecosystem dynamics. *Computational Biology and Chemistry, 27*(6), 523–530.

Van der Maarel, E. (1993). Some remarks on disturbance and its relations to diversity and stability. *Journal of Vegetation Science, 4,* 733–736.

Vitousek, P. M., Ehrlich, P. R., Ehrlich, A. H., et al. (1986). Human appropriation of the products of photosynthesis. *BioScience, 36*(6), 363–373.

Vranken, I., Baudry, J., Aubinet, M., et al. (2015). A review on the use of entropy in landscape ecology: Heterogeneity, unpredictability, scale dependence and their links with thermodynamics. *Landscape Ecology, 30,* 51–65.

Wilson, J. B. (1990). Mechanisms of species coexistence: Twelve explanations for Hutchinson's 'paradox of the plankton': Evidence from New Zealand plant communities. *New Zealand Journal of Ecology, 13,* 17–42.

Wilson, J. B. (1994). The 'intermediate disturbance hypothesis' of species coexistence is based in on patch dynamics. *New Zealand Journal of Ecology, 18,* 176–181.

Wrbka, T., Erb, K.-H., Schulz, N. B., et al. (2004). Linking pattern and process in cultural landscapes. An empirical study based on spatially explicit indicators. *Land Use Policy, 21*(3), 289–306.

## Author Biographies

**Joan Marull** is senior researcher and director of the Department of Ecology and Territory at the Barcelona Institute of Regional and Metropolitan Studies (Autonomous University of Barcelona). His main field of activity is land management and urban planning from a systemic approach. Currently, his research focuses on the study of social metabolism, ecological economics, urban ecology, landscape ecology, and territorial efficiency, as well as the application of satellite technology in the analysis of land use change. His transdisciplinary proposal integrates the study of socio-ecological processes to address research and policy issues at multiple time and space scales.

**Carme Font** works as Research Technician in Environmental Modelling at the Catalan Institute for Water Research (ICRA) in Girona, Spain. Her research focuses on mathematical models and the statistical analysis of natural systems. She has developed indicators for agroecosystem energy flows. Currently, she is working on inland water modelling.

# Part II
# Case Studies and Empirical Evidence

# Chapter 5
# Does Your Landscape Mirror What You Eat? A Long-Term Socio-metabolic Analysis of a Local Food System in Vallès County (Spain, 1860–1956–1999)

**Roc Padró, Inés Marco, Claudio Cattaneo, Jonathan Caravaca and Enric Tello**

**Abstract** We assess the social metabolism of very different farm systems that existed in Vallès County, along the socio-ecological transition from organic to industrial agriculture at three different time points from 1860 to 1999. This allows us to analyze these contrasting food systems by focusing on four perspectives: agricultural labour productivity in relation to regional diets, the importance of multi-functionality in agroecosystems, the loss of landscape diversity and species richness, and the impacts of the current food regime at global and local scales. The socio-metabolic profiles obtained show that (1) winegrowing specialization co-existed with sustenance-oriented organic farming in 1860; (2) in 1956, the resumption of grain growing, combined with incipient use of industrial fertilizers, led to a more diverse agroecosystem where greater dependence on external inputs was countered by an increased productivity, providing more balanced diets and producing minor impacts on landscape ecology; (3) by 1999, a specialization in feedlots had disconnected local diets from a linear agro-industrial feed-meat chain based on huge feed imports from the Global South, leading to highly polarized

R. Padró (✉) · I. Marco · E. Tello
Department of Economic History, Institutions, Policy and World Economy, University of Barcelona, Diagonal 690, 08034 Barcelona, Spain
e-mail: roc.padro@gmail.com

I. Marco
e-mail: ines.marco@ub.edu

E. Tello
e-mail: tello@ub.edu

C. Cattaneo
Barcelona Institute of Regional and Metropolitan Studies (IERMB), Autonomous University of Barcelona, MRA Building, Bellaterra, 08193 Barcelona, Spain
e-mail: claudio.cattaneo@uab.cat

J. Caravaca
University of Barcelona, Barcelona, Spain
e-mail: jonathancaravaca@gmail.com

 
E. Fraňková et al. (eds.), *Socio-metabolic Perspectives on the Sustainability of Local Food Systems*, Human-Environment Interactions 7,
https://doi.org/10.1007/978-3-319-69236-4_5

socio-ecological impacts. Whereas unequal ecological exchange affects peasant communities and agroecosystems in feed-exporting countries, local landscapes suffer from the accumulation of dung waste poured into flatlands and from forest abandonment in steeper areas.

**Keywords** Social metabolism · Local food systems · Labour productivity · Land-use integration · Political ecology

## 5.1 Introduction

Agriculture is the economic activity that maintains a biophysical interaction between society and nature with the greatest spatial extent on Earth. Over the last 150 years, this socio-ecological interaction has been transformed dramatically, driven by deep changes in food production and consumption. Due to the disconnection between urban consumers and the rurality brought about by the socio-metabolic changes undertaken in agricultural practices through the Green Revolution, current agriculture is far from being sustainable. Global changes in food trade, along with new dietary patterns and food demands from consumers, have shaped the regional specialization of farm producers. Purchasers of these globalised food baskets have become increasingly alienated from foodscapes that farmers cultivate in ever more distant locations, and vice versa (Leguizamón 2016).

Under this prevailing globalized system, cheap agro-industrial food has been transformed into a highly branded, packaged and de-spatialized commodity severed from time (e.g. season), space (e.g. landscape) and culture (e.g. meaning) (Weis 2010). Indeed, cheap oil subsidizes the huge amount of biomass globally traded that experienced a 5-fold increase from 1962 to 2010 (Mayer et al. 2015). Most people in the Global North buy their food in supermarkets, often without knowing anything about where it comes from or which social and ecological impacts it entails. Meanwhile, most of the information and the decision-making power is kept at the headquarters of transnational agribusiness, far away from the actual labour needed to produce this food and far from its consumption (Friedmann 2016). For these corporations, the only relevant link between both extremes of the food chain is the price paid to the producer, and the one received from the consumer.

New social movements, like La Via Campesina, have brought this issue to the forefront, putting food sovereignty in research programmes and decision-making agendas worldwide (Edelman et al. 2014). They want a relocation of agri-food chains, so as to empower peasant producers as well as urban and rural consumers, and to raise collective awareness about the socio-ecological consequences that the prevailing food regime carries. From a research standpoint, this emerging social demand can be met with the study of social metabolism, a way to overcome current cultural alienation by carefully accounting the material and energy flows moved across food chains, from farmers working in agro-ecosystems to the tables of consumers and their kitchens' dustbins (Gliessmann 1998; González de Molina and Toledo 2014). This socio-metabolic scanning may help us realize that what we eat, wear, and burn is always generating socio-ecological imprints on the environment (Infante Amate et al. 2014). Given that agroecosystems are the result of the complexity of both ecological and social systems interacting (Kay and Schneider 1994), a socio-metabolic analysis of food regimes cannot be generated from a single dimension. The multiscale character of food chains (Tello and González de Molina, Chap. 2 in this volume) demands a new holistic approach in natural and social sciences (Tilman and Clark 2014).

Food sovereignty is aimed not only at criticizing current agri-food systems, but also to open the way for other, more sustainable ones. This chapter has a unique approach to show this, as it offers a historical perspective from which we can learn for the future. Studying the socio-ecological transition from a past organic agri-food system to the current industrial ones is an opportunity to perform a natural experiment aimed at identifying crucial points, linkages and limits from a systemic point of view (González de Molina and Guzmán 2006).

We are going to use a case study located in Vallès County (Barcelona Metropolitan Region, Catalonia) as a bench test. The area covers 12,400 ha located in a Mediterranean plain between the coastal and pre-coastal Catalan mountain ranges, with an average rainfall from 600 to 800 mm and a mean annual temperature around 14 °C. During the last 150 years, its population density has increased 5-fold from 66 to 336 inhab./km$^2$.

Starting from an organic local food system with a vineyard specialization, we study the evolution into a highly globalized industrial one. We are going to analyse the agroecosystems of these four municipalities c.1860, and in 1956 and 1999, from a multidimensional and multi-scalar perspective to draw some general conclusions from the socio-ecological transition carried out in food production and consumption. The socio-metabolic patterns at the start and the end of this period were extremely different. In 1860, winegrowing specialization coexisted with a subsistence-oriented organic farming that kept a significant level of self-reliance (Garrabou et al. 2007). In 1956, the resumption of grain growing, combined with an incipient use of industrial fertilizers, led to a more diversified agroecosystem where greater dependence on external inputs was countered by increased productivity; this provided a more balanced diet, and produced lesser impacts on the local landscape ecology. In 1999, a specialization in feedlots disconnected local diets from a linear agro-industrial feed-meat chain, based on huge feed imports from the Global South which entail strong socio-ecological impacts both locally and worldwide.

By comparing these very different agri-food systems four key dimensions are brought to light: (1) the evolution of agricultural labour productivity and its connection or disconnection with regional diets; (2) the importance and increasing loss of multi-functionality in agroecosystems; (3) a significant loss of landscape ability to host biodiversity; and (4) the strong socio-ecological impacts of the current food regime at global and local scales. Combining them, we gain a multidimensional perspective of the socio-ecological functioning that links production and consumption with the satisfaction or dissatisfaction of human needs, and with socio-ecological patterns and processes at different scales. Throughout the period studied all these factors were modified following nutritional transition and agroecological change from organic to industrial farm systems. Behind them there lay the overall energy transition to fossil fuels, a growing socio-metabolic rift in the circulation of nutrients in and out of soils (Clark and Foster 2009), and a vanishing of former landscape mosaics (González-Bernáldez 1981; Levers et al. 2015) which kept a great deal of farm-associated biodiversity (Altieri 1999).

## 5.2 Materials and Methods

By using material-energy-flow accounting (MEFA), we develop an analysis of the Vallès agroecosystem over three time periods in order to profile the socio-metabolic transition from organic to industrial agriculture. Previous works (Cussó et al. 2006; Tello et al. 2015, 2016) have compared c.1860 and 1999 time points, but they missed an intermediate one which we mention in this approach, as well as in a fund-flow analysis developed by Marco et al. (2017). This allows us to identify different patterns and drivers of transformation along the socio-metabolic transition from traditional organic to industrial farming.

So far, energy accounting with MEFA is introducing wider perspectives going beyond a simple input-output energy analysis (Galán et al. 2016; Guzmán and González de Molina 2015; Tello et al. 2016), but the scale of our analysis remained at the agroecosystem level. Indeed, there is a growing body of research linking ecological economics with political ecology (Martinez-Alier et al. 2010). Therefore, we propose a forward step towards a multi-scalar analysis (Giampietro et al. 2008) considering the local food system embedded in the global food systems. As this article focuses on the dimension of local food systems, and how socio-ecological transition shaped the external links and internal synergies of the agroecosystems, we perform an analysis through four different perspectives: population, land uses, landscape, and the global food system. In doing so, we refer to three theoretical frameworks: analysis of social metabolism, landscape ecology, and political ecology.

### *5.2.1 Social Metabolism in Agroecosystems: Population and Land-Use Perspectives*

Our analysis regarding social metabolism in agroecosystems takes two different elements into account: on the one hand, satisfaction of human needs; on the other hand, we acknowledge the multifunctional role that different land uses and livestock may or may not have. We consider the agricultural active population as the main fund of an agroecosystem,[1] together with other basic funds such as the farmland, the livestock, and the farm-associated biodiversity.

In order to get an adequate profile of the agroecosystem functioning, the analysis has to take into consideration that the labour cost to provide food to society is as important as the ability of the agroecosystem to keep providing the required biotic materials over time. Section 5.3.1 assesses the evolution of the agricultural labour energy productivity (ALEP), i.e. agricultural produce obtained per unit of labour, and its linkages with the potential capability of the agroecosystem to cover local human needs of subsistence through food and fuel '*satisfiers*'. This potential capacity would depend on (i) the composition of agricultural produce, (ii) food and fuel requirements per capita according to the prevailing diet, (iii) the percentage of the workforce engaged in agriculture, and (iv) population density. Following these steps, we first compare agricultural labour productivity with food and fuel biomass requirements per capita, thus assessing which was the surplus or deficit per farm worker at different levels of aggregation. We then compare if this fits the ratio of non-agricultural population per farm worker, observing whether local food and fuel produce was adapted to local food and fuel demand.

In order to assess the energy performance of these contrasting farm systems in a way that does not conceal their internal agroecological reproduction (Tello et al. 2016; Galán et al. 2016; Marco et al. 2017), we account for the Gross Calorific Value (GCV) of all biomass produced in a year using the energy converters given by Guzmán et al. (2014). For any external biomass flow entering the agroecosystem boundaries, the embodied energy cost of transport, processing and packaging is added to its own GCV—indeed, given the predominance of feed imports in our 1999 case study, we used as references Pérez Martínez and Monzón de Cáceres (2008) for transportation costs, and Cooperativas Agro-alimentarias (2010) for feed processing. For non-renewable materials used as external inputs—such as tractors and other farm implements, fossil fuels, or agrochemicals—all the embodied energy required for their production and delivery is accounted based on Aguilera et al. (2015). Animal feeding requirements are taken from Soroa (1953) and Church (1984), and applied to pre-industrial livestock husbandry or to industrial animal fattening in feedlots. Then the data obtained in these energy balances of the three analyzed farm systems will be used to assess the multifunctional capacity of their underlying funds to keep reproducing these flows over time. Agricultural

[1]When referring to funds, we assume the distinction between stock and fund made by Georgescu-Roegen (1971). A biophysical fund provides a flow while either maintaining itself or being maintained (Faber et al. 1995), thus remaining as such within the time span adopted to account for a specific process (Mayumi 1991). Our flowchart of an agroecosystem differentiates among four principal funds: farming community, farmland, livestock, and associated biodiversity (Tello et al. 2016). For more details, see Marco et al. (forthcoming).

produce will only include those products suitable for human consumption, excluding reused biomass for animal feeding or soil fertilizers, and marketable products such as grain or fodder for animal consumption. Agricultural labour productivity is estimated through the agricultural produce per farm worker, thus dividing the agricultural produce by the total number of farm workers at each time point. Units of labour are defined as full-time Agricultural Working Units (AWU) a year. Diet composition for the mid-19th century is based on a thorough research made by Cussó and Garrabou (2001) supplemented by contemporary historical sources such as the *Estudio Agrícola del Vallès* (1874) (Garrabou and Planas 1998). For the 1954 and 1999 diets, we have used Catalan averages gathered in the Household Budget Survey made in 1963–1966 (INE 1969), and statistics of consumption (DARP 1998).

In order to avoid a bias in our results due to the effect of winegrowing specialization c.1860, given its very low energy content, and considering it a cash crop linked to its exchange value more than to its use value, we propose a slightly different approach to assess it. We transform wine surplus into an equivalent of food (such as bread or legumes) through market prices taken from regional cadastres (*amillaramientos*).

Regarding household fuelwood consumption patterns, we have reviewed historical and current data (Sancho et al. 1885; FAO 1983; Reddy 1981; Wijesinghe 1984). Based on these references, we propose an average daily consumption of 1.56 kg of firewood for heating and cooking, adapting the estimation to climate and seasonality (Giampietro and Pimentel 1990; Colomé 1996; Bhatt and Sachan 2004).

In Sect. 5.3.2 we evaluate the loss of multifunctionality through energy accounting. Going beyond a purely efficiency analysis, the methodology followed in this chapter also allows us to show how the energy flows from a specific fund contribute to different types of services. As has already been assessed, either from a physiological or from a socio-metabolic perspective (Krausmann 2004), the metabolizable energy incorporated by a fund is distributed into different energy carriers of different qualities (e.g. livestock can produce milk, but also manure, draft power as well as heat). We will then take these different services of the agroecosystem (fertilization, food, power for tillage) to analyse the share of them that are provided by livestock at the three time points.

### 5.2.2 Landscape Ecology Indicators as a Proxy for Capabilities to Host Biodiversity

A further dimension analysis refers to landscape ecology. We adopt a land-sharing approach and assume that intermediate levels of human intervention in agroecosystems can benefit *beta* biodiversity—related to differentiation among habitats—compared to non-intervention (Gliessmann 1998; Marull et al. 2015, 2016). We define farm-associated biodiversity as communities of non-domesticated species that are a part of, and play a relevant role in, the reproduction of an agroecosystem. They provide ecosystem services, but are not the focus of farming activity, and depend on the dispersal ability of a landscape to provide proper material conditions for the survival of these and other non-farm related communities (Loreau et al. 2003; Tscharntke et al. 2012).

In order to assess these conditions for farm-associated biodiversity, we will take into account the interaction among social metabolism and landscape patterns, by using four main indicators: the amount of biomass left to non-domesticated food chains, as well as the ecological disturbance exerted by the flows of social metabolism (NPPEROI; AFEROI); habitat differentiation (L); and landscape fragmentation (EMS). Similar approaches have been used in Marull et al. (2010, 2016), but have only focused on a landscape ecology perspective.

The first indicator is the energy left for farm-associated biodiversity. We start with the methodological proposal on NPPEROI (Galán et al. 2016), which is the ratio of the Actual Net Primary Production ($NPP_{act}$ as defined in Vitousek et al. 1986) to the External Inputs and Biomass Reused (as a proxy of human ecological disturbance).[2] We are also interested in the Agro-ecological FEROI proposed by Guzmán and González de Molina (2015), which accounts for the land cost of producing an amount of final produce regarding the total investment made, which also includes the unharvested biomass.

The third indicator relies on the Shannon index to account for landscape heterogeneity (Vranken et al. 2014), a key mechanism for biodiversity maintenance (Loreau et al. 2003). Spatial habitat differentiation is also associated to margins management, which reinforces ecosystem services such as plague and disease control, in turn enhancing biodiversity (Holland et al. 2012). Finally, we also account for other aspects of social metabolism that are beyond agrarian activity, such as fragmentation of habitats due to linear transport infrastructures as well as patch dimension, using the Effective Mesh Size (EMS; Jaeger 2000). This latter indicator accounts for the probability that any two random points in a region may be connected, i.e. not separated by barriers. Both indicators are landscape metrics that can be calculated through GIS analysis of digital land cover maps. To this aim, we divided the whole surface of the four municipalities into 95 squared cells of $1 \times 1\ km^2$. Calculating the Shannon Index requires assessment of the ratio of land uses in the total surface, while for the Effective Mesh Size only the surface of each

---

[2]These indicators are derived from the Energy Return on Investment (EROI) that is calculated through energy analysis: External Inputs mean those flows coming from outside the agroecosystem boundaries; Biomass Reused is the share of $NPP_{act}$ devoted to maintain the livestock, or farmland soil fertility; Unharvested Biomass is the share of $NPP_{act}$ that remains available for the associated biodiversity; Final Produce is the total amount of $NPP_{act}$ that is available to be consumed by the farming community or that goes outside the agro-ecosystem. For a deeper definition of these concepts, see Tello et al. (2016). Once the flows have been calculated, the indicators are the following: $NPPEROI = \frac{NPP_{act}}{BR + EI}$ where BR is the Biomass Reused and EI the External Inputs; $AFEROI = \frac{FP}{BR + EI * UB}$ where FP is the Final Produce and UB the Unharvested Biomass.

patch is needed.[3] Then, using these four indicators, we can gauge how socio-metabolic transition has affected landscape functionality to host biodiversity.

### 5.2.3 Global and Local Effects, Political Ecology

The interpretation of our results is based on a political ecology perspective, assessed in terms of who gets the environmental benefits of a foodscape and who has to bear the environmental costs. As we have outlined in previous sections, current regional specialization in Vallès County in industrial livestock fattening is supported by the global food regime via massive feed imports (McMichael 2016). We therefore focus on the agroecosystem's ability to host livestock density, and its political ecological implications. Our aim is to assess to what extent an unequal exchange of environmental benefits and impacts might be at stake. Thus, to implement this approach, we estimate the carrying capacity of the agroecosystem studied to host different livestock densities, taking into account local productivities and the total metabolizable energy required by current animal diets (Church 1984; Flores and Rodríguez-Ventura 2014; Instituto Nacional de Estadística 1999). Then, imports have to be considered for all the animal feed not supplied from local sources. In order to identify the source of the different feed imported, we use international trade statistics at state level to calculate the apparent consumption, that is, the share that is produced in Spain, the non-consumed part, and the imports from any other countries.[4] To approach the virtual land cost of industrial livestock fattening with feed imports, we calculate the hectares required using FAO's average land yields of each crop. This allows us to identify hot spots in countries specialized in supplying this feed and then, through a bibliography review, to refer to the ensuing impacts in their landscapes.

Finally, we focus on the internal impacts on the local agroecosystem functioning that these feed imports generate. We gauge how livestock breeding specialization is overproducing dung that exceeds the carrying capacity assessed through nutrient balances (González de Molina et al. 2010) and, again by reviewing bibliography, how this pollution raises concerns that require new researches and policies addressed to relocate and downsize the sources of these impacts.

---

[3]The mathematical expression of the Shannon Index, modified for agrarian metabolism, is shown in Marull et al. (2016) as $L = \left(-\sum_{i=1}^{k} p_i \log_k p_i\right)(1 - p_u)$ where p is the share of surface for each land use, k is the number of land covers not considering the urban ones, and $p_u$ the share of urban area over the total. On the other hand, the formula for the Effective Mesh Size, using the definition given by Jaeger (2000), is $EMS = \frac{1}{A_t}\sum_{i=1}^{n} A_i^2$ being $A_t$ the total surface, n the number of patches, and $A_i$ the surface of each patch.

[4]In Catalonia, vertical integration on pig feeding accounts for around 75% of the feedlots, and the greatest share measuring it in total weight. So it seems reasonable to estimate that its' consumption of feed will have a similar pattern in international sources as the Spanish one (Observatori del Porcí 2009).

## 5.3 Results and Discussion

### *5.3.1 Agroecosystem as a Human Needs Satisfier*

#### 5.3.1.1 Modern Nutritional Transition in Vallès County

We assume that, either historically or at present, the main aim of society-nature interaction by means of agro-ecosystems is to satisfy *subsistence* needs, mainly through food and fuel. No doubt these *satisfiers* are not static, but shift over time. Although dietary needs have not changed throughout time, as endosomatic energy requirements are fairly constant (Lotka 1956), dietary composition as well as heating and cooking practices have changed dramatically between 1860 and the present. A modern nutrition transition (Popkin 1993; Smil 2000; Cussó and Garrabou 2007) took place in Europe from the beginning of the 19th century to the end of the 20th century, with different paths between Northern and Southern Europe. This process entailed a change from a local, seasonal, eminently vegetarian and often monotonous diet, to a diversified, excessive, unbalanced and globalized diet (Cussó and Garrabou 2010).

However, at the same time, the substitution of traditional renewable energy carriers (manpower, firewood, wind and water) by modern fossil-fuel ones (coal, gas and oil) has occurred. Gales et al. (2007) found that the contribution of traditional biomass energy carriers to total energy input was less than 50% in the Netherlands and the United Kingdom in 1864, but not at this level in Italy and Spain until the 1940s. Both nutritional and energy transitions involved a disconnection of food and fuel consumption from farming communities and their agroecosystems (Kander et al. 2013).

Changes in Catalan diet (Table 5.1) are in line with the Spanish nutritional transition (Cussó and Garrabou 2010). From a Mediterranean diet, based on a great consumption of bread, potatoes and legumes accompanied by vegetables, fruits and fresh or salted fish, including some pork and mutton as main contributions of animal origin; to an increasingly globalized diet with a greater prominence of animal products (Nicolau and Pujol 2005). Circa 1860, 70% of the total dietary intake in ME were cereals, legumes and potatoes. This percentage was reduced to 40% in 1956 and 25% in 1999. Consumption of animal products has experienced a 4-fold increase, from 12% of total intake c.1860 to 15 and 30% in 1999, which equals the peak reached at Spanish scale (Marrodán et al. 2012; Infante Amate and González de Molina 2013; Infante Amate et al. 2015; Soto et al. 2016). Whereas in the first interval, from 1860 to 1956, dairy products and eggs were the items that increased the most, from the 1950s it was meat, as happened in other European countries (Teuteberg and Flandrin 1999). Current protein intake comes from animal products which have replaced legumes, despite the fact that the latter are much more energy efficient to produce. On average 6 kg plant protein is required to yield 1 kg meat protein (Pimentel and Pimentel 2003; Smil 2000).

**Table 5.1** Change in diet composition for Vallès County c.1860, 1956 and 1999

| Products | 1860 | | 1956 | | | 1999 | | |
|---|---|---|---|---|---|---|---|---|
| | Fresh weight | Metabolizable energy | Fresh weight | | Metabolizable energy | Fresh weight | | Metabolizable energy |
| | $gr \cdot day^{-1} \cdot cap^{-1}$ | $kcal \cdot day^{-1} \cdot cap^{-1}$ | $gr \cdot day^{-1} \cdot cap^{-1}$ | Variation rate (%) | $kcal \cdot day^{-1} \cdot cap^{-1}$ | $gr \cdot day^{-1} \cdot cap^{-1}$ | Variation rate (%) | $kcal \cdot day^{-1} \cdot cap^{-1}$ |
| Bread | 437 | 1149 | 302 | −31 | 793 | 192 | −36 | 504 |
| Olive oil | 15 | 135 | 72 | 380 | 643 | 59 | −18 | 425 |
| Wine | 214 | 130 | 165 | −23 | 100 | 287 | 74 | 167 |
| Other cereals | 92 | 325 | 32 | −65 | 112 | 19 | −41 | 69 |
| Pasta | – | – | 15 | – | 55 | 15 | 0 | 54 |
| Legumes | 74 | 41 | 33 | −55 | 18 | 15 | −55 | 54 |
| Potatoes | 460 | 327 | 238 | −48 | 170 | 119 | −50 | 86 |
| Vegetables | 293 | 75 | 223 | −24 | 57 | 300 | 35 | 107 |
| Fresh fruits and nuts | 52 | 83 | 225 | 333 | 239 | 200 | −11 | 72 |
| Fish | 30 | 34 | 66 | 120 | 76 | 90 | 36 | 103 |
| Meat | 88 | 306 | 83 | −6 | 258 | 192 | 131 | 580 |
| Eggs, milk and cheese | – | – | 223 | – | 179 | 394 | 77 | 265 |
| Others | 10 | 5 | 30 | 200 | 119 | 524 | 1647 | 382 |
| Total | 1765 | 2610 | 1707 | −3 | 2820 | 2406 | 41 | 2869 |

*Source* Our own through Cussó and Garrabou (2001, 2007, 2012) for 1860, and average Catalan food consumption from INE (1969) for 1956 and DARP (1998) for 1999

Variation means the difference in fresh weight from the previous time point to the one in the column

Moreover, population growth in the Vallès adds pressure on local agroecosystems. If meat produce per person and year has increased 4-fold throughout the period studied, increase of total meat requirements has grown by a factor of 19 when we include the effect of an increased population density. Given the lower bioconversion efficiency, and the astonishing land requirements to feed this huge livestock increase, strong environmental pressures have ensued due to this dietary change (Smil 2002; de Boer et al. 2006), as discussed in greater detail in Sect. 5.3.4. Finally, processed food, which involves high levels of embodied energy through its industrial transformation (Infante Amate et al. 2014), has also boomed during this period, from 0.2 to 13%. All this points to the fact that globalized diets are clearly no longer linked to natural resource endowments at local and regional scales, as age-old practices were.

#### 5.3.1.2 Evolution of Labour Productivity, Agricultural Surplus, Agricultural Population and Total Population

From the beginning of the 20th century, but particularly in the second half of the century, the introduction of chemical fertilizers has increased yields. Rainfed wheat yields rose from 1135 kg/hectare (fresh weight) to 1231 in 1956 and 2795 in 1999. Mechanization boosted total power capacity from 449 kW (only human and animal power) to 780 kW in 1956 (with mechanization on 25% of cropland) and to 12,065 kW in 1999.

Labour productivity increased, on the one hand due to the abandonment of traditional fertilizing techniques,[5] but on the other because machinery replaced human and animal labour. Final Produce (FP)[6] per farm worker and year rose from 128 to 204 GJ, and then to 1249 GJ in 1999 (Table 5.2), that is, 67, 106 and 650 MJ/h respectively. This increase has more to do with the decrease in the total number of farm workers (−87%) than to an increase of FP as such (this only grew by 19%). In turn, farm workers were substituted by capital, with a dramatic growth in external inputs and consequent decrease in EFEROI.[7] At the same time FP in Vallès County shifted through the whole period, from a pretty balanced composition between cropland (34%) and woodland produce (65%), and a residual share of animal FP (1%), to an animal-produce specialized system (76% of FP) (Marco et al. 2017). But how have these changes in agricultural production, in terms of size and

[5]The most labour intensive fertilizing activities abandoned were burying fresh biomass and burning *formiguers*, a series of small charcoal kilns incorporated into the soil (Garrabou and Planas 1998; Olarieta et al. 2011).

[6]This flow refers to that part of an agroecosystem's products and services (from agriculture, livestock and forestry) that is destined to final use or consumption, as explained in Tello et al (2016).

[7]The External Final EROI (EFEROI) is calculated as follows: $EFEROI = \frac{FP}{EI}$ where FP is the Final Produce and EI the External Inputs (Tello et al. 2016); Biomass Reused is not included.

**Table 5.2** Comparison between Final Produce per farm worker and adult food and fuel requirements for Vallès County c.1860, 1956 and 1999

| Products | | 1860 | | | 1956 | | | 1999 | | |
|---|---|---|---|---|---|---|---|---|---|---|
| | | Produced | Required | Self-sufficiency[b] | Produced | Required | Self-sufficiency | Produced | Required | Self-sufficiency |
| | | GJ/L·farm worker$^{-1}$ ·year$^{-1}$ | | % | GJ/L·farm worker$^{-1}$ ·year$^{-1}$ | | % | GJ/L·farm worker$^{-1}$ ·year$^{-1}$ | | % |
| Vegetables, Fresh fruits, nuts | | 2.1 | 0.9 | 60 | 10.7 | 0.8 | 130 | 28.9 | 0.6 | 31 |
| Cereals | | 6.1 | 2.5 | 65 | 16.7 | 1.7 | 96 | 19.0 | 0.9 | 13 |
| Livestock food produce | | 1.3 | 1.0 | 35 | 6.0 | 0.8 | 74 | 666.6 | 1.4 | 295 |
| Wine | | 11.0 | 0.2 | 1239[c] | 5.0 | 0.2 | 281 | 1.5 | 0.3 | 3 |
| Olive oil | | 1.8 | 0.2 | 212 | 0.3 | 1.0 | 3 | 2.5 | 0.8 | 2 |
| Total food, wine and olive oil | | 22.4 | 4.2[d] | 137 | 38.7 | 4.5 | 83 | 718.4 | 4.0 | 114 |
| Woody biomass | Fresh fruits and nuts | 1.1 | 11.0 | 253 | 8.9 | 11.0 | 109 | 24.6 | – | – |
| | Vineyard | 19.1 | | | 1.1 | | | 6.1 | | |
| | Olive trees | 6.8 | | | 3.1 | | | 15.3 | | |
| | Woodland | 80.6 | | | 111.6 | | | 87.1 | | |
| Total woody biomass | | 107.6 | 11.0 | 253 | 124.7 | 11.0 | 109 | 133.3 | – | – |
| Total food and woody biomass | | 129.9 | 15.2 | 221 | 163.4 | 15.5 | 101 | 851.7 | – | – |

| Population data | Inhabitants | Ratio[a] | Inhabitants | Ratio | Inhabitants | Ratio |
|---|---|---|---|---|---|---|
| Farm workers | 2057 | 3.9 | 1154 | 10.4 | 250 | 156.8 |
| Total population | 7941 | | 12,047 | | 39,189 | |

*Source* Our own

[a]Inhabitants per farm worker, which includes non-agricultural population and also non-working agricultural population (mainly children). [b]Self-sufficiency refers to the percentage of the total requirements covered with total produce. [c]Note that as we transformed wine surplus into food equivalents (see Sect. 5.2), this ratio would be lower in terms of wine surplus over wine requirements per capita. Without this transformation, wine produce per farm worker would be 7.3, thus the ratio would be reduced to 32. [d]Note that total GJ do not coincide with those appearing in Table 5.1, as here we are not including the item Others, and we are referring to Gross Calorific Value while in Table 5.1 we account for Metabolizable Energy

composition, and agricultural productivity been related to local food and fuel consumption, agricultural labour and demographic dynamics?

In the mid-19th century, we estimate that FP per farm worker and year was 22 GJ of food and wine, equivalent to the dietary needs of more than 5 male adults, and 108 GJ of woody biomass, equivalent to the fuel needs of nearly 10 people: with a farmer every 4 people, local food and fuel needs could be more than satisfied. Notwithstanding, if we distinguish between different food typologies, unbalances emerge: winegrowing specialization (64% of cropland area) implied a shortage of cereals, which were imported from inner Spain (Garrabou et al. 2007; Garrabou and Tello 2008). Therefore, although farmers' surplus was just enough to satisfy the dietary needs of the local population, a balanced diet could only be achieved through commercial exchange, thus c.1860 the Vallès was not entirely a local food system. But simply growing cereals instead of having vineyards would have not been the same. The multifunctionality of woody crops (Infante-Amate and Parcerisas 2013; Infante Amate et al. 2016) and unequal access to woodland can explain why poor farmers relied on their small vineyards as a source of fuel that could not be supplied otherwise. Moreover, the wealthiest landowners, who had woodland in excess, destined part of woodland production to the fuel-intensive brick industry and lime furnaces (Garrabou and Planas 1998: 121).

In 1956, food and fuel requirements had not changed a great deal from the 19th century. Particularly, regarding household fuel consumption, modern energy carriers had not yet spread broadly, and less so in rural areas. Arroyo (2006) estimates that in 1956 only 13% of households in Barcelona used gas for cooking and heating; this percentage drops to 11.4 and 8% for the nearby industrial towns of Sabadell and Terrassa. Gas cylinders only started to be widely used from the 1960s onwards. The increase in labour productivity—food surplus grew by 60%, allowing each farm worker to supply produce to nearly 9 male adults—was more than offset by an increase in the total population density (+51.7%), and a decrease in agricultural workers (−43.9%). In contrast to both 1860 and 1999, in 1956 there is no evidence of any outstanding specialization, and agroecosystem produce was sufficient in quantity and diverse enough to supply practically the whole population's requirements, including cereals and vegetables. The end of winegrowing specialization, after the *Phylloxera* plague (Badia-Miró et al. 2010) and the partial substitution by herbaceous and vegetable crops, along with the beginning of livestock dairy specialization in the Vallès, may explain this change. It still appears a very tight, albeit reduced, adjustment between farmers produce and population needs, with no need to import. Indeed Spain was just leaving behind the autarchic period imposed by Franco's dictatorship.

In 1999, we observe an increase on per farmer productivity—food produce per farm worker was multiplied by 18—going hand in hand with the combined effect of less farm workers and an even higher population. This implies that at aggregate level more than enough food calories were produced for the local population, however, 78% of this was animal produce. On the one hand, local animal produce per farm worker could supply the dietary requirement of 463 male adults, and with a farmer every 157 inhabitants, we can expect that Vallès is exporting most of its meat; on the other hand, this huge increase is offset by a fall in the agroecosystem's capacity to supply enough food for the local population. Wood biomass comparison makes no sense in this period due to the diffusion of heat obtained from fossil sources. New consumption patterns linked to supermarket chains and large groceries broke the link with local vegetable and cereal produce. The main purpose of FP in 1999 was not to cover local food needs, but to provide livestock produce to markets. This pork, once slaughtered, and some parts processed as inlay, was mainly distributed in Spain and only around 11% went to consumers abroad, mostly in France, Portugal and Germany (INE 1999).

Finally, some limitations of this approach need to be highlighted as the results are aggregated and averaged. First, we are accounting for production without considering that land and livestock distribution was strongly influenced by inequality. Second, labour productivities differ slightly, depending on the kind of agricultural work (e.g. cropland, woodland and livestock). Third, individual requirements refer to an adult male, not to average requirements that would depend on demographic structure. And fourth, we did not include here the need to generate a surplus to cover the local population's other needs, such as clothing, housing, the building of infrastructures and tool repairs.

## 5.3.2 *The Loss of Multifunctionality*

By assessing the energy balances of these very different farm systems and activities, we capture the loss of their original multifunctionality. In turn, we also identify the unsustainable processes of land-use specialization, as well as the ensuing impacts on emergent properties such as farm-associated biodiversity at a landscape scale. Below we detail the effects of this loss of multifunctionality on two different agroecosystem funds (land and livestock).

### 5.3.2.1 Disturbed Fields, Silent Forests: Land-Use Polarization

Before the socio-metabolic transition to industrial farming, food and fuel produce had to be in balance with feed production and the maintenance of ecosystem services that minimized crop diseases or poor harvests. In advanced organic agricultures, only some valuable cash crops were profitable enough to allow for imports of nutrients from outside agroecosystems.[8] In turn, land productivity was maintained in these traditional organic farm systems by local nutrient transfers from forests and livestock: fertilization was one of the main drivers of biophysical

[8]A good example could be the case of biomass imports of *toxo* (*Ulex europaeus*) for vineyard fertilization; these were imported in carts during the 19th century from the interior to the coast in Galiza (North-West of Iberia), as described in Corbacho-González (2015).

internal loops, which required a high labour intensity and tight land-use integration. The progressive introduction of synthetic fertilizers, which occurred throughout the 20th century, broke the integration between the basic funds of agroecosystems through these internal loops, and fostered the functional disconnection of agrarian activities. The use and abuse of industrial fertilizers also contributed to the fall in Final and External EROI[9] for the high energy cost of their production (Aguilera et al. 2015).

In addition to cropland intensification, crop homogenization also occurred. If we measure land cover diversity with the Shannon Index, we first observe an increase of spatial heterogeneity from 0.59 c.1860 to 0.84 in 1956, mainly because vineyards ceased to dominate the cultural landscape following the *Phylloxera* plague in the 1880s and 1890s. Afterwards, as a result of the new specialization in livestock fattening in industrial feedlots, heterogeneity was reduced again, to 0.65 in 1999.

At the same time, landscape homogenization has given rise to a reduction in the ecosystem services of plague and disease regulation and led to a greater need for biocide consumption, from 3.47 kg/ha in the 1960s up to 4.66 kg/ha currently. As shown in other case studies, loss of land cover heterogeneity increases consumption of pesticides (Jonason et al. 2013; Landis et al. 2000). Therefore getting rid of biocides requires the restoration of ecosystem services, which in turn means recovering the multifunctionality of integrated land-use management and animal husbandry (Foley et al. 2005).

Synthetic fertilizers and cropland homogenization explain the process of linearization of agricultural and livestock energy flows, where external inputs have replaced internal recirculation of biomass and services. In the same vein, biocides have decreased the rate of unharvested biomass with weed management and the ecological services it can provide. Complex agroecosystems have been converted into a simplified soil (Pollan 2008).

In 1860, 44.7% of the Actual Net Primary Production grown in cropland was reused to maintain soil fertility, an energy loop that entailed high labour requirements. By 1956 this share had decreased to 10%, and then to 9% in 1999 (Marco et al. 2017). This is an example of substitution among productive factors, where labour and land—in their mainstream economic meaning—seem to be replaced by external biophysical inputs, which in turn hide embodied labour and natural resources. Thus, the stronger the functional disconnection among agroecological funds is, the higher the required external inputs are, and as we will see in Sect. 5.3.4, the deeper the socio-ecological impacts on other agroecosystems will be.

Paradoxically, while human disturbance on agricultural land increased with the advance of an agro-industrial system, the transition in household energy carriers to fossil fuels which boomed in the 1960s (Soto et al. 2016) relegated forests to

[9]Again, this indicator emerges from the proposal of the so-called Energy Return of Investment (Tello et al. 2016). The Final EROI (FEROI) accounts for the energy efficiency of the whole agroecosystem and is calculated with the following formula: $FEROI = \frac{FP}{BR + EI}$, where FP is the Final Product, BR the Biomass Reused, and EI the External Inputs.

practically providing only aesthetic and conservational functions. The effects of this no-management strategy provided forests with a low quality endorsement and lower internal diversity (Cervera et al. 2016). While woodland area has increased from 35.5 to 56.7%, mainly in the steepest marginal areas, where agriculture could not be industrialized, its contribution to Final Produce has decreased from 36 to 5% of FP due to the collapse of charcoal making and wood extraction.

Forests have been the greatest source of final product per unit of labour spent thanks to the lower human intervention required in comparison to agriculture. In 1860, FP per farm worker was 59 GJ in cropland, 968 GJ in woodland and 9 GJ in livestock, respectively. The widespread access of households to gas cylinders, natural gas and oil fuel tanks, which began in the 1950s (Tello et al. 2014), resulted in forest extraction rates dropping from 41.1 GJ/ha in 1860 to 21 in 1956, and to only 3.5 in 1999. Although historically woodland has offered the greatest amount of Final Produce per unit of labour, market prices do not incorporate the positive environmental externalities of sustainable woodland management, or the negative externalities of fossil fuel consumption. Market signals have cancelled society's ability to supply cooking and heating fuel from local renewable sources.

Forest transition was reinforced by the collapse of extensive livestock grazing in meadows and wood pastures, a source of animal feeding that covered 16% of the total animal feed intake in 1860, 22% in 1956, and an insignificant 0.1% in 1999. From a landscape ecology perspective, this has also entailed the loss of a wide network of open spaces, leading to an important decrease of landscape diversity and the loss of habitats for many species, e.g. orchids or butterflies (Marull et al. 2014, 2015; Tello et al. 2014; Otero et al. 2015).

#### 5.3.2.2 Pork Is no Longer the Greater Profiteer: Losing the Functions of Livestock

Despite the low energy efficiency of feed-meat animal bioconversion, in traditional organic agroecosystems livestock had been a key factor in closing the greatest part of metabolic cycles: draught power, fertilization, heat emanating from stables, by-products revalorization and, as a complementary source, food. Traditional organic farming kept livestock breeding carefully integrated with cropland, pasture and forestland (Krausmann 2004)—as well as with peasant housing. The multi-functionality of livestock husbandry in the mid-19th century was lost during the 20th century. Mechanization and synthetic fertilizers on the one hand, and nutritional transition and agribusiness on the other, reduced livestock functions in the Vallès to only one: animal food (principally meat, milk and eggs). Livestock densities[10] have increased dramatically from 7 LU500/km$^2$ of farmland in 1860 to 187 in the present, 74% of which are constituted by swine.

[10]We express livestock density with LU500/km$^2$, meaning the number of equivalent animals of 500 kg per km$^2$.

Pigs have also experienced a nutritional transition, from being the best profiteer of residues to becoming responsible for massive grain imports. From being fed with domestic garbage, horticultural and vineyard by-products, to being fed mainly with corn, soybean meal and barley. Today 70% of swine intake is barley, which entails both a high land cost for its production and a strong competition with human food requirements (Haberl 2015). Not in vain, is the change in livestock density viewed as the main reason for the fall in the whole agroecosystem energy efficiency (Final EROI) from 1.03 to 0.22. Animal produce constitutes 76% of Final Produce, and animal intake 85.3% of Total Inputs Consumed (Marco et al. 2017). Conversely, c.1860 ruminants were fed in traditional organic farm systems, only one third with grains, because of their ability to degrade fibrous feed that does not compete with human demand of food. This become even more relevant, if we take into account that ruminants were the largest and most diverse source of livestock services in those traditional agroecosystems. Thus, looking at livestock diets can help to understand the role animal husbandry plays in agroecosystems, either integrating energy flows across different funds or contributing to a further linearization of food chains within an input-output, simplified and over-specialized, industrial production system in feedlots.[11]

We can also analyse the different services provided by livestock as a share of the total service in the agroecosystem. As stated before, we consider here the functions of fertilization, the power of tillage, and food provision. Heat obtained from stables —not accounted for in our balances—has been used for centuries for warming farms and houses, thus contributing to minimizing fuelwood demand.[12] Organic fertilization through manure still played an important role in 1956, as can be seen in Fig. 5.1, but was almost anecdotal four decades later. Livestock contribution to tillage power was reduced earlier, and in 1956 covered less than 10% of total installed tillage power despite its high weight in applied tillage power figures. On the contrary, meat production doubled its contribution to the overall production and, as mentioned, had a huge contribution in 1999.

Low livestock densities in the Vallès area c.1860 required integrated land-use management and diversity of fertilizing techniques. As shown in Fig. 5.1, at that time the replenishment of nutrients required, besides animal manure, the reintroduction of human dung in the soils and also *formiguers*, which represented one of the most labour-intensive ways to maintain soil fertility (Olarieta et al. 2011) as well as an opportunity for nutrient catchment from forestlands. All these diverse biocultural managements were abandoned with the introduction of synthetic fertilizers, which in 1956 accounted for 64% of total fertilization requirements, increasing to 92% in 1999. Yet, as we have highlighted in Sect. 5.3.2.1, current fertilization

[11]In turn, this entails an associated contradiction: Farmers are giving sodium bicarbonate to ruminants to prevent the acidity produced by the excessive consumption of grains (Ferre and Baucells 2009).

[12]The isolated Catalan farms (*masies*) usually included the stable on the ground floor, where animals stayed during the night, while the chambers were on the upper floor, taking advantage of the animal heat that flowed from downstairs (Closa 2012).

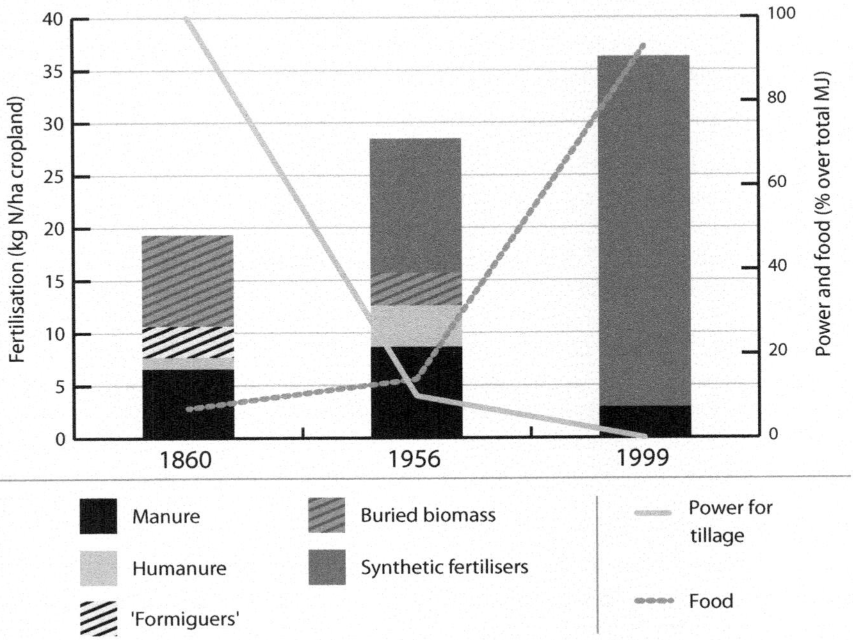


**Fig. 5.1** Nitrogen supplies, share of animal power for tillage and food provided by livestock among the Vallès transition. *Source* Our own from the sources listed in Marco et al. (2017)

patterns in most industrialized agroecosystems conceal the new role that pig slurry plays in the feedlots' specialized regions, which is being converted from resource to waste.

### 5.3.3 *Emergent Properties of Cultural Landscapes: Farm-Associated Biodiversity*

The progressive linearization of flows interlinking farmland and livestock funds has implied a partial disconnection of agrarian activities from their endowment of natural resources and an impoverishment in land covers. An emergent property of landscapes rich in land covers is habitat differentiation, enhancing farm-associated biodiversity (Loreau et al. 2003; Tscharntke et al. 2012; Marull et al. 2016).

The distribution patterns of farm-associated biodiversity not only depend on the flows that agrarian activity voluntarily or involuntarily devotes to them, but on how land covers are managed by farmers, along with other site-specific physical and biological characteristics. It is, in fact, through habitat heterogeneity at the landscape scale that the best advantage can be taken from ecosystem services for farmers (Power 2010). Therefore, we combine land-use changes with landscape

**Table 5.3** Dynamics of constraints for biodiversity

| | | 1860 | 1956 | 1999 |
|---|---|---|---|---|
| Unharvested biomass | $GJ \cdot ha^{-1}$ | 24.5 | 44.8 | 60.2 |
| External inputs | $GJ \cdot ha^{-1}$ | 1.0 | 7.8 | 106.8 |
| NPP-EROI | Dimensionless | 3.27 | 3.08 | 0.55 |
| A-FEROI | Dimensionless | 0.49 | 0.25 | 0.16 |
| L | Dimensionless | 0.70 | 0.73 | 0.33 |
| EMS | $Km^2$ | 142.5 | 135.0 | 89.0 |

*Source* Our own

ecology analysis to consider four indicators constituting the landscape—social metabolism interface: *NPPEROI, AFEROI, L* and *EMS*. We have presented them in Sect. 5.2.2, and the values found in the study area are shown in Table 5.3.

Unharvested Biomass (UB) is a measure of the amount of Actual Net Primary Production ($NPP_{act}$) that is left for the associated biodiversity: the higher its amount, the better for biodiversity. On the contrary, External Inputs (EI) can be understood as a measure of anthropic ecological disturbance, so the lower the EI remains, and the lesser the human perturbation is. NPP-EROI and A-FEROI show how these, and other flows, interact in the whole agroecosystem. Higher values of both are desirable, as the first accounts for the $NPP_{act}$ that an agroecosystem has per unit of farming disturbance exerted, and the second for the amount of final production that goes outside the system boundaries per unit of the overall anthropic effort made to keep the agroecosystem functioning.[13] Finally, higher values of L and EMS show better habitat conditions to host different ecological niches.

Our results show an increase in farming disturbance from 1860 to 1956, expressed in the magnitude of EI per unit of farmland, compensated by an increase in UB per surface unit due to a less intense forestry activity which has allowed the retaining of a certain amount of undisturbed biomass available for biodiversity. In the same period, 1860–1956, we observe a slight increase in land cover heterogeneity (here taken as a proxy for habitat diversification)—mainly associated with the abandonment of vineyards—followed by a sharp decline from 1956 to 1999. This decrease in land cover richness has gone hand in hand with the growing presence of roads and other linear infrastructures whose barrier effects are affecting the ability of wild animals and plants communities to connect with each other.

The most dramatic change occurred during the last time period. From 1956 to 1999, there was a 6-fold drop in *NPPEROI, L* decreased by 55%, and *EMS* by 34%. What the *NPP-EROI* and Agroecological Final EROI reveal is to what extent the

---

[13]It is important to remember that the first motivation of an agroecosystem is to provide biotic materials for a society. Yet the continuous extraction of this final produce involves an ecological disturbance that needs to be kept below a certain level compatible with the reproduction of the agroecosystem funds. Therefore, in order to have sustainable farm systems, production must be balanced to the ecological disturbance exerted through the investment made to keep the agroecosystem functioning.

proportion of Final Produce has decreased with respect to the sum of UB, EI and the Biomass Reused. The combined effect of forest abandonment and feed imports have strikingly increased both UB and EI with respect to the previous time points. This socio-metabolic shift has translated into a landscape polarization between abandoned woodlands in the steepest lands, and intensification in flatter ones. This land-use polarization entailed a loss of landscape complexity, and a vanishing of land cover mosaics, which reduced habitat differentiation and species richness—as the values of their proxy indicators [Shannon (L) and EMS] show (Marull et al. 2014, 2015; Tello et al. 2014; Otero et al. 2015).

These results reveal that through the time period studied, and particularly in keeping with the Green Revolution from 1956 to 1999, the farming social metabolism gave up its efforts to increase the overall share of biomass harvested, in an unintended land sparing effect, while the disturbance exerted in the remaining cropland increased. Both opposite changes entail agroecosystem degradation, which becomes apparent with the loss of agro-forestry mosaics, and Unharvested Biomass accumulation in woodland that make them more fire-prone, and give rise to homogeneous landscapes whose niches grow out of control, leading to plagues—as happens in Catalonia with the wild boar (Bosch et al. 2012). This could explain the lesser variety of ecological niches, and the loss of complexity in food chains, when UB is accumulated only in some specific habitats. Besides these agroecosystem metabolism impacts, the barrier effect of linear infrastructures, such as highways, high-speed railway lines, and a significant urban sprawl, have added a strong habitat fragmentation, and a decrease of ecological connectivity between landscape patches that reinforces biodiversity loss.

In short, the role played by agro-forestry mosaics was very important, not only because of the agroecosystem multifunctionality they entailed, but also in terms of the farm-associated biodiversity they provided. The socio-metabolic transition towards industrialized farming systems has led to a loss of habitat differentiation, tied to a higher level of disturbance, a greater fragmentation, and a lower ecological connectivity that has grown even deeper in the last decades.

## *5.3.4 Expelling Socio-environmental Unsustainability*

### 5.3.4.1 Agroecosystems' Carrying Capacity

Besides the hazard that external inputs represent in terms of linearization of agroecosystems functioning, there are impacts that are hidden when a local scale of analysis is adopted: External Inputs (EI) are not a cybernetic issue, they imply biophysical flows proceeding from elsewhere with all the energy embodied in their production and transport processes (Tello et al. 2016). In 1999 71% of the EI energy flow in the Vallès area was animal feeding. In addition to 71% of local cropland already being devoted to feed products, imports account for 87% of all the biomass required for livestock maintenance. To put it simply, livestock density in 1999 cannot be supported by the carrying capacity of the local agroecosystem.

Despite the increase of meat consumption per capita, and the higher population densities in Vallès County, this huge livestock density is linked to regional economic specialization in meat production. According to the estimated diets of this area, pork production is 17-fold greater than the requirement for average diet types at present. Considering all types of meat, the study area produces 3 times the dietary requirement of its local population (if they were all male adults; even more if we differentiate by age and gender). Moreover, if all cropland area in 1999 had been devoted to livestock feed, 56.8% of animal intake would still have to be imported from abroad. In fact, the livestock density that could be carried within agroecosystems boundaries, assuming a complete local specialization of cropland to animal feeding, is 35.4 LU500/km$^2$, that is 5.3 times less than the actual one in 1999.

#### 5.3.4.2 A Global Land Sprawl? The Appropriation of the Land of Others

To gain a better understanding of the footprint of local food systems, it is necessary to expand the scale of analysis. Since social metabolism is no longer closed at local scale, information tends to be hidden at higher scales (González de Molina and Toledo 2014). We have analysed the likely origins of these feed imports. As can be seen in Fig. 5.2, corn represents the largest feed import item, mainly coming from other Spanish regions, but also from France, Argentina and Brazil. Indeed, nearly 10.000 tons of soybean meal were imported from South America.

Setting aside the sea surface required for fish meal, livestock breeding in the four municipalities need an 8.3 greater area than the one devoted to cropland within the Vallès study area region. The metabolic rift[14] generated in terms of surfaces of fertile land is enormous, as Lassaletta et al. (2014) have already pointed out in terms of nitrogen flows. Moreover, this estimated food imprint disregards the degradation of agroecosystem funds in feed-supplying countries (Guzmán et al. 2011).

The virtual land cost of the intensive meat production performed in the study area is not only associated with local diets being disconnected from their local territory. Actually, only 32% of the land devoted to agriculture in the mid-19th century in the Vallès is cropped nowadays. This also refers to a question of the concentration of decision-making power and ejection of unsustainability from the global North to global South; as can be seen in Fig. 5.2, the main providers of feed are peripheral countries. Agrarian activities and food consumption baskets in the open economies of the Global North trace impacts on global environment due to

[14]We take the definition of metabolic rift from Schneider and McMichael (2010) as "a social, ecological, and historical concept describing the disruption of natural cycles and processes and ruptures in material human-nature relations under capitalism".

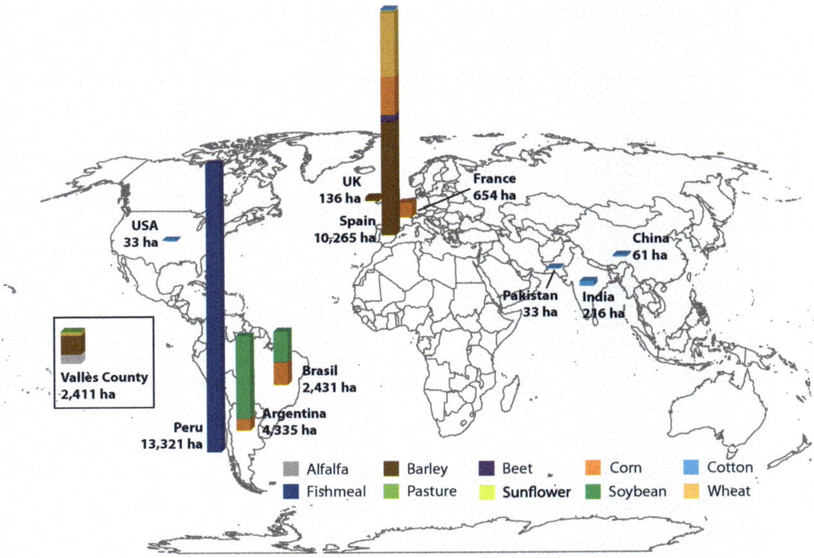


**Fig. 5.2** Average surface required to maintain the Vallès feed imports for 1999. *Source* Our own [The area of fishmeal referred to is the total sea surface required considering the total amount of fish products gathered from the sea in Peru (the main provider of fishmeal) divided by its marine are. The surface required from the rest of Spain excludes the production inside the Vallès county (which is located in the box)]

market power asymmetry (Garmendia et al. 2016), which underlies unequal economic and ecological exchange (Foster and Holleman 2014).

Soybean imports, which constitute some 20% of the total feed imports, come mainly from Argentina. Its growth is supplied by an agro-industrial production system whose environmental and social impacts would not be allowed under current Spanish legislation. Agribusiness consolidated in Argentina with the introduction of Roundup Ready™ soybeans in 1996 (Leguizamón 2016). Almost all soya bean produced is GMO, and 90% is exported. National food security risks, the high volatility of the commodity market, and a handful of large corporations controlling the whole production chain, are only one part of the negative economic and social effects of this food regime. The disappearance of family farms, the displacement of indigenous communities, deforestation, biodiversity loss and health problems associated with the aerial spraying of glyphosate have to be accounted too (Teubal 2008). As Leguizamón (2016) argues, agribusiness chains create a geographic and cultural distance between farmers and consumers that hides its socio-ecological negative impacts. The socio-ecological distance increases further once this agro-industrial feed is consumed indirectly through livestock bioconversion.

From a farm-associated biodiversity perspective, the impact emerges with the loss of management diversity (Brookfield and Stocking 1999): in Argentina monoculture is conducted by big estates, larger than 5000 ha, doubling the current agricultural surface of the Vallès (Catacora-Vargas et al. 2012). Obviously, this also entails a landscape homogenization and the ensuing decline of farm-associated biodiversity, as well as deforestation. Moreover, behind these biophysical changes there are also social consequences, such as the impact on the livelihoods of populations affected through the violent process of peasants' expulsion from the land and the loss of subsistence farming (Magdoff 2013). This is what a rough analysis of soybean meal production in Argentina reveals. Similar societal and environmental impacts have been reported with cotton production in India, fish meal in Peru, and soybeans in Brazil (Temper et al. 2015).

#### 5.3.4.3 Local Socio-ecological Costs of Global Trade

The importation and bioconversion of such an enormous amount of foreign corn and soybeans is not an inert process for the local territory. From a socio-metabolic perspective, the negative externalities associated to the farming of the imported feed in the producing countries go hand in hand with strong agroecological impacts at local scale. The overall estimated volume of dung produced by animal husbandry in Vallès County is up to 221.700 $m^3$ of slurry per year, equivalent to a cubic pool with a 60-m side. Because of the difference in its concentration of nutrients between grain and dung, the energy cost of returning to the original soils all this biomass, in order to close the nutrient cycles, would be 4.68 times the transport cost of feed imports. It is simply entropy that makes it impossible in monetary terms to close such a nutrient cycle.

This animal excreta, which in former historical periods would have been considered a precious resource in a region with a structural scarcity of manure (Galán et al. 2012), is now treated as an economic problem that leads to serious environmental pollution. In other words, within the current globalized food regime it becomes an out of place resource (Odum 1993). This volume of dung, after being composted, could fertilize 7.8 times the nitrogen requirements of all the cropping area in the four municipalities of Vallès County. Yet, due to economic decisions adopted in the context of a great atomization of agricultural activities, farmer instead use significant amounts of chemical fertilizers. Pig slurry is thus applied as a pre-planting fertilizer, with rates that exceed 400 kg N/ha·year (Sisquella et al. 2004), while recommendations are not to trespass a limit of 170 kg N/ha·year on organic amendments (DOGC 2009). Therefore, only 37% of all this dung is actually devoted to soil fertility while the rest is being leached, or wasted if not applied to cropland soil. Replacing chemical fertilizers with dung would reduce by 10,971 GJ the External Inputs required, as well as 37,600 GJ in waste flows, 15% of the total biophysical wastage of this industrial farm system. This nitrogen leaching is polluting aquifers with a widespread diffuse impact. Indeed, over-fertilization is increasing the eutrophication risk in agricultural soils due to the excess of phosphorus and potassium (Penuelas et al. 2009).

Summing up, the metabolic rift driven by cheap fuel prices does not only allow the breaking of nutrient cycles; it also damages the environmental quality of the local aquifers and river streams, as well as being responsible for the opportunity cost of not closing nutrient cycles by recovering multifunctionality of agroecosystem funds both in exporting and importing countries. Not without reason, has part of the scientific community long been calling for a downscaling of livestock densities to the real carrying capacity of soils at a municipal or regional scale, in order to avoid pollution (Teira-Esmatges and Flotats 2003). However, as long as this problem remains a consequence of massive feed imports at global scale, any local assessment will only partially tackle the problem. The actual solution means devising and implementing agroecological strategies oriented towards local sustainable food systems.

## 5.4 Conclusions

During the end of the 19th, and the first half of the 20th centuries, the agroecosystem of Vallès County was tightly connected with the food and fuel requirements of the local population. This was so despite the fact that vineyard specialization c.1860 implied a higher dependence on market exchanges. In other words, people were not only living *within* a territory but they lived *of* the territory. Interestingly, in the mid-20th century, increases in labour productivity allowed for a lower share of an agriculturally active population and higher population densities. This could be explained partially by the abandonment of labour-intensive vegetable fertilizing techniques; these were gradually replaced by chemical fertilizers at a time when they were devised to supplement, but never replace, organic manure. During the second half of the 20th century, nutritional and energy transitions, the massive spread of the Green Revolution, and industrial livestock fattening in feedlots, broke all these linkages. Agricultural produce was no longer defined by the local population's food and fuel needs. Labour productivity rose steeply, while the role of agricultural labour within the whole economy shrunk. At the same time, animal produce went on to dominate Final Produce composition. New dietary patterns explain both livestock specialization in Vallès County, and the ensuing disconnection between local food requirements and local food produce.

Disconnection with local people went hand in hand with a disconnection of agro-industrial meat produce with the surrounding territory. The abandonment of integrated farmland management during the socio-metabolic transition to industrial farming, and livestock fattening in feedlots, also supposed the atomization and linear behaviour of agroecosystem funds due to the end of many multipurpose farm activities. While production is focused only on maximizing short-term economic profit, disregarding the positive externalities of closing biophysical cycles at local level, the ensuing imbalance among funds (e.g. between livestock and farmland) derived from an ever greater need to rely on external inputs which, in turn, increased anthropic ecological disturbance. Particularly interesting is the case of livestock, whose ecosystem services have been reduced to only meat production, while their animal diets have lost their former reusing ability and are increasingly competing with human food production. The ensuing disappearance of the former complex agro-forestry mosaics has resulted in a loss of farm-associated biodiversity. Hence, socio-metabolic transition to agro-industrial food systems has led to less variety in food-chains available for non-domesticated species, together with a loss of habitat differentiation. To this, a greater landscape fragmentation has been added as a result of an increasingly polarized land-use pattern between a highly disturbing industrial farming of flatter lands, and the abandonment of steeper ones to forest encroachment.

While in former organic agricultures commercial specializations were somehow adjusted to the local or regional agroecosystem's carrying capacity, in the current industrialized capitalist ones specializations depend on the massive imports of external inputs. We have seen in the Vallès case study that meat produce would have had to be 5.3 times lower in 1999, if it had been adjusted to the local capacity to grow animal feed. This disconnection implies a global footprint that appropriates the land yields of a cropland surface 8.3 times greater than agricultural land in the Vallès, mainly from Spain but also coming from the Global South. These global trade relations involve a power asymmetry of agri-business corporations that entails relevant societal and environmental impacts in the exporting countries which cannot be kept in check by current social and legal constraints in Spain. In turn, this unequal ecological exchange also damages the importing area, where dung accumulation is harming water and soils' environmental quality. These local impacts of a global trading chain cannot be countered with just local analysis and action, but require a changing of the food regime as a whole. To give but one example, the nutrients extracted from export regions in the Global South cannot close their nutrient cycles in an organic, sustainable manner. Agri-food systems can only become sustainable if their biophysical and socio-economic flows are relocated, along with their cultural practices and political decision-making processes.

Our study also teaches us some lessons from a methodological point of view. Carrying out a multi-dimensional and multi-scalar analysis of social metabolism is a useful tool that allows us to comprehend the diffuse impacts of policy making in food regimes at different dimensions: diets, land uses, landscapes and international trade. This permits, on the one hand, going beyond single-sided analysis to enhance the complexity of agrarian systems and food regimes through its social and ecological linkages; from our perspective, this is the only way to face current challenges to global food systems. On the other hand, performing a long-term dynamic analysis of how local food systems have changed over time helps us address an applied history task: to recover peasants' bio-cultural knowledge and expertise that had been employed in managing society-nature relations for such a long time, as a key resource to devise new relocated and resilient agri-food systems for a more sustainable future.

# References

Aguilera, E., Guzmán, G. I., Infante-Amate, J., et al. (2015). *Embodied energy in agricultural inputs. Incorporating a historical perspective*. DT-SEHA 1507.

Altieri, M. A. (1999). The ecological role of biodiversity in agroecosystems. *Agriculture, Ecosystems & Environment, 74*(1–3), 19–31. doi:10.1016/S0167-8809(99)00028-6.

Arroyo, M. (2006). Los cambios en el proceso de producción y de distribución de gas en Barcelona y su hinterland (1930–1969). Entre el gas de hulla y el gas natural. *Scripta Nova: Revista Electrónica de Geografía y Ciencias Sociales, 10,* 29.

Badia-Miró, M., Tello, E., Valls, F., et al. (2010). The grape phylloxera plague as a natural experiment: The unkeep of vineyards in Catalonia (Spain), 1858–1935. *Australian Economic History Review, 50*(1), 39–61. doi:10.1111/j.1467-8446.2009.00271.x.

Bhatt, B. P., & Sachan, M. S. (2004). Firewood consumption along an altitudinal gradient in mountain villages of India. *Biomass and Bioenergy, 27*(1), 69–75. doi:10.1016/j.biombioe.2003.10.004.

Bosch, J., Peris, S., Fonseca, C., et al. (2012). Distribution, abundance and density of the wild boar on the Iberian Peninsula, based on the CORINE program and hunting statistics. *Folia Zoologica, 61*(2), 138–151.

Brookfield, H., & Stocking, M. (1999). Agrodiversity: Definition, description and design. *Global Environmental Change, 9*(2), 77–80. doi:10.1016/S0959-3780(99)00004-7.

Catacora-Vargas, G., Galeano, P., Agapito-Tenfen, S., et al. (2012). *Soybean production in the Southern Cone of the Americas: Update on land and pesticide use*. Cochabamba: Virmegraf.

Cervera, T., Pino, J., Marull, J., et al. (2016). Understanding the long-term dynamics of forest transition: From deforestation to afforestation in a Mediterranean landscape (Catalonia, 1868–2005), *Land Use Policy (in press)* . doi:10.1016/j.landusepol.2016.10.006.

Church, D. C. (1984). *Alimentos y alimentacion del ganado*. Tomo I. Hemisferio sur, Buenos Aires.

Clark, B., & Foster, J. B. (2009). Ecological imperialism and the global metabolic rift unequal exchange and the guano/nitrates trade. *International Journal of Comparative Sociology, 50*(3–4), 311–334. doi:10.1177/0020715209105144.

Closa, E. (2012). *Els valors artístics, arquitectònics i constructius de la masia*. Barcelona.

Colomé, J. (1996). *L'especialització vitícola a la Catalunya del segle XIX. La comarca del Penedès*. Dissertation, Universitat de Barcelona.

Cooperativas Agro-alimentarias. (2010). *Manual de ahorro y eficiencia energética del sector*. Madrid.

Corbacho-González, B. (2015). *La transición socioecológica de una agricultura atlántica europea: El caso de la agricultura gallega, 1750–1900*. Paper presented at the III Seminario de la Sociedad Española de Historia Agraria, Ministerio de Agricultura, Alimentación y Medio Ambiente, Madrid, November 28, 2014.

Cussó, X., & Garrabou, R. (2001). Alimentació i nutrició al Vallès Oriental en les darreres dècades del segle XIX. *Lauro: revista del Museu de Granollers, 21,* 26–34.

Cussó, X., & Garrabou, R. (2007). La transición nutricional en la España contemporánea: Las variaciones en el consumo de pan, patatas y legumbres (1850–2000). *Investigaciones de Historia Económica, 3*(7), 69–100.

Cussó, X., & Garrabou, R. (2010). *La globalización de la dieta en España en el siglo XX*. Paper presented at the X Congreso Español de Sociología, Public University of Pamplona, June 1–3, 2010.

Cussó, X., & Garrabou, R. (2012). *Alimentacio i nutricio al Vallès Occidental. Un segle i mig de canvis i permanències: 1787–1936*. Unitat d'Història Econòmica 2012_05.

Cussó, X., Garrabou, R., Olarieta, J. R., et al. (2006). Balances energéticos y usos del suelo en la agricultura catalana: Una comparación entre mediados del siglo XIX y finales del siglo XX. *Historia Agraria, 40,* 471–500.

DARP Departament d'Agricultura, Ramadería i Pesca. (1998). *Dades bàsiques de l'agricultura, la ramaderia i la pesca a Catalunya*. Gabinet Tècnic.

de Boer, J., Helms, M., & Aiking, H. (2006). Protein consumption and sustainability: Diet diversity in EU-15. *Ecological Economics, 59*(3), 267–274. doi:10.1016/j.ecolecon.2005.10.011.

Diari Oficial de la Generalitat de Catalunya DOGC. (2009). *DR 136/2009 Zones vulnerables en relació amb la contaminació de nitrats que procedeixen de fonts agràries i de gestió de les dejeccions ramaderes*. Barcelona.

Edelman, M., Weis, T., Baviskar, A., et al. (2014). Introduction: Critical perspectives on food sovereignty. *The Journal of Peasant Studies, 41*(6), 911–931. doi:10.1080/03066150.2014.963568.

Faber, M., Manstetten, R., & Proops, J. L. R. (1995). On the conceptual foundations of ecological economics: A teleological approach. *Ecological Economics, 12*(1), 41–54.

FAO. (1983). *Métodos simples para fabricar carbón vegetal*. Roma: Estudio Montes. FAO.

Ferre, D., & Baucells, J. (2009). El bicarbonato sódico como aditivo insustituible en dietas de alta producción. http://dialnet.unirioja.es/servlet/articulo?codigo=2877973. Accessed November 15, 2016.

Flores, M., & Rodríguez-Ventura, M. (2014). Curso de nutricion animal. http://www.webs.ulpgc.es/nutranim/index.html. Accessed November 15, 2016.

Foley, J. A., Defries, R., Asner, G. P., et al. (2005). Global consequences of land use. *Science, 309*, 570–574. doi:10.1126/science.1111772.

Foster, J. B., & Holleman, H. (2014). The theory of unequal ecological exchange: A Marx-Odum dialectic. *The Journal of Peasant Studies, 41*(2), 199–233. doi:10.1080/03066150.2014.889687.

Friedmann, H. (2016). Food regime analysis and agrarian questions: Widening the conversation. *The Journal of Peasant Studies, 43*(3), 671–692. doi:10.1080/03066150.2016.1146254.

Galán, E., Padró, R., Marco, I., et al. (2016). Widening the analysis of Energy Return on Investment (EROI) in agro-ecosystems: Socio-ecological transitions to industrialized farm systems (the Vallès County, Catalonia, c.1860 and 1999). *Ecological Modelling, 336*, 13–25. doi:10.1016/j.ecolmodel.2016.05.012.

Galán, E., Tello, E., Cussó, X., et al. (2012). Métodos de fertilización y balance de nutrientes en la agricultura orgánica tradicional de la biorregion mediterránea: Cataluña (España) en la década de 1860. *Historia Agraria, 65*, 95–119.

Gales, B., Kander, A., Malanima, P., et al. (2007). North versus South: Energy transition and energy intensity in Europe over 200 years. *European Review of Economic History, 11*(2), 219–253.

Garmendia, E., Urkidi, L., Arto, I., et al. (2016). Tracing the impacts of a northern open economy on the global environment. *Ecological Economics, 126*, 169–181. doi:10.1016/j.ecolecon.2016.02.011.

Garrabou, R., Cussó, X., & Tello, E. (2007). La persistència del conreu de cereals a la província de Barcelona a mitjan segle XIX. *Estudis d'Història Agrària, 20*, 165–221.

Garrabou, R., & Planas, J. (1998). *Estudio Agrícola del Vallès 1874*. Granollers: Impremta de Granollers.

Garrabou, R., & Tello, E. (2008). L'especialització vitícola catalana i la formació del mercat blader espanyol: Una nova interpretació a partir del cas de la província de Barcelona. *Recerques, 57*, 91–134.

Georgescu-Roegen, N. (1971). *The entropy law and the economic process*. Cambridge: Harvard University Press.

Giampietro, M., Mayumi, K., & Ramos-Martin, J. (2008). *Multi-Scale Integrated Analysis of Societal and Ecosystem Metabolism (MUSIASEM): An outline of rationale and theory*. Barcelona: Elsevier.

Giampietro, M., & Pimentel, D. (1990). Assessment of the energetics of human labor. *Agriculture, Ecosystems & Environment, 32*(3), 257–272.

Gliessmann, S. (1998). *Agroecology: Ecological processes in sustainable agriculture*. London: Lewis Publishers.
González-Bernáldez, F. (1981). *Ecología y paisaje*. Madrid: Editorial Blume.
González de Molina, M., García Ruiz, R., Guzmán, G. I., et al. (2010). *Guideline for constructing nutrient balance in historical agricultural systems (and its application to three case-studies in southern Spain)*. DT-SEHA 1008.
González de Molina, M., & Guzmán, G. I. (2006). *Tras los pasos de la insustentabilidad agricultura y medio ambiente en perspectiva histórica (siglos XVIII–XX)*. Barcelona: Icaria.
González de Molina, M., & Toledo, V. (2014). *The social metabolism*. New York: Springer.
Guzmán, G. I., Aguilera, E., Soto, D., et al. (2014). *Metodología y conversores para el cálculo de la biomasa total producida en los agroecosistemas*. DT-SEHA.
Guzmán, G. I., & González de Molina, M. (2015). Energy efficiency in agrarian systems from an agroecological perspective. *Agroecology and Sustainable Food Systems, 39*(8), 924–952. doi:10.1080/21683565.2015.1053587.
Guzmán, G. I., González de Molina, M., & Alonso, A. M. (2011). The land cost of agrarian sustainability. An assessment. *Land Use Policy, 28*(4), 825–835. doi:10.1016/j.landusepol.2011.01.010.
Haberl, H. (2015). Competition for land: A sociometabolic perspective. *Ecological Economics, 119*, 424–431. doi:10.1016/j.ecolecon.2014.10.002.
Holland, J. M., Oaten, H., Moreby, S., et al. (2012). Agri-environment scheme enhancing ecosystem services: A demonstration of improved biological control in cereal crops. *Agriculture, Ecosystems & Environment, 155*, 147–152. doi:10.1016/j.agee.2012.04.014.
Infante Amate, J., Aguilera, E., & González de Molina, M. (2014). *La gran transformacion del sector agroalimentario español. Un análisis desde la perspectiva energetica (1960–2010)*. DT-SEHA 1403.
Infante Amate, J., & González de Molina, M. (2013). 'Sustainable de-growth' in agriculture and food: An agro-ecological perspective on Spain's agri-food system (year 2000). *Journal of Cleaner Production, 38*, 27–35. doi:10.1016/j.jclepro.2011.03.018.
Infante Amate, J., & Parcerisas, L. (2013). Nuevas hipótesis sobre la especialización agraria en el Mediterráneo español. El olivar y la viña en perspectiva comparada (1850–1935). *XIII Congreso de la Sociedad Española de Historia Agraria, Badajoz, 7–9 de Noviembre*.
Infante Amate, J., Soto, D., Aguilera, E., et al. (2015). The Spanish transition to industrial metabolism: Long-term material flow analysis (1860–2010). *Journal of Industrial Ecology, 19*(5), 866–876. doi:10.1111/jiec.12261.
Infante Amate, J., Villa, I., Aguilera, E., et al. (2016). The making of olive landscapes in the south of Spain. A history of continuous expansion and intensification. In M. Agnoletti & F. Emanueli (Eds.), *Biocultural Diversity in Europe* (pp. 157–179). New York: Springer.
Instituto Nacional de Estadística. (1969). *Encuesta de presupuestos familiares (Marzo 1963–Marzo 1965)* (pp. 37–58). Madrid.
Instituto Nacional de Estadística. (1999). *Censo agrario 1999*. Madrid.
Jaeger, J. A. G. (2000). Landscape division, splitting index, and effective mesh size: New measures of landscape fragmentation. *Landscape Ecology, 15*(2), 115–130. doi:10.1023/A:1008129329289.
Jonason, D., Smith, H. G., Bengtsson, J., et al. (2013). Landscape simplification promotes weed seed predation by carabid beetles (Coleoptera: Carabidae). *Landscape Ecology, 28*(3), 487–494. doi:10.1007/s10980-013-9848-2.
Kander, A., Warde, P., & Malanima, P. (2013). *Power to the people: Energy in Europe over the last five centuries*. Princeton: Princeton University Press.
Kay, J., & Schneider, E. D. (1994). Embracing complexity, the challenge of the ecosystem approach. *Alternatives, 20*(3), 32–38. doi:10.1017/CBO9781107415324.004.
Krausmann, F. (2004). Milk, manure, and muscle power. Livestock and the transformation of preindustrial agriculture in Central Europe. *Human Ecology, 32*(6), 735–772. doi:10.1007/s10745-004-6834-y.

Landis, D. A., Wratten, S. D., & Gurr, G. M. (2000). Habitat management to conserve natural enemies of arthropod pests in agriculture. *Annual Review of Entomology, 45,* 175–201.

Lassaletta, L., Billen, G., Romero, E., et al. (2014). How changes in diet and trade patterns have shaped the N cycle at the national scale: Spain (1961–2009). *Regional Environmental Change, 14*(2), 785–797. doi:10.1007/s10113-013-0536-1.

Leguizamón, A. (2016). Disappearing nature? Agribusiness, biotechnology and distance in Argentine soybean production. *The Journal of Peasant Studies, 43*(2), 313–330. doi:10.1080/03066150.2016.1140647.

Levers, C., Müller, D., Erb, K., et al. (2015). Archetypical patterns and trajectories of land systems in Europe. *Regional Environmental Change*, 1–18. http://doi.org/10.1007/s10113-015-0907-x.

Loreau, M., Mouquet, N., & Gonzalez, A. (2003). Biodiversity as spatial insurance in heterogeneous landscapes. *Proceedings of the National Academy of Sciences of the United States of America, 100*(22), 12765–12770. doi:10.1073/pnas.2235465100.

Lotka, A. J. (1956). *Elements of mathematical biology*. New York: Dover Publications.

Magdoff, F. (2013). Twenty-first-century land grabs: Accumulation by agricultural dispossession. *Monthly Review, 65,* 1–12.

Marco, I., Padró, R., Cattaneo, C., Caravacca, C., & Tello, E. (2017). From vineyards to feedlots: A fund-flow scanning of sociometabolic transition in the Vallès County (Catalonia) 1860–1956–1999. *Regional Environmental Change*. doi:10.1007/s10113-017-1172-y.

Marrodán, M. D., Montero, P., & Cherkaoui, M. (2012). Transición Nutricional en España durante la historia reciente. *Nutrición clínica y dietética hospitalaria, 32*(2), 55–64.

Martinez-Alier, J., Kallis, G., Veuthey, S., Walter, M., & Temper, L. (2010). Social metabolism, ecological distribution conflicts and valuation languages. *Ecological Economics, 70*(2), 153–158.

Marull, J., Font, C., Padró, R., et al. (2016). Energy-Landscape Integrated Analysis: A proposal for measuring complexity in internal agroecosystem processes (Barcelona Metropolitan Region, 1860–2000). *Ecological Indicators, 66,* 30–46. doi:10.1007/s13398-014-0173-7.2.

Marull, J., Otero, I., Stefanescu, C., et al. (2015). Exploring the links between forest transition and landscape changes in the Mediterranean. Does forest recovery really lead to better landscape quality? *Agroforestry Systems, 89*(4), 705–719. doi:10.1007/s10457-015-9808-8.

Marull, J., Pino, J., Tello, E., et al. (2010). Social metabolism, landscape change and land-use planning in the Barcelona Metropolitan Region. *Land Use Policy, 27*(2), 497–510. doi:10.1016/j.landusepol.2009.07.004.

Marull, J., Tello, E., Wilcox, P. T., et al. (2014). Recovering the landscape history behind a Mediterranean edge environment (The Congost Valley, Catalonia, 1854–2005): The importance of agroforestry systems in biological conservation. *Applied Geography, 54,* 1–17. doi:10.1016/j.apgeog.2014.06.030.

Mayer, A., Schaffartzik, A., Haas, W., et al. (2015). Patterns of global biomass trade. Implications for food sovereignty and socio-environmental conflicts. *EJOLT Reports, 20*.

Mayumi, K. (1991). Temporary emancipation from land: From the industrial revolution to the present time. *Ecological Economics, 4,* 35–56.

McMichael, P. (2016). Bernstein-McMichael-Friedmann dialogue on food regimes. Commentary: Food regime for thought. *The Journal of Peasant Studies, 43*(3), 648–670. doi:10.1080/03066150.2016.1143816.

Nicolau, R., & Pujol, J. (2005). El consumo de proteínas animales en Barcelona entre las décadas de 1830 y 1930: Evolución y factores condicionantes. *Investigaciones de Historia Económica, 1*(3), 101–134.

Observatori del Porcí. (2009). *B3 Anàlisi simplificat de l'estructura de la cadena*. Barcelona.

Odum, E. P. (1993). *Ecology and our endangered life-support systems*. Massachussets: Sinauer Associates.

Olarieta, J. R., Padró, R., Masip, G., et al. (2011). "Formiguers", a historical system of soil fertilization (and biochar production?). *Agriculture, Ecosystems & Environment, 140*(1–2), 27–33. doi:10.1016/j.agee.2010.11.008.

Otero, I., Marull, J., Tello, E., et al. (2015). Land abandonment, landscape, and biodiversity: Questioning the restorative character of the forest transition in the Mediterranean. *Ecology & Society, 20*(2), 7. http://dx.doi.org/10.5751/ES-07378-200207.

Penuelas, J., Sardans, J., Alcaniz, J. M., et al. (2009). Increased eutrophication and nutrient imbalances in the agricultural soil of NE Catalonia, Spain. *Journal of Environmental Biology, 30*(5), 841–846.

Pérez Martínez, P., & Monzón de Cáceres, A. (2008). Consumo de energía por el transporte en España y tendencias de emisión. *Observatorio Medioambiental, 11,* 127–147.

Pimentel, D., & Pimentel, M. (2003). Sustainability of meat-based and plant-based diets and the environment. *The American Journal of Clinical Nutrition, 78*(3), 660S–663S.

Pollan, M. (2008). *In defence of food: An eater's manifesto.* New York: Penguin.

Popkin, B. M. (1993). Nutritional patterns and transitions. *Population and Development Review, 19*(1), 138–157.

Power, A. G. (2010). Ecosystem services and agriculture: Tradeoffs and synergies. *Philosophical Transactions of the Royal Society of London. Series B, Biological Sciences, 365*(1554), 2959–2971. doi:10.1098/rstb.2010.0143.

Reddy, A. K. N. (1981). An Indian village agricultural ecosystem case study of Ungra village. Part II. Discussion. *Biomass, 1,* 77–88. doi:10.1016/0144-4565(81)90016-0.

Sancho i Puig. (1885). *Ahorra aunque sean lágrimas* (pp. 113–116). Certamen del Ateneo de Villanueva i Geltrú.

Schneider, M., McMichael, P. (2010) Deepening, and repairing, the metabolic rift. *The Journal of Peasant Studies 37*(3), 461–484. doi: 10.1080/03066150.2010.494371.

Sisquella, M., Lloveras, J., Álvaro, J., et al. (2004). *Técnicas de cultivo para la producción de maíz, trigo y alfalfa en los regadíos del valle del Ebro.* Lleida.

Smil, V. (2000). *Feeding the world: A challenge for the 21st century.* Cambridge, MA: MIT Press.

Smil, V. (2002). Worldwide transformation of diets, burdens of meat production and opportunities for novel food proteins. *Enzyme and Microbial Technology, 30*(3), 305–311. doi:10.1016/S0141-0229(01)00504-X.

Soroa, J. (1953). *Prontuario del agricultor y el ganadero* (8th ed.). Madrid: Dossta.

Soto, D., Infante-Amate, J., Guzmán, G. I., et al. (2016). The social metabolism of biomass in Spain, 1900–2008: From food to feed-oriented changes in the agro-ecosystems. *Ecological Economics, 128,* 130–138. doi:10.1016/j.ecolecon.2016.04.017.

Teira-Esmatges, M. R., & Flotats, X. (2003). A method for livestock waste management planning in NE Spain. *Waste Management, 23*(10), 917–932. doi:10.1016/S0956-053X(03)00072-2.

Tello, E., Galán, E., Cunfer, G., et al. (2015). A proposal for a workable analysis of Energy Return On Investment (EROI) in agroecosystems. Part I : Analytical approach. *Social Ecology Working Papers, 156.*

Tello, E., Galán, E., Sacristán, V., et al. (2016). Opening the black box of energy throughputs in agroecosystems: A decomposition analysis of final EROI into its internal and external returns (the Vallès county, Catalonia c. 1860 and 1999). *Ecological Economics, 121,* 160–174. doi:10.1016/j.ecolecon.2015.11.012.

Tello, E., Valldeperas, N., Ollés, A., et al. (2014). Looking backwards into a Mediterranean edge environment: Landscape changes in El Congost Valley (Catalonia), 1850–2005. *Environment & History, 20*(3), 347–384. doi:10.3197/096734014X14031694156402.

Temper, L., del Bene, D., & Martinez-Alier, J. (2015). Mapping the frontiers and front lines of global environmental justice: The EJAtlas. *Journal of Political Ecology, 22,* 255–278.

Teubal, M. (2008). Soja y agronegocios en la Argentina: La crisis del modelo. *Lavboratorio, 22,* 5–7.

Teuteberg, H. J., & Flandrin, J. L. (1999). The transformation of the European diet. In J. L. Flandrin, M. Montanari, & A. Sonnenfeld (Eds.), Food: A culinary history from antiquity to the present (Histoire de l'alimentation) (C. Botsford et al., Trans.) (pp. 442–456). New York: Columbia University Press (Original work published in 1996).

Tilman, D., & Clark, M. (2014). Global diets link environmental sustainability and human health. *Nature, 515,* 518–522. doi:10.1038/nature13959.

Tscharntke, T., Tylianakis, J. M., Rand, T. A., et al. (2012). Landscape moderation of biodiversity patterns and processes—Eight hypotheses. *Biological Reviews, 87*(3), 661–685. doi:10.1111/j.1469-185X.2011.00216.x.

Vitousek, P. M., Ehrlich, P. R., Ehrlich, A. H., et al. (1986). Human appropriation of the products of photosynthesis. *BioScience, 36*(6), 363–373.

Vranken, I., Baudry, J., Aubinet, M., et al. (2014). A review on the use of entropy in landscape ecology: heterogeneity, unpredictability, scale dependence and their links with thermodynamics. *Landscape Ecology*. doi:10.1007/s10980-014-0105-0.

Weis, T. (2010). The accelerating biophysical contradictions of industrial capitalist agriculture. *Journal of Agrarian Change, 10*(3), 315–341. doi:10.1111/j.1471-0366.2010.00273.x.

Wijesinghe, L. C. A. de S. (1984). A sample study of biomass fuel consumption in Srilanka households. *Biomass, 5,* 261–82. doi:10.1016/0144-4565(84)90073-8.

## Author Biographies

**Roc Padró** is a forestry engineer and has a Master's degree in Soil and Water Management from the University of Lleida. He participates in social and political movements mainly focused on access to land and towards food sovereignty. He is currently working on his Ph.D. on socio-ecological transitions from organic to industrial agriculture at the University of Barcelona. The foundations of his thesis are on the historical impacts of food regimes on environmental and social sustainability. His research is part of the *Sustainable Farm Systems* international project, in which he participates on the theoretical development of the *Energy Return On Investment* framework, as well as setting an *Energy-Landscape Integrated Analysis* in order to analyse the impacts of social metabolism on energy efficiency, information, and landscape functionality. His main contribution is the development of an agroecosystem socio-ecological modelisation called *Sustainable Farm Reproductive Analysis* (SFRA), in order to analyse how the use of a specific territory could be optimised under its biophysical and social constraints regarding its prevailing social goals.

**Inés Marco** is an ecofeminist economist. She holds a Bachelor's degree in Economics and a Master's degree in Economic History from the University of Barcelona. She is currently completing her Ph.D. on the socio-ecological transition of agricultural systems, focused on a case study from Catalonia. After having assessed the socio-metabolic structure of three different stages of the transition, her current research interests include the analysis of the immaterial elements of the agricultural metabolism structure and dynamics. For this purpose, she has worked on the theoretical framework to understanding the links between social inequalities (from a gender and class perspective) and social metabolism. Indeed, she proposes a methodology to assess how these links worked in traditional organic agriculture. She has also been involved in many social movements to denounce the social and ecological impacts of the capitalist system, especially in the Global South, such as the effects of transnational corporations and illegitimate external debt.

**Claudio Cattaneo** works as a postdoctoral researcher at the Barcelona Institute of Regional and Metropolitan Studies and as associate professor of ecological economics at the Department of Economics and Economic History at the Autonomous University of Barcelona, Catalonia, and at the School of Economics of LIUC University, Castellanza, Italy. He also teaches MSc Ecological Economics at the School of Geosciences, University of Edinburgh, Scotland. He is senior researcher at the Research and Degrowth association in Barcelona and a member of the informal Squatting Europe Kollective network. His research interests and passions include the socio-metabolic analysis of agro-ecosystems and landscape ecology, degrowth, the squatters' movement, and alternative communities and grass-root practices.

**Jonatan Caravaca** is an Environmental Sciences' student. He conducted his degree between Cordoba University, Institut de Géographie Alpine in Grenoble and the University of Barcelona, where he gained a research placement on the local team of the international "Sustainable Farm Systems: Long-Term Socio-Ecological Metabolism in Western Agriculture" (SFS) project. There he worked on Energy Flow Accounting in Agroecosystems in a Catalonian case study focused on the socio-ecological transition from organic to industrial agriculture. He is also interested in the application of landscape ecology approaches to agrarian systems, food and technological sovereignty, and grass-root solutions to global change challenges.

**Enric Tello** is an Agricultural and Environmental Historian at the Department of Economic History, Institutions, Policy and World Economy at the University of Barcelona, where he leads a multidisciplinary research team included in the international project on "Sustainable Farm Systems: Long-Term Socio-Ecological Metabolism in Western Agriculture" (SFS) funded by the Social Sciences and Humanities Research Council of Canada from 2012 to 2018. Along with the SFS Partnership Grant, this team has developed ground-breaking methodologies like a multi-EROI approach to Energy Flow Accounting of Agroecosystems, an Energy-Landscape Integrated Analysis (ELIA), and a Sustainable Farm Reproduction Analysis (SFRA) which aim at carrying out Environmental History research as a transdisciplinary meeting point where some of the main research topics on Sustainable Science can be addressed from a long-term perspective—e.g. assessing how bio-economically circular farm systems are, which ecosystem services they can offer, and which environmental impacts they cause.

# Chapter 6
# Sustainability Challenges of Pre-industrial Local Food Systems—Insights from Long-Term Socio-Ecological Research in Austria

**Michael Gizicki-Neundlinger, Simone Gingrich and Dino Güldner**

**Abstract** This chapter digs into the history of local food systems (LFS), arguing that we need to understand pre-industrial, *organic* forms of agriculture to learn about its potentially sustainable futures. A close look back in time provides insights into the structure and functioning of LFS under the conditions of an agrarian socio-metabolic regime. We use socio-ecological concepts and methods derived from the "Vienna School" of Social Ecology to address three sustainability challenges of pre-industrial local food systems (PILFS)—the maintenance of soil fertility in the long run, the stable provision of food for the farming community and the equal distribution of critical resources. We draw upon a rich tradition of Long-Term Socio-Ecological Research (LTSER), applying it to Austria for a systemic understanding of PILFS and their sustainability potentials and constraints. The results and lessons learnt from pre-industrial agriculture are used to explore specific solutions and "leapfrogging" opportunities in recent discussions on the sustainability of local food systems.

**Keywords** Long-Term Socio-Ecological research · Pre-Industrial local food systems · Soil fertility · Food provision · Social inequality

## 6.1 Introduction

In recent years, local and organic food systems have been advocated as viable and sustainable alternatives to intensive agriculture associated with the dominant globalized and industrial food regime. Advantages of local food systems (LFS) are

M. Gizicki-Neundlinger (✉) · S. Gingrich · D. Güldner
Institute of Social Ecology, Schottenfeldgasse 29, A-1070 Vienna, Austria
e-mail: michael.gizicki.neundlinger@gmail.com

S. Gingrich
e-mail: simone.gingrich@aau.at

D. Güldner
e-mail: dino.gueldner@aau.at

 
E. Fraňková et al. (eds.), *Socio-metabolic Perspectives on the Sustainability of Local Food Systems*, Human-Environment Interactions 7,
https://doi.org/10.1007/978-3-319-69236-4_6

manifold: Short food supply chains can reduce environmental impact caused by trade and transport (Coley et al. 2009; Iles 2005). Also, LFS may support a higher share of added value, to be retained locally and distributed more equally among producers and consumers (Kneafsey et al. 2013). Finally, mixed and integrated farming methods would allow for better nutrient management within relatively closed nutrient cycles, avoiding the large scale application of synthetic fertilizers and fostering both biodiversity and agroecological stability (Pretty 2008; Tilman 1998). In the context of the current food crisis and growing environmental impact related to conventional food production, the promotion of LFS has gained momentum among scientists, civil society movements and policy makers (Holt-Gimenez and Altieri 2012; Martinez-Alier 2011; Seyfang 2006)—even though some argue that the productivity of organic systems is drastically overestimated (Connor 2008) or that LFS still have to hold their promises against empirical evidence (Born and Purcell 2006; Edwards-Jones et al. 2008; DeLind 2011).

The integration of "traditional", "old" or "historic" agricultural knowledge and practices into today's LFS has played a central part in the discourse since its early beginnings (Feenstra 1997). For example, Pimentel et al. (2005: 580) argue that "conventional agriculture can be made more sustainable and ecologically sound by adopting some traditional organic farming technologies, [which] have been used for about 6000 years to make agriculture sustainable while conserving soil, water, energy, and biological resources." In his widely acknowledged contribution in 2009, Miguel Altieri states that the "new models of agriculture [...] will need to include forms of farming that are more ecological, biodiverse, local, sustainable, and socially just. They will be rooted in the ecological rationale of traditional small-scale agriculture, representing long established examples of successful community-based local agriculture." Traditional organic agriculture constituted the single, most important supplier of food for millennia in industrial Europe—and continues to be of relevance for food security in developing countries today. However, pre-industrial food systems were not entirely sustainable per definition, but were facing severe structural constraints and sustainability problems (Diamond 1997; Netting 1981). By analysing and comparing them to their modern counterparts, we can learn about the challenges of traditional—truly organic—local food systems, and derive implications for a transition towards more sustainable food systems. This will be the prime objective of our chapter.

We take the perspective of Long-Term Socio-Ecological Research (LTSER) (Haberl et al. 2006; Singh et al. 2013) in discussing local food systems. LTSER integrates long-term modelling of ecosystem processes and patterns with more qualitative, historical research. Doing so, it brings together experts from different disciplinary backgrounds—in our example social ecologists, anthropologists and historians—to assess the sustainability of coupled human-environment (or socio-ecological) systems. Taking an LTSER-perspective, we argue that the dominant notion of "traditional agriculture" in the discourse on sustainable LFS is misleading in two ways: (1) So far, no single, agreed-upon definition of "traditional" can be found in the literature. The term may refer to a vague, unspecified reference condition of agriculture in the past, as we have seen above, or it may refer to a variety of local, indigenous practices of land use and food distribution in the Global South (Johns et al. 2013; Kuhnlein et al. 2013). Accordingly, we believe that the concept of "traditional agriculture" needs some elaboration and clarification. (2) Furthermore, perspectives on "traditional" LFS neglect historical realities. In the historical sources we encounter traces of complex and evolving agricultural systems, which were highly variable across time and space. In many cases they were successful in providing sufficient food for multiple generations, but often they were not. Rather, those systems faced sustainability problems, as indicated by contemporary accounts on crop failures, problems with soil fertility, malnutrition and associated diseases. Thus, empirical case studies from Austria will help to critically assess this diagnosis of "sustainable" LFS.

To contextualize and empirically investigate sustainability issues of past LFS, we need a robust definition from an LTSER-perspective. In the following, we define historical "traditional" food systems using their main socio-ecological characteristic. We understand "Pre-Industrial Local Food Systems" (PILFS) as characterized by the absence of inputs from industrial sources (for example synthetic fertilizers, electricity or machinery). Accordingly, our notion of PILFS refers to any land use practice operating under the conditions of a solar-energy regime (Fischer-Kowalski and Haberl 2007; Sieferle 1997). As sustainability is a very broad and abstract concept, we provide a narrower understanding using a variation of the "sustainability triangle" (see Fig. 6.1). The concept has been used in LTSER before (Gaube and Haberl 2013), and refers to an intricate balance of ecological, economic and social dimensions. Hereby, the three spheres are not embedded within a hierarchical structure, but are mutually interacting and evolving as coupled socio-ecological systems. To assess the sustainability of PILFS we translate the triangle framework into three central dimensions: (1) maintenance of soil fertility of the area under cultivation, (2) provision of sufficient quantities of food for a given population and (3) access and equal distribution of food and other critical resources in the agrarian system. In the following, we refer to these three aspects as the most important "sustainability challenges" of PILFS.

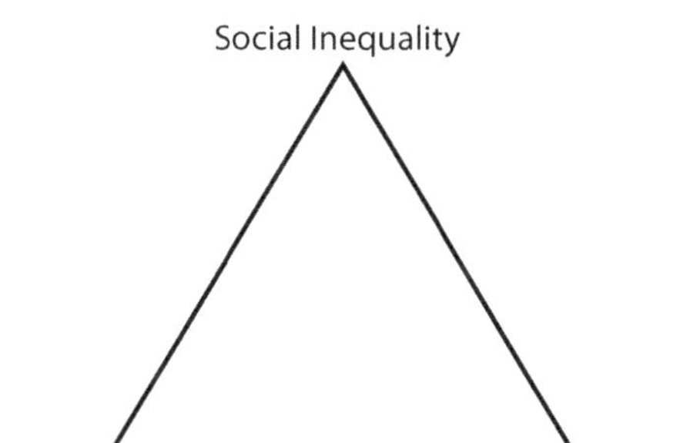


**Fig. 6.1** The sustainability triangle in the context of PILFS

The chapter is structured as follows: In the next section, we briefly introduce the context of 19th century Austria, where our empirical case studies are situated. Next, we introduce the archival material used to reconstruct PILFS and the specific, socio-metabolic reading required to systematically assess them. The third part is devoted to the three sustainability challenges of pre-industrial agriculture—soil fertility, food provision and social inequality. Finally, we provide some preliminary reflections on the practical implications for contemporary discussions on LFS from the perspective of Austrian LTSER.

## 6.2 Austrian Agriculture and Food Systems at the Dawn of Industrialization

Within LTSER, agricultural and food systems have been studied under the auspices of "socio-ecological metabolism" (Krausmann 2004; Sieferle et al. 2006; Gingrich et al. 2013, 2015). Socio-ecological metabolism is a conceptual framework that understands social systems as constantly interacting with their natural environment in order to biophysically reproduce themselves. Within socially organized processes, they extract, process and consume natural resources (materials, energy and nutrients) from the ecosystems (inputs) and return them as waste flows or other emissions (outputs).[1] External Outputs (EO) From this perspective, *pre-industrial* local food systems—not only in Austria, but everywhere on the planet—are characterized by the absence of industrial inputs and, therefore, outputs. In those systems, solar radiation is the prime source of energy. The amount of energy available depends on (1) the regional intensity of solar radiation, (2) climatic conditions (in terms of precipitation and temperature), (3) soil properties of the respective land area, (4) the photosynthetic efficiency of the plant communities that convert solar radiation into biomass and (5) the efficiency of the agricultural tools employed to harness biomass energy (Gliessman 2007). To utilize solar energy fixed in plant tissue, farmers invest their community labour. They breed favourable plant and animal species, extinguish non-favourable ones and harvest biomass from cropland, grassland or woodland, transforming pristine ecosystems into cultivated agro-ecosystems. Farmers only tap into the energy flows within a given agro-ecosystem in order to render (and maintain) it more useful according to their specific needs of production and reproduction—a process which has been referred to as "colonization of natural systems" (Fischer-Kowalski and Haberl 2007). Over time, different colonization strategies form complex landscape mosaics, channelling flows of energy and nutrients through highly diverse agricultural landscapes. Therefore, access to land suitable for agriculture plays a crucial role in PILFS, as it ultimately limits biomass production and the availability of food energy for the farming community (Mazoyer and Roudart 2006).

[1]See also Tello and González de Molina, Chap. 2 in this volume.

To investigate the sustainability of LFS in a truly organic socio-ecological context, Austrian food systems of the early nineteenth century can serve as ideal case studies. Well until the end of the century, agriculture was the predominant socio-economic sector in Austria. Compared to other European countries, the Habsburg Monarchy may be considered a relative "latecomer" to agricultural and also industrial modernization (Sieferle et al. 2006). Accordingly, plant biomass obtained within agricultural systems largely dominated Austria's energy metabolism until the beginning of the twentieth century (Krausmann and Haberl 2002). A broad array of different land uses co-existed. The Alps, covering approximately 30% of the territory today, offered relatively unfavourable topographic and climatic conditions. Here, systems of transhumance, where cattle were seasonally shifted to pastures above the treeline, and mixed farming with crop production extending to high elevations were predominantly practised. In contrast, fertile soils characterized the lowland ecosystems, for example along the river Danube or in the vast Austro-Hungarian plains, allowing for more intensive agriculture. Here, three-field rotation systems were combined with livestock keeping. In the hilly transition zones, permanent cultivation of orchards and vineyards added to the land use portfolio. And finally, Austria's rich forest resources (approximately 45% of the whole agricultural landscape were covered with woodlands) were exploited in the pre-alpine uplands and along alluvial floodplains. Woodlands were not only used for the production of timber and fuel wood, but also as source for litter and other bedding material and they gave grazing animals access to abundant nutrients stockpiled in the fertile forest soils.

Looking at the socio-economic context, we find that Austria was severely polarized between ruling landlords and subjected peasants. Until the Liberal Reforms in 1848, landlords formally owned the land and leased it to their subjects ("manorialism"). In return for this land tenure, peasants were bound to deliver a fraction of their harvest as tithes or other forms of taxes. Commonly, peasants gave around 10% of their annual produce directly to the lords, or compensated for their obligations in money. Additionally, peasants had to perform regular forced (courveé) labour duties on their landlord's demesne (domestic lands), whereas labour requirements regularly peaked during the harvest season. Until they were finally abolished in the middle of the nineteenth century, tithes and courveé obligations drastically influenced peasant land use and regularly led to fierce disputes throughout the Monarchy (Gizicki-Neundlinger et al. 2017).

Our LTSER case studies represent this diversity of Austria's different agroecological zones and its large socio-economic spectrum, from smallholder subsistence family farming in the village of *Kamp*, to early proto-industrial areas (*Grünburg*), to large, market-oriented manorial enterprises (*Breuner* demesne). Also, the Austrian examples may be considered representative for the broader Central European context (Krausmann 2004). Table 6.1 gives an overview on the agrarian structure of the selected case studies, ranging from entire regions (R), to villages (V) and specific agrarian agents (for example manorial farmers, smallholders) (A). Also, it provides some first socio-ecological indicators to shed more

**Table 6.1** Agrarian structure and socio-ecological indicators of selected Austrian PILFS in the first half of the nineteenth century

| Region (R) Village (V) Agent (A) | St. Florian (R) | Grünburg (R) | Möll Valley (R) | Theyern (V) | Grafen-wörth (V) | Kamp (V) | Breuner Demesne (A) | Kamp Small-holder (A) |
|---|---|---|---|---|---|---|---|---|
| Socio-economic features | Big monastery Intensive agriculture | Small urban centres Metal manufacture | Transhumance Silver mining Lowland cropping | Intensive cropping Communal grazing areas | Forestry, Artisans Communal grazing areas | Manorial structures Vineyard Lowland cropping | Feudal crops for market Manufacture Forestry | Family farms Subsistence crops Wage labourer |
| Agroecology | Pre-Alpine hills Fertile lowland soils | Fringes of the Alps Mountains to lowlands | Alpine pastures Bad climate and topo-graphy for cropping | Fertile loess soil with high clay content | Cropland on fertile chernozem and humid black soils Vineyards on podsol loam soil uphill Riparian forest and wet Danube meadows | | | |
| Pop. Density (cap/km$^2$) | 77 | 87 | 19 | 45 | 68 | 147 | 2 | 119 |
| Area (km$^2$) | 52 | 61 | 234 | 3.6 | 11.6 | 2.4 | 47 | 0.9 |
| Cropland (%) | 68 | 30 | 9 | 20 | 43 | 79 | 14 | 90 |
| Grassland (%) | 15 | 32 | 38 | 2 | 24 | 5 | 3 | 0 |
| Gardens & Vineyards (%) | 0 | 0 | 0 | 17 | 5 | 9 | 1 | 9 |
| Woodland (%) | 17 | 28 | 32 | 35 | 29 | 6 | 82 | 1 |
| Farm Size (ha) | 17 | 5 | 4 | 8 | 4 | 4 | 278 | 1 |
| Cereal Yields (t/ha) | 1.2 | 0.9 | 0.8 | 0.8 | 0.7 | 0.8 | 1.1 | 1.2 |
| Livestock Density (LSU/ha) | 31 | 28 | 8 | 24 | 5 | 6 | 11 | 7 |

(continued)

**Table 6.1** (continued)

| Region (R) Village (V) Agent (A) | St. Florian (R) | Grünburg (R) | Möll Valley (R) | Theyern (V) | Grafen-wörth (V) | Kamp (V) | Breuner Demesne (A) | Kamp Small-holder (A) |
|---|---|---|---|---|---|---|---|---|
| Food Provision (GJ/cap) | n.d. | n.d. | 4.2 | 9.4 | 5.7 (R) | | 17 | 1.5 |
| External Final EROI (GJ/GJ) | 7 | 10 | n.d. | 6 | n.d. | n.d. | n.d. | n.d. |
| Internal Final EROI (GJ/GJ) | 0.6 | 0.6 | n.d. | n.d. | n.d. | n.d. | n.d. | n.d. |
| Total nitrogen balance (kg N/ha) | n.d. | n.d. | 0.9 | −2 | −3.6 | −2.2 | 5.5 | n.d. |

*Sources* Own calculations, Cunfer and Krausmann (2009), Gingrich et al. (2015), Gingrich et al. (forthcoming), Gizicki-Neundlinger et al. (2017), Krausmann (2004); n.d. = no data light on the sustainability challenges of the Austrian PILFS, which will be discussed in the following sections.

The nineteenth century may be considered a decisive turning point, as the socio-economic context of Austrian agriculture changed significantly. Accelerated urbanization posed new challenges, as a growing urban population increased the demand for agricultural products. The population of the capital city of Vienna, for example, increased by a factor of ten during the 19th century, with large implications for both urban supply and discharge (Gierlinger et al. 2013; Krausmann 2013). The integration of city-hinterland commodity flows via the introduction of a large-scale railway network, as well as the expansion of market relations throughout Austria, further exacerbated pressures on rural areas (Gingrich et al. 2012; Sandgruber 2005). The redirection of biomass from the rural producers towards urban consumers made it increasingly difficult to maintain agroecological nutrient cycles and contributed to growing nutrient deficits in the hinterlands—a process described as "metabolic rift" (Foster 1999, Güldner and Krausmann 2017). And finally, growing quantities of agricultural products from the Americas were imported to Europe during the first wave of agricultural globalization in the late nineteenth century, creating additional pressures on food markets all over the continent (Cunfer and Krausmann 2009; McMichael 2009). How did PILFS respond to this new situation? In Austria (as in most parts of 19th century Europe), the possibilities to put additional land under cultivation ("expansion") were strictly constrained (Krausmann 2001; Gingrich et al. 2007). Consequently, land use and food production were intensified during the course of the century. In the absence of any industrial inputs, the only way of raising productivity was to invest more labour time and to use local resources more efficiently, but without pushing the resource base towards agroecological disequilibrium. So, in the period under investigation, PILFS faced many structurally similar challenges compared to modern LFS. This specific historical situation provides another ideal prerequisite to learn lessons for today.

## 6.3 Historical Sources in Austria and Their Socio-Metabolic Reading

Rich archival sources on 19th century agriculture are an additional reason why Austrian case studies can provide fruitful insights into pre-industrial agriculture and food systems. The archives contain abundant information on past manorial and peasant agriculture and food production. In particular,19th century sources provide detailed quantitative and qualitative information. We have focussed on the two most promising sources in Austria's historiography: the Franciscean cadastre and manorial accounting records. The former is one of the most important and best studied sources in the historiography of the Habsburg Empire (Moritsch 1970; Sandgruber 1979; Marquart 2006; Sieferle et al. 2006). From 1817 until 1856,

Habsburg expert commissions systematically surveyed the entire Monarchy (530,000 km$^2$ in total) to undertake a comprehensive land tax survey for each village at the plot level. Unlike earlier surveys, the Franciscean cadastre may be considered very accurate (Bauer 2014). Over the years, the commissions created numerous documents covering many details of agricultural life. The core documents (cadastral elaborates) contain information on agricultural yields, livestock numbers, the demographic and occupational structure of the village, peasant diets, feeding practices, land use, etc. Additionally, the "parcel protocols" provide insights into land tenure and ownership. And finally, colourful and very detail-rich maps were created, showing landscape elements down to a 1 m$^2$ resolution.

Recently, manorial accounting records have been used to investigate pre-industrial land use on the demesne lands (Gizicki-Neundlinger et al. 2017). These books were issued by the manorial administration, and give a comprehensive overview on the whole demesne economy of landlords throughout the entire agricultural year, and for continuous time periods. Even though Campbell (1983) has pointed out their value for economic, food and agricultural history research, manorial accounting records still remain largely under-used. The books comprise all relevant inputs and outputs to and from manorial farmsteads, and cover the most important agricultural activities in the demesne economy: the production of cereals, legumes, hay and wine, the harvest of fodder crops, livestock production, the application of seeds, export of commodities, collection of tithes and other taxes, and the importation of goods such as salt, tools and candles, etc. Commonly, multiple issues are available in the archives, allowing a reconstruction of the dynamics of manorial resource flows over decades or even centuries.

Each of the two sources has its specific benefits and structural shortcomings. The Franciscean cadastre provides more detailed information on land use and livestock, but only for one point in time. Also, the cadastral sources lack precise information on the processing of agricultural goods, local and regional trade relations, and on consumption patterns. In contrast, accounting records allow for a closer look into a large farm's internal processes (from production to processing and distribution), and also on their change over time. In addition, aspects of socio-economic distribution are covered, as the books report on tithes, taxes and rents in physical units. Still, some important information remains fragmentary or missing: the accounting books contain no data on monetary flows, labour relations, land use and population numbers. Additionally, aspects of land ownership and land rent are not explicitly addressed in the accounting records. Finally, accounting may be very selective and biased against its specific socio-economic context (Planas and Saguer 2005). These biases, inherent to all types of historical sources, need to be considered carefully when interpreting results obtained from accounting books as well as from the Franciscean cadastre.

Within the Austrian LTSER case studies, the data obtained from the historical sources was processed as follows: In a first step, we used agricultural handbooks (Hitschmann 1920) and agricultural statistics (Sandgruber 1978, 2005) to convert historic data into modern metric units. Considering the relative biases of the data, as well as the specific historically embedded context of data generation, this is a very sensitive step. Accordingly, we critically evaluated and, when possible, cross-checked cadastral and accounting book information using site-specific qualitative and quantitative archival sources or relevant literature. Rough estimations on the plausibility of production and consumption, as well as data from structurally similar regions and villages, also helped to close data gaps. Depending on the respective research question, data was converted into metric units of mass (tonnes of fresh weight and dry matter), energy (in gross calorific and nutritional value) and nutrients (content of nitrogen). In a second step, we identified material and energy flows that were relevant for our accounting procedures, but not covered in the historical sources. Missing material and energy flows were modelled using assumptions from historical and modern agroecology. Krausmann (2008), Güldner (forthcoming) and Guzman et al. (2014) give detailed descriptions of relevant conversion factors and modelling assumptions. For example, written sources usually do not contain data on livestock grazing. Using information on the metabolic requirements of farm animals, along with some quantitative data on feedstuff and local or regional feeding practices, we were able to estimate the share of grazed biomass in the animal diet. Similarly, manure availability was modelled based on gross feed intake, retention rates for the specific livestock species, and average stable keeping days in the respective region. Wherever possible, we used additional sources to test the assumptions we derived from this socio-metabolic understanding of local agricultural systems. In a third step, we used the comprehensive dataset to calculate a broad array of socio-ecological indicators to assess the three relevant sustainability dimensions.

We used different indicators for each of the sustainability challenges: (1) To assess soil fertility we used socio-ecological nutrient accounting (Garcia-Ruiz et al. 2012; Güldner forthcoming; Güldner et al. 2016). Here, we reconstructed nutrient balances at the soil-surface scale, i.e. we assessed the amount of nitrogen inputs and outputs to and from agriculturally used soils. Natural inputs (for example rainfall, symbiotic and asymbiotic fixation) and outputs (for example denitrification, leaching, losses due to erosion) were quantified, as well as socio-economic flows to and from the soil system (for example harvest, manure, seeds). Positive nutrient balances represent fertile agricultural soils and favourable yields, whereas negative balances indicate possible overexploitation of soil resources. (2) To reconstruct aspects of food provision, we used metabolic profiles indicating the total biomass resource availability per capita. To obtain net food available for human consumption, we considered the supply of plant- and animal-based foods in terms of their respective nutritional energy values. Vegetal food production was assessed by deducing seed output and processing losses from gross crop harvest, and food of animal origin was calculated based on livestock numbers and typical productivities. These reconstructions allow the comparison of resource access on the level of different regions or villages. Additionally, we have estimated the energy efficiency of Austrian PILFS. Following the "Energy Return on Investment (EROI)" accounting procedures applied in Tello et al. (2015, 2016), we compared Final Produce from agro-ecosystems to external energy inputs ("External Final EROI") and locally recycled biomass flows ("Internal Final EROI").[2] This agroecological EROI accounting framework enabled us to measure the relative energy efficiency of PILFS, the relation of external and internal energy inputs, and its change over time. (3) And finally, we disaggregated data on biomass resource use and final food consumption on the level of individual households, which helped us to unravel issues of social inequality in terms of the uneven distribution of crucial resources. Here, we used information on tithes, taxes and rents to arrive at an approximation on final food available per household type and per capita, providing a fine-grained picture of resource distribution within the local land use system. These three perspectives allow a comprehensive and systematic socio-ecological assessment of the sustainability of past PILFS, which may assist a better understanding of the local food systems of today and tomorrow.

## 6.4 The Challenge of Soil Fertility

The maintenance of soil fertility over the long run may be considered one of the fundamental challenges for all types of agricultural and food provisioning systems. In fertile agricultural soils, the amount of nutrients extracted (or lost) does not exceed the amount of nutrients replenished in the course of an agricultural year (Tello et al. 2012). Among the major plant nutrients, nitrogen (N) proved to be the most limiting factor of agricultural productivity in pre-industrial times (Allen 2008). Consequently, the availability of N in agricultural soils is one of the most significant determinants of crop yields and—therefore—food (and feed) provision for the farming community (Connor et al. 2011). So it is imperative for farmers to warrant a steady backflow of N into the respective soil systems. LFS are regularly associated with high rates of N replenishment and therefore hypothesized to warrant agro-ecological stability and community resilience (Badgley et al. 2007; Gomiero et al. 2011; Tilman 1998).

Various natural and socio-economic variables determine the N content of agricultural soils. Regarding output, farmers extract N from soils by harvesting biomass (or when livestock feed on grazed biomass). In addition, denitrification, leaching and losses via volatilization contribute to total soil N depletion. In turn, nutrients are replenished by natural processes, as well as through cultivation measures. Highly variable biogeographical conditions determine the relative share of natural N inputs into the soils. The amount of N deposited via rainfall may vary significantly between different regions. For example, in Upper Austria, high annual precipitation of approximately 1100 $l/m^2$ heavily contributed to positive N balances (Gingrich

[2]External Final EROI relates final produce (i.e. all biomass products either used and consumed locally or sold at markets) to external inputs (i.e. human labour, non-local biomass bought on markets and industrial inputs such as machinery or fossil fuels). Internal Final EROI relates the final produce to local biomass reuses (i.e. livestock feed and litter, seeds, stubble ploughed into fields or biomass burned).

et al. 2015), whereas Lower Austria or the Austro-Hungarian plains only benefitted from 670 to 550 $l/m^2$ (Güldner and Krausmann 2017). In these regions, other natural N inputs were of greater importance. In Lower Austria, asymbiotic fixation accounted for up to 20% of the annual N replenishment, and on the Austro-Hungarian Plains symbiotic fixation of atmospheric N also accounts for 20% of total N replenished. This means that important N flows may be considered beyond the scope of socio-economic management practices and were determined primarily by biogeographical conditions. Accordingly, the different local land use strategies and extraction regimes of PILFS aimed at efficiently channelling and re-directing the remaining N flows. For example, seed input was not only essential to plant next year's crop, but also constituted an important socio-economic flow of N (up to six per cent of total N input into the soil). In quantitative terms, though, animal manure (and in some cases also human excreta) may be considered the most important N input to the soil systems. Accordingly, the prime socio-economic strategy of Austrian PILFS to tackle the universal problem of nutrient replenishment in agricultural soils was the efficient re-direction of nutrients from the livestock system towards agriculturally used lands.[3]

Livestock keeping may be considered an important way to convert biomass that cannot be directly digested or used by humans (e.g. grass vegetation, bushes, stubbles from cropland, fallen leaves) into the following: (1) useful energy in the form of draught power, used to pull ploughs and carts or to run mills (2) livestock produce, mostly milk, meat, eggs and wool and, finally (3) into nutrients richly concentrated in animal manure. So livestock—and its' ability to mobilize N via digestion—played an integral role within pre-industrial agriculture. Livestock keeping and feeding may be used to transfer vitally important nutrients throughout the whole agricultural landscape, locally integrating different types of land uses and providing a major socio-ecological nutrient backflow. Normally, livestock was used to transfer nutrients from extensively used land (for example grassland or woodland) to more intensively used land (cropland, vineyards or kitchen gardens). But the necessity to integrate favourable livestock densities into pre-industrial agriculture had important implications for local land use. To support the necessary livestock numbers, sufficient land area was needed to either feed the animals directly (summer grazing) or indirectly from crops (e.g. oats, potatoes or fodder legumes) and their residues (mostly straw or bran). So sufficient land resources were required to produce enough manure to ensure cropland productivity. This "land cost of sustainability" may be considered an important determining factor of both Central European and Mediterranean PILFS (Guzmán and González de Molina 2009). Therefore, collecting the maximum and avoiding losses of manure was of prime importance in PILFS. Depending on the livestock density in the respective region, the potential application of human excreta and the efficiency of the manure

[3]In Catalonia, burning and burying of biomass replaced a significant amount of N lost in soils (Padro et al., Chap. 5 in this volume). In Austria, those strategies were of minor quantitative importance.

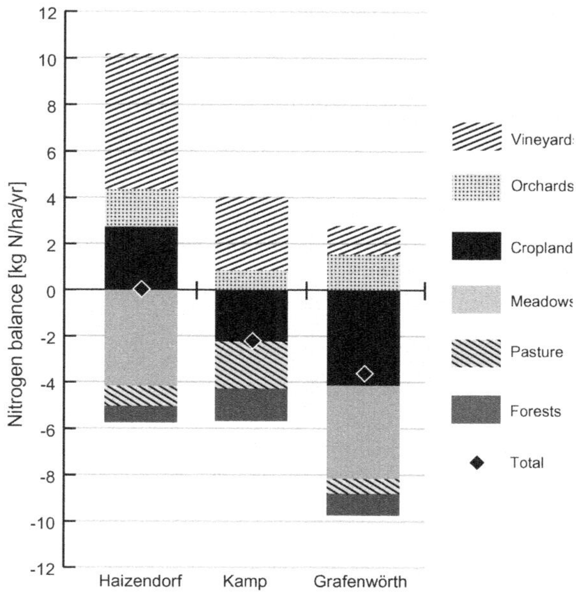


**Fig. 6.2** Nitrogen balances of the three Lower Austrian villages of *Haizendorf*, *Kamp* and *Grafenwörth* in 1830

management practices, 40–90% of N extracted via harvest may be replenished (Güldner and Krausmann 2017; Krausmann 2004). Figure 6.2 shows N balances for three pre-industrial villages. In *Haizendorf* (left bar), we can see N transfers from meadows, pasture and forests towards cropland, economically important vineyards, and smallholder orchards, allowing for a balanced N budget at the village level.

During the nineteenth century, urbanization and population growth increased demand on local food production. But under the conditions of a solar-energetic regime, N was not available in abundance. Accordingly, measures had to be found to increase the overall N availability for PILFS and to manage N flows more efficiently. So now clover (or other leguminous plants) were cultivated on previously unmanaged fallow lands. Their ability to fix atmospheric N drastically increased overall N availability in the local land use systems. Along with clover, other fodder crops were introduced all over the Habsburg Monarchy. This lead to a significant increase in livestock density, again raising the total manure available for food production. In addition, stable keeping gained significance all over the country (Sandgruber 1982). Stables were a decisive innovation, as livestock keeping in barnyards allowed for a more efficient collection and utilization of organic fertilizers derived from animal manure. In the stables, a relevant fraction of manure that was previously "lost" on grazing areas could now be collected and redirected towards cropland to raise agricultural yields, food and feed production. The interplay of intensified forage and livestock production induced a significant expansion of agricultural production, as nutrient availability and management were optimized within the conditions and limitations of the solar-energetic regime. This new land use regime was so important that it has been called the "first agricultural revolution of modern times" (Mazoyer and Roudart 2006). But the more intensive use of biomass resources may have other, unintended side consequences for the soils. On the one hand, integration of clover and other leguminous crops into agricultural rotations may lead to depletion of other soil nutrient resources in the long run. The atmospheric N, organically bound in clover forage, could easily be locally recycled via the livestock system. However, phosphorous (P) contained in the forage did not originate from organic sources but mainly from slowly depleting, inorganic bedrock soil stocks. So, once mobilized via the livestock system, P could not be easily recovered (Güldner and Krausmann 2017). On the other hand, intensified production was only possible until a certain threshold and may lead to over-exploitation of local N resources. Figure 6.2 illustrates N deficits in the villages of *Kamp* and *Grafenwörth*. Here, N extraction via grazing (and timber extraction) exceeded local natural N replenishment via deposition or fixation. This led to N deficits not only on the extensively used land, but also on cropland, posing challenges for the food (and feed) provision of the farming community.

## 6.5 The Challenge of Food Provision

To sustain a certain agricultural and non-agricultural population, PILFS needed to secure a relatively stable amount and favourable composition of food (Boserup 1965; Netting 1993; Wolf 1966). Also, in the discourse on local systems, various aspects of food provision are at the core of academic interest. For example, yield gaps between conventional and organic more local forms of farming (and how to technically close them) remain a matter of open dispute (Connor 2008; Ponisio et al. 2014; Ponti et al. 2012; Seufert et al. 2012). Some empirical evidence from Austrian PILFS may help to better understand the intricate relationship between food resources, population numbers and nutrition at the local level. Under the conditions of a solar-energy based agricultural system, transportation opportunities and long distance trade with bulk commodities were restricted (Sieferle et al. 2006). As a consequence, most inputs to the respective food system had to be of local origin. Local animal manure (and human excreta) accounted for a major fraction of biomass input into the land use system, as we have seen in the previous section. The same holds true for other crucial agrarian resources, for example crop seeds needed in the subsequent growing season. Also, the local farm family household was the prime source of labour power within PILFS. The agrarian population may be considered relatively immobile, and monetary opportunities to employ additional farmhands were often restricted for the poor smallholder communities (Gizicki-Neundlinger et al. 2017).

In sum, a relatively large share of agrarian resources got used and reused at the local level in pre-industrial agriculture. To better understand the energy relations of PILFS, we followed Tello et al. (2015, 2016) and calculated "Internal Final EROI", i.e. the ratio of final produce (the sum of crops, livestock produce and wood) to biomass reused at the local level (stubble, feed, litter and seeds). Table 6.1 gives values for the two Upper Austrian regions of *St. Florian* and *Grünburg*, indicating that the final produce accounted for only c. 60% of local biomass reuse, i.e. agricultural productivity was constrained by the availability of local resources. In both regions, livestock served to transfer energy (and nutrients, see above) from grasslands and forests to croplands in the form of manure, and provided draught power required for cultivation. So, also in terms of energy, the different compartments of agroecosystems (livestock, forests, and agricultural land) were tightly integrated (Gingrich et al. forthcoming). This pronounced "locality" of energy flows corresponds with relatively low yields. In our sample, cereal yields ranged between 0.7 and 1.2 tonnes of dry matter per hectare cropland (see Table 6.1). Compared to modern conventional and organic standards—around 4.2 tonnes per hectare in organic systems (Pimentel et al. 2005) and between 5.5 and 6.5 tonnes per hectare achieved in conventional agriculture (Gingrich et al. forthcoming, Krausmann 2016)—the area productivity of Austrian PILFS may be considered very low.

But, despite the "local input—low output" regime, PILFS were characterized by relatively high energy efficiency. In pre-industrial times, agriculture was the most important means of energy provision for any given community (Sieferle et al. 2006). Consequently, at some level, the energy withdrawn from agroecosystems, and in particular nutritional energy contained in food, needed to be higher than external energy inputs that farmers invested to produce this final energy. If we calculate "External Final EROI", i.e. the ratio of final produce to external inputs (energy embodied in labour, household wastes, and non-local biomass) we find significantly positive energy relations. Table 6.1 also shows some quantifications of the energy input-output ratio in the regions of *St. Florian* and *Grünburg*, and in the village of *Theyern*. All of the three cases show a positive energy balance, ranging from 6 to 10 energy units obtained per energy unit expended.[4] In pre-industrial Europe, where large-scale agricultural expansion was not an option, intensification through more labour and biomass reuse per unit of area was the sole means to raise productivity and food production (Gingrich et al. forthcoming). Both of these options faced local constraints. Reinvesting more biomass locally in order to raise yields (for example by keeping more livestock to provide manure) could directly impede final production of food (for example land used for feed production was not available for food production). On the other hand, investing more labour (for example by shifting to more labour-intensive crops, such as from cereals to potatoes) required the feeding of more agricultural workers locally, and produce yet more food.

Reconstructions of per-capita net food availability show that the average final food production varied greatly between 1.5 and 17 gigajoule nutritional value per capita and year on the household level, and between 4.2 and 9.4 GJ nutritional value per capita and year for different villages and regions (see Table 6.1).[5] As reported in the cadastral elaborates, diets within PILFS were largely plant-based.

---

[4]EROI of Theyern was calculated in Krausmann (2004), following the methodology in Hall et al. (1986).

[5]Taxes and tithes were considered at the household level only.

From an energetic perspective, crop production was significantly more efficient than livestock management. Crop production provided more food per unit of area and per unit of labour input than livestock production. According to the historical sources, meat was consumed once a week or less (for example, in *Kamp* we have reconstructed that only 5% of total food consumed was of animal origin, Gizicki-Neundlinger et al. 2017). Following Freudenberger (1998) and Smil (2001) we consider 3.4 GJ per capita as the minimum annual food requirement needed to sustain the individual metabolism—even though this value may vary substantially according to age, weight and occupation. Still, this minimum metabolic requirement provides opportunity for a first approximation on aspects of (mal-)nutrition in PILFS. We can see that, at the per capita level, the amount of surplus food available was very limited—except in the heavily export-oriented demesne economies of the earliest agrarian capitalists, for example the Lower Austrian *Breuner* family.

In the other cases, the marginal surplus achieved served multiple purposes and needs. First, the additional food could be transferred to landlords as payment of manorial rents and other tax obligations. Second, the food surplus could allow modest participation at local markets or may have been sold to growing urban centres in the vicinity, or to Vienna. Still, under the conditions of high transportation costs within the solar-energetic regime, trade relations remained of minor importance. And finally, the surplus food could be stored to increase the resilience of the PILFS, i.e. to minimize risk in times of bad harvests or extreme weather events. So our findings on food provision suggest that in most of the cases, enough food was produced to sustain a certain agricultural and non-agricultural population. But in some instances we found that the minimum metabolic requirements could not be met. For example, in the Upper Austrian *Enns* valley only 74%, and in the village of *Nußdorf* only 83%, of the regional or village demand was covered by local production (Krausmann 2004; Gingrich et al. 2015), indicating that in some regions, supra-regional trade relations were the only way to abate malnutrition, even under pre-industrial conditions. At the household level, we found similar examples of subsistence pressure (see Table 6.1 for the example of smallholders in *Kamp*).

## 6.6 The Challenge of Social Inequality

The diagnosis of the huge potential for subsistence pressure within pre-industrial agriculture brings us to the last sustainability challenge—social inequality. In the vast literature on current LFS, we find many advocates of the idea that those systems are more equitable and socially just compared to their conventional, industrialised counterparts (DuPuis and Goodman 2005; Holt-Gimenez and Altieri 2012). But under the conditions of manorialism (as discussed above), we found an unequal distribution of crucial resources (foremost land, biomass, manure and labour) in many of the PILFS under investigation. Traditional land tenure excluded most of the peasant population from the critical resource land. Table 6.1 shows a tremendous polarization regarding farm size. On the one hand, landlord demesnes occupied extremely large areas, enjoying significant economies of scale and large per capita surpluses. On the other hand, smallholder families cultivated farms no larger than one hectare of cropland, or small kitchen gardens, lacking livestock and grazing areas to compensate for their land costs of sustainability (Gizicki-Neundlinger and Güldner forthcoming). Consequently, this land exclusion created a situation of "structural scarcity" (Homer-Dixon 1999) among the peasant population, fostering sharp agrarian inequality and subsistence pressure. For example, in the village of *Kamp*, located in the centre of one of the manorial land use systems under investigation (*Breuner* demesne), the local smallholder families yielded only 45% of the minimum metabolic requirements of their farm household. Considering that an average farm size of five to six hectares sufficed to warrant an average family household metabolism (Gizicki-Neundlinger et al. 2017), the *Kamp* case seems to represent a more general trend of pre-industrial subsistence conditions within Austrian PILFS. For example, in the nearby village of *Untersebarn* we found that under the conditions of rigid manorial land tenure, only 5% of the total peasant population enjoyed farm sizes greater than 10 hectares, 20% accessed 5 hectares of land, and 75% cultivated land no larger than one hectare.[6] In addition to the exclusive land tenure relations, landlords exerted comprehensive feudal rights to extract a fraction of the local produce from their subjected peasants. In *Kamp,* peasants delivered approximately 10% of their total crop produce to the landlords. On a regional level, these tithes and tax relations were of lesser quantitative importance—in two of the *Breuner* administrative districts we found 5 to 6% of agrarian surplus accumulated by the landlords (Gizicki-Neundlinger and Güldner forthcoming).

Under the conditions of manorial land scarcity and surplus extraction, peasants had to find strategies to compensate for the food deficits. One was to participate in the local labour markets. As their small land plots needed less labour time than the large demesnes, most of the population worked in the service of the local landlords to generate income. But competition on the local labour market was very high, as the labour force was available in abundance. Landlords benefitted from this surplus situation, keeping wages relatively low (Gizicki-Neundlinger et al. 2017). In some regions with favourable climate and soil conditions the cultivation of vineyards or other permanent crops created additional family income (Bauer and Landsteiner 2006). Most of the Austrian PILFS case studies showed this mixed strategy of family farm subsistence agriculture in combination with varying degrees of market participation. The process of agricultural intensification may be considered another strategy used to comply with subsistence pressure. Regarding land use per household, we found that the lesser land was accessible to the family farm, the more intensively it was used. Figure 6.3 shows an analysis of early 19th century households in three Upper Austrian villages (*St. Florian, Untergrünburg,*

[6]Further research would be required to better understand how different soil qualities were distributed between lords and peasants. The Franciscean cadastre would entail information on different types of land use classes (e.g. arable I = highly productive, arable IV = poorly productive).

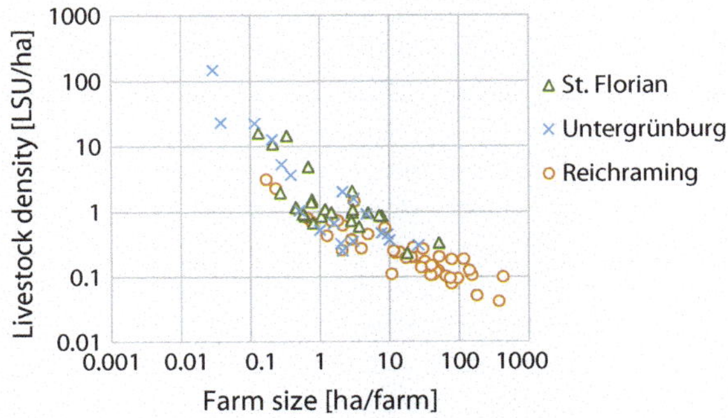


**Fig. 6.3** Correlation between farm size and livestock density in the three Upper Austrian villages of *St. Florian, Untergrünburg* and *Reichraming* in 1830

*Reichraming*), which reveals a negative correlation between farm size (hectare land area per household) and livestock density (livestock units of 500 kg per unit of area available to the respective household). There was no communal pasture in any of the villages, so the higher livestock density of smaller farms hints at higher land-use intensity. Smaller farms needed to provide more feed per unit of land area (or were forced to buy it), and produced more manure per unit of land area (if they didn't sell it). Below a threshold of c. 0.1 ha per household, no livestock was kept. On the other hand, large land owners tended to have higher shares of grassland and/or forest, where extensive land management was possible, producing less feed and requiring less manure per unit of land area, i.e. the larger land owners had enough agrarian capital at their disposal (e.g. access to manure) to compensate for their land costs of sustainability. On the contrary, smallholders tended to erode their agrarian resource base in the middle or long run. They were not able to pay the land costs of agricultural intensification, but exploited stockpiled soil resources.

We can see these unintended side effects of smallholder intensification in our nutrient budgets. The central bar of Fig. 6.2 shows the reconstructed N balance for the *Kamp* case. Here, the more intensive land use strategies under the conditions of subsistence pressure led to negative N balances on cropland (−2.7 kg N/ha/yr), pastures (−2.6 kg N/ha/yr) and in the village forests (−1.7 kg N/ha/yr). The negative balance indicates that annual N reservoirs were exploited and that no soil N was accumulated, which is reflected in the relatively low cereal yields in the village (see Table 6.1). Conversely, N budgets on orchards and vineyards were slightly positive, ranging from 0.8 to 3.2 kg N/ha/yr, underlining the prime economic importance of permanent cultures for the smallholder household budgets. In total, the *Kamp* peasant economy exceeded their annual N capital for −2.2 kg N/ha/yr, threatening the sustainability of the local land use system.

On the regional or village level, an important way of alleviating socio-ecological pressures exerted on the local land were common pool resources (CPR). The high significance of CPR institutions for the sustainability of complex socio-ecological systems has been discussed intensively (Hardin 1968; Ostrom 1990, 2009). In PILFS, we may also find many examples of communal resource use, which was the subject of regular disputes between landlords and peasants (Soto 2014;

Winkelbauer 1986). Consequently, complex legal statuses and access rights evolved around the pre-industrial CPRs. In many European countries including Austria, peasant livestock was regularly led to graze on communal grassland or on open fallow fields, either within village territory or on adjacent lands. This strategy of communal grazing transferred vitally important nutrients from the CPRs to smallholder plots, contributing to the fertility of their agricultural soils. As Austria is also rich in forest resources (c. 45% of today's territory is covered with forests), woodland areas provided another opportunity for communal grazing. In some instances, we even found that forest grazing was a major source of N input into cropland. For example, in one of the administrative districts of the *Breuner* demesne positive N balances were only achieved as long as forest was open to peasant livestock grazing. But, as the landlords exerted comprehensive manorial rights over forest resources, they could have closed woodland areas for grazing very easily, pushing the peasant land use systems towards lower N budgets and agroecological disequilibrium (Gizicki-Neundlinger and Güldner forthcoming). Another case study shows that CPRs may restore the soil fertility of cropland only up to a certain threshold. The right bar of Fig. 6.2 shows a drastically negative N balance in the village of *Grafenwörth,* where 30% of the total grassland areas of the region were located. The peasant livestock of neighbouring villages were regularly led to graze in *Grafenwörth*, exhausting grassland areas (−5 kg N/ha/yr on meadows, −0.8 kg N/ha/yr on pasture and forests (−1.1 kg N/ha/yr). The amount of N removed via communal grazing even led to a negative balance, on both cropland (−5.1 kg N/ha/yr) and on the village level (−3.6 kg N/ha/yr). The scarcity of fertile agricultural soils drove the peasant population of *Grafenwörth* towards forestry or the secondary sector, mostly artisanry and proto-manufacture.

## 6.7 Lessons Learned for Sustainable Local Food Systems

In this chapter, we have elaborated on the structure and functioning of pre-industrial agriculture in order to learn lessons for the future of local food systems. We have used our Austrian case studies to understand sustainability dimensions in a pre-industrial, organic context, and to identify thresholds and bottlenecks of PILFS operating under these conditions. Our LTSER studies on Austrian agriculture have shown that most of the PILFS had to face all three challenges simultaneously, only within varying degrees and nuances. For example, cereal yields and food output could not be increased above a certain critical threshold without threatening soil fertility. Or enough food was produced to feed the local population, but was distributed unevenly under the conditions of manorialism. Accordingly, the prime objective of the PILFS was to navigate these trade-offs within the sustainability triangle of soil fertility—food provision—social inequality. Under the conditions of a solar-energetic regime, the margins for increasing agricultural productivity were constrained. Local production was determined by biogeographical conditions (e.g. N deposition or fixation), and only a few wealthy landlords and peasants were able to compensate for the increased land costs of agricultural intensification. Only the modernization and industrialization of agriculture provided practical solutions at the local level. The widespread diffusion of industrial inputs fundamentally transformed the Austrian PILFS. With the introduction of chemically synthesized N, the bottleneck of local nutrient availability was finally resolved, rendering the land costs of sustainability quasi-obsolete. Whereas efficient manure management played an integral role in the pre-industrial context, utilization and application of animal (and human) manure was almost completely abandoned after the industrialization of agriculture. The abundancy of synthetic N led to a tremendous growth in food production and shifted asymmetries in resource distribution away from the local production systems to the global scale (McMichael 2009). Evidently, externalization of the sustainability challenges on to a global level created the manifold sustainability problems associated with today's dominant conventional food regime. This means that the industrialization of agriculture not only triggered the disintegration of local energy and nutrient cycles, and producers and consumers, but also led to shifting the three important sustainability challenges of agricultural and food-provisioning systems to the global level. This transition, from the local navigation of challenges towards problem solving via externalization and globalization, blurred the picture on local options, solutions and operating spaces. We want to conclude with a close look at this historical bifurcation at the dawn of industrialization: was there a more sustainable option to "leapfrog" to and what did it look like?

Empirical evidence from the second half of the nineteenth century allows a glimpse of possible pathways for more sustainable intensification. Sustainable land-use intensification in pre-industrial agriculture may have occurred under specific circumstances, when yields increased without diminishing soil fertility. In one of our case studies, on the *Breuner* demesne economy, we found that cereal yields were raised from approximately 1000 kg DM/ha to almost 3500 kg DM/ha cropland. Innovations enabling such yield increases included the introduction of clover on fallow land, which was combined with the cultivation of turnips and/or potatoes, vitally contributing to food and feed provision (the "first agricultural revolution", see above). In our case study, the overall fodder availability was raised by a factor of 8 during the second half of the century. Also, livestock management played an important role—a shift from extensive sheep rearing to more intensive cattle farming is an example. The multi-functionality of cattle had a positive effect on cropland productivity. More cattle induced a significant leap in the availability of manure, livestock products and draught power compared to sheep, which predominantly provided wool. Finally, the increasing practice of stable-keeping enabled higher crop yields through more effective nutrient transfers from extensively-used grasslands to cropland. Table 6.2 indicates these production increases on the *Breuner* demesne. As the nutrient budgets were very positive at the beginning of the century (we reconstructed a net surplus of approximately 5 kg N/ha/yr in the late 1830s), we assume that this increase in cropland productivity still occurred within relatively sustainable limits (Gizicki-Neundlinger and Güldner

**Table 6.2** Production dynamics of cereals, roots and tubers and clover, as well as livestock numbers, indexed for the year 1870

| Year | Cereals | Roots & Tubers | Clover | Livestock |
|---|---|---|---|---|
| 1870 | 100 | 100 | 100 | 100 |
| 1875 | 112 | 182 | 444 | 77 |
| 1880 | 132 | 114 | 437 | 156 |
| 1885 | 108 | 337 | 763 | 140 |
| 1890 | 189 | 374 | 1184 | 189 |
| 1895 | 234 | 362 | 1108 | 210 |
| 1900 | 226 | 416 | 932 | 183 |
| 1905 | 257 | 470 | 1024 | 237 |

Data taken from *Breuner* demesne in Lower Austria. *Source* Naturalhauptbuch

forthcoming).[7] We have to be very cautious when interpreting the results from this single case study, as demesne economies do not represent the broader Austrian picture. Also, we have to be cautious about the land costs of sustainability. Presumably, the increase in livestock has created additional need for land, again at the costs of smaller peasants, pushing them even more towards unsustainable agroecological limits. Still, we believe that demesne economies and their sustainability strategies may provide some fruitful insights, a fact that was pointed out earlier (Pretty 1990). A thorough socio-ecological analysis of demesne agriculture can contribute significantly to the debate on how to close the gaps, not only between conventional and organic system, but also between current organic and pre-industrial ones.

What are the practical implications of our findings? (1) Within PILFS in Central Europe, the integration of livestock into the local land use regime was of prime importance. Not just for meat, milk and egg consumption, but also to cover the huge land costs of sustainability. With its multi-functionality (supply of draught power, animal products and manure), every LFS significantly benefitted from favourable livestock density and composition. Today, livestock is primarily used for milk, meat or egg provision, creating food systems adjusted to this single dimension. A dietary transition towards more vegetal-based consumption preferences would help recover some of the multi-functional aspects of livestock. In close correlation, animal and human manure management could be significantly improved (see Simon-Rojo and Duží, Chap. 11 in this volume), inducing a change in perception from "unwanted by-product" and "waste" to a vital source for soil fertility (Winiwarter 2001).[8] (2) An increase in local biomass reused led to higher energy efficiency of the agricultural systems, but to a decrease in agricultural

[7]Further research would be required to understand implications for P availability in the soils (see Sect. 6.4).

[8]Since 2010, the International Fund for Agricultural Development (IFAD) and the FAO division on Integrated Crop-Livestock Systems (ICLS) have put forward similar ideas under the auspices of "Integrated crop-livestock farming systems", providing knowledge and toolkits for smallholders on how to efficiently integrate livestock into their farming systems.

output. But low yields do not necessarily lead to a malnourished population, as we have seen in *Theyern* and *Grafenwörth*. On the contrary, our reconstructions suggest that in most of the cases, enough food was produced to sustain the local population. Rather, an unequal distribution contributed to sustainability problems (in the case of the *Kamp* smallholders). Surely, pre-industrial population densities were below today's demographic trajectories. Still, strategies of regional specialization (Gingrich et al. forthcoming) or demesne patterns of sustainable intensification (see above) may help to close yield gaps even under the conditions of population growth. (3) Finally, access to land played an integral role within PILFS. Farmers needed to be endowed with enough land to compensate for their land costs of sustainability. This means that peasants needed access not only to cropland, but also to grassland and/or forests. In recent years, land grabbing has led to large-scale dispossessions of peasants and enclosures of common lands in the Global South (Borras and Franco 2012; McMichael 2012). But without the type of integrated farming (a favourable mix of different land uses in combination with livestock keeping) we found in PILFS, peasant livelihoods and efficient nutrient transfers to intensively used lands are both severely threatened. Also, effective nutrient management is imperative to avoid degradation and over-exploitation of extensively used lands (grassland and forests). Again, animal and human manure could be used to replenish N extracted (and lost) on these lands, which are normally depending only on natural N inputs. Optionally, communal institutions need to be implemented to pay for the land costs within the whole farming community, but not above certain agroecological limits (see the case of *Grafenwörth*). So, through the lens of the Austrian LTSER integration of livestock, an increase in biomass reuse and options to pay for pre-industrial land costs could help to build local food systems with greater resilience and socio-ecological sustainability.

## References


Allen, R. C. (2008). The nitrogen hypothesis and the English agricultural revolution: A biological analysis. *Journal of Economic History, 68*(1), 182–210.

Altieri, M.A. (2009). Agroecology, small farms, and food sovereignty. *Monthly Review, 61.*

Badgley, C., Moghtader, J., Quintero, E., Zakem, E., Chappell, M. J., Aviles-Vazquez, K., et al. (2007). Organic agriculture and the global food supply. *Renewable Agriculture and Food Systems, 22*(2), 86–108.

Bauer, M. (2014). Agrarsysteme in Niederösterreich im frühen 19. Jahrhundert: Eine Analyse auf Basis der Schätzungsoperate des Franziszeischen Katasters. Institut für Geschichte des ländlichen Raumes, St. Pölten.

Bauer, M., & Landsteiner, E. (2006). Der Weinbau der Waldviertler. In H. Knittler (Ed.), *Wirtschaftsgeschichte des Waldviertels* (pp. 195–216). Horn, Waidhofen an der Thaya: Waldviertler Heimatbund.

Born, B., & Purcell, M. (2006). Avoiding the local trap. Scale and food systems in planning research. *Journal of Planning Education and Research, 26*(2), 195–207.

Borras, S. M., & Franco, J. C. (2012). Global land grabbing and trajectories of agrarian change: A preliminary analysis. *Journal of Agrarian Change, 12*(1), 34–59.

Boserup, E. (1965). *The conditions of agricultural growth: The economics of agrarian change under population pressure*. London: Allen & Unwin.
Campbell, B. (1983). Agricultural productivity in Medieval England: Some evidence from Norfolk. *Journal of Economic History, 43,* 26–46.
Coley, D., Howard, M., & Winter, M. (2009). Local food, food miles and carbon emissions: A comparison of farm shop and mass distribution approaches. *Food Policy, 34*(2), 150–155.
Connor, J. (2008). Organic agriculture cannot feed the world. *Field Crops Research, 106*(2), 187–190.
Connor, D. J., Loomis, R., & Cassman, K. G. (2011). *Crop ecology: Productivity and management in agricultural systems*. Cambridge: Cambridge University Press.
Cunfer, G., & Krausmann, F. (2009). Sustaining soil fertility. Agricultural practices in the old and new worlds. *Global Environment, 4,* 8–47.
DeLind, L. B. (2011). Are local food and the local food movement taking us where we want to go? Or are we hitching our wagons to the wrong stars? *Agriculture and Human Values, 28*(2), 273–283.
Diamond, J. (1997). *Guns, germs, and steel: The fates of human societies*. London: Cape.
DuPuis, E. M., & Goodman, D. (2005). Should we go "home" to eat? Toward a reflexive politics of localism. *Journal of Rural Studies, 21*(3), 359–371.
Edwards-Jones, G., Mila i Canals, L., Hounsome, N., Truninger, M., Koerber, G., Hounsome, B., et al. (2008). Testing the assertion that 'local food is best': The challenges of an evidence-based approach. *Trends in Food Science and Technology, 19* (5), 265–274.
Feenstra, G. W. (1997). Local food systems and sustainable communities. *American Journal of Alternative Agriculture, 12*(1), 28–36.
Fischer-Kowalski, M., & Haberl, H. (Eds.). (2007). *Socioecological transitions and global change: Trajectories of social metabolism and land use*. Cheltenham: Edward Elgar.
Foster, J. B. (1999). Marx' theory of metabolic rift. Classical foundations for environmental sociology. *American Journal of Sociology, 105*(2), 366–405.
Freudenberger, H. (1998). Human energy and work in a European village. *Anthropologischer Anzeiger, 56,* 239–249.
Garcia-Ruiz, R., Gonzalez de Molina, M., Guzman, G., Soto, D., & Infante-Amate, J. (2012). Guidelines for constructing nitrogen, phosphorus, and potassium balances in historical agricultural systems. *Journal of Sustainable Agriculture, 36*(6), 650–682.
Gaube, V. & Haberl, H. (2013). Using integrated models to analyse socio-ecological system dynamics in long-term socio-ecological research: Austrian experiences. In: S. J. Singh, H. Haberl, M. Chertow, M. Mirtl. & M. Schmid (Eds.), *Long term socio-ecological research: Studies in society-nature interactions across spatial and temporal scales* (p. 53–76). Dordrecht: Springer.
Gierlinger, S., Haidvogl, G., Gingrich, S., & Krausmann, F. (2013). Feeding and cleaning the city: The role of the urban waterscape in provision and disposal in Vienna during the industrial transformation. *Water History, 5*(2), 219–239.
Gingrich, S., Erb, K. H., Krausmann, F., Gaube, V., & Haberl, H. (2007). Long-term dynamics of terrestrial carbon stocks in Austria: A comprehensive assessment of the time period from 1830 to 2000. *Regional Environmental Change, 7*(1), 37–47.
Gingrich, S., Haidvogl, G., & Krausmann, F. (2012). The Danube and Vienna: Urban resource use, transport and land use 1800–1910. *Regional Environmental Change, 12*(2), 283–294.
Gingrich, S., Haidvogl, G., Krausmann, F., Preis, S., & Garcia-Ruiz, R. (2015). Providing food while sustaining soil fertility in two pre-industrial Alpine agroecosystems. *Human Ecology, 43* (3), 395–410.
Gingrich, S., Schmid, M., Gradwohl, M., & Krausmann, F. (2013). How material and energy flows change socio-natural arrangements: The transformation of agriculture in the Eisenwurzen Region, 1860–2000. In S. J. Singh, H. Haberl, M. Chertow, M. Mirtl, & M. Schmid (Eds.), *Long term socio-ecological research: Studies in society-nature interactions across spatial and temporal scales* (pp. 297–313). Dordrecht: Springer.

Gingrich, S., Theurl, M. C., Erb, K. H. & Krausmann, F. (forthcoming). Regional specialization and market integration: Agroecosystem energy transitions in Upper Austria. *Regional Environmental Change*. online first. doi:10.1007/s10113-017-1145-1.
Gizicki-Neundlinger, M., & Güldner, D. (forthcoming). Surplus, scarcity, and soil fertility in pre-industrial Austrian agriculture: The sustainability costs of inequality. *Journal of Peasant Studies*.
Gizicki-Neundlinger, M., Gingrich, S., Güldner, D., Krausmann, F., & Tello, E. (2017). Land, food, and labour in pre-industrial agro-ecosystems: A socio-ecological perspective on early 19th century seigneurial systems. *Historia Agraria.*
Gliessman, S. (2007). *Agroecology: The ecology of sustainable food systems*. Boca Raton: CRC Press.
Gomiero, T., Pimentel, D., & Paoletti, M. G. (2011). Is there a need for a more sustainable agriculture? *Critical Reviews in Plant Sciences, 30*(1–2), 6–23.
Güldner, D. (forthcoming). *Zur Umweltgeschichte der Schießpulverherstellung in der Habsburgermonarchie: Die Folgen der Salpeterherstellung auf die Fruchtbarkeit von Böden*. Working Papers in Social Ecology. Institute of Social Ecology. Vienna.
Güldner, D., & Krausmann, F. (2017). Nutrient Recycling and Soil Fertility Management in the Course of the Industrial Transition of Traditional, Organic Agriculture: The Case of Bruck Estate, 1787–1906. *Agriculture, Ecosystems & Environment, 249*, 80–90.
Güldner, D., Krausmann, F., & Winiwarter, V. (2016). From farm to gun and no way back. Habsburg gunpowder production in the eighteenth century and its impact on agriculture and soil fertility. *Regional Environmental Change, 16*(1), 151–162.
Guzman Casado, G., Aguilera, E., Soto, D., Cid, A., Infante, J., García Ruiz, R., Herrera, A., Villa, I., & González de Molina, M. (2014). *Methodology and conversion factors to estimate the net primary productivity of historical and contemporary agroecosystems*. Sociedad Española de Historia Agraria, Documentos de Trabajo 1407. Pablo de Olavide University, Sevilla.
Guzmán, G., & González de Molina, M. (2009). Preindustrial agriculture versus organic agriculture: The land cost of sustainability. *Land Use Policy, 26*, 502–510.
Haberl, H., et al. (2006). From LTER to LTSER. Conceptualizing the socioeconomic dimension of long-term socioecological research. *Ecology and Society, 11*(2), 13.
Hall, C. A. S., Cleveland, C. J., & Kaufmann, R. (1986). *Energy and resource quality*. New York: Wiley.
Hardin, G. (1968). The tragedy of the commons. *Science, 162*(3859), 1243–1248.
Hitschmann, R. (1920). Vademekum für den Landwirt. Wien.
Holt-Gimenez, E., & Altieri, M. A. (2012). Agroecology, food sovereignty and the New Green Revolution. *Agroecology and Sustainable Food Systems, 37*(1), 90–102.
Homer-Dixon, T. (1999). *Environment, scarcity, and violence*.New Jersey: Princeton University Press.
Iles, A. (2005). Learning in sustainable agriculture: food miles and missing objects. *Environmental Values, 14*, 163–183.
International Fund for Agricultural Development. (2010). *Integrated crop-livestock farming system*. Rome (https://www.ifad.org/topic/resource/overview/tags/livestock).
Johns, T., Powell, B., Maundu, P., & Eyzaguirre, P. B. (2013). Agricultural biodiversity as a link between traditional food systems and contemporary development, social integrity and ecological health. *Journal of the Science of Food and Agriculture, 93*(14), 3433–3442.
Kneafsey, M., Venn, L., Schmutz, U., Balázs, B., Trenchard, L., Eyden-Wood, T., et al. (2013). *Short food supply chains and local food systems in the EU: A state of play of their socio-economic characteristics*. Sevilla: European Commission, Joint Research Centre, Institute for Prospective Technological Studies.
Krausmann, F. (2001). Land use and industrial modernization: An empirical analysis of human influence on the functioning of ecosystems in Austria 1830–1995. *Land Use Policy, 18*, 17–26.
Krausmann, F. (2004). Milk, manure, and muscle power: Livestock and the transformation of preindustrial agriculture in Central Europe. *Human Ecology, 32*(6), 735–772.

Krausmann, F. (2008). *Land use and socio-economic metabolism in pre-industrial agricultural systems: Four nineteenth century Austrian villages in comparison*. Working Papers in Social Ecology 72. Institute of Social Ecology, Vienna.

Krausmann, F. (2013). A city and its hinterland: Vienna's energy metabolism 1800–2006. In S. J. Singh, H. Haberl, M. Chertow, M. Mirtl, & M. Schmid (Eds.), *Long term socio-ecological research: Studies in society-nature interactions across spatial and temporal scales* (pp. 247–268). Dordrecht: Springer.

Krausmann, F. (2016). From energy source to sink: Transformations of Austrian agriculture. In H. Haberl, M. Fischer-Kowalski, F. Krausmann, & V. Winiwarter (Eds.), *Social ecology: Analysing society-nature interaction and its history* (pp. 433–446). New York: Springer.

Krausmann, F., & Haberl, H. (2002). The process of industrialization from the perspective of energetic metabolism: Socioeconomic energy flows in Austria 1830–1995. *Ecological Economics, 41*(2), 177–201.

Kuhnlein, H. V., Erasmus, B., Spigelski, D., & Burlingame, B. (Eds.). (2013). *Indigenous Peoples' food systems and well-being: Interventions and policies for healthy communities*. Rome: FAO.

Marquart, E. (2006). Grundlagen für eine umwelthistorische Bearbeitung des Franziszeischen Katasters. Universität Wien.

Martinez-Alier, J. (2011). The EROI of agriculture and its use by the Via Campesina. *Journal of Peasant Studies, 38*(1), 145–160.

Mazoyer, M., & Roudart, L. (2006). *A world history of agriculture: From the Neolithic age to the current crisis*. New York: Monthly Review Press.

McMichael, P. (2009). A food regime genealogy. *Journal of Peasant Studies, 36*(1), 139–169.

McMichael, P. (2012). The land grab and corporate food regime restructuring. *Journal of Peasant Studies, 39*(3–4), 681–701.

Moritsch, A. (1970). Der Franziszeische Kataster und die dazugehörigen Schätzungsoperate als wirtschafts- und sozialhistorische Quellen. *East European Quarterly, 3*, 438–448.

Netting R. (1981*). Balancing on an Alp: Ecological change and continuity in a Swiss mountain community*. Cambridge: Cambridge University Press.

Netting, R. (1993). Smallholders, householders: *Farm families and the ecology of intensive, sustainable agriculture*. Palo Alto: Stanford University Press.

Olarieta, J. R., Padro, R., Masip, G., Ochoa-Rodriguez, R., & Tello, E. (2011). 'Formiguers', a historical system of soil fertilization (and biochar production?) *Agriculture. Ecosystems and Environment, 140*(1–2), 27–33.

Ostrom, E. (1990). *Governing the commons: The evolution of institutions for collective action*. Cambridge: Cambridge University Press.

Ostrom, E. (2009). A general framework for analysing sustainability of socio-ecological systems. *Science, 325*(5939), 419–422.

Pimentel, D., Hepperly, P., Hanson, J., Douds, D., & Seidel, R. (2005). Environmental, energetic, and economic comparisons of organic and conventional farming systems. *BioScience, 55*(7), 573–582.

Planas, J., & Saguer, E. (2005). Accounting records of large rural estates and the dynamic of agriculture in Catalonia (Spain), 1850–1950. *Accounting, Business and Financial History, 15*, 171–185.

Ponisio, L. C., M'Gonigle, L. K., Mace, K. C., Palomino, J., de Valpine, P., & Kremen, C. (2014). Diversification practices reduce organic to conventional yield gap. *Proceedings of the Royal Society B: Biological Sciences, 282*, 20141396.

Ponti, T., Rijk, B., & van Ittersum, M. K. (2012). The crop yield gap between organic and conventional agriculture. *Agricultural Systems, 108*, 1–9.

Pretty, J. (1990). Sustainable agriculture in the Middle Ages: The English manor. *Agricultural History Review, 38*(1), 1–19.

Pretty, J. (2008). Agricultural sustainability: Concepts, principles and evidence. *Philosophical Transactions of the Royal Society B: Biological Sciences, 363*(1491), 447–465.

Sandgruber, R. (1978). *Österreichische Agrarstatistik 1750-1918*. München: Oldenbourg.

Sandgruber, R. (1979). Der Franziszeische Kataster als Quelle für die Wirtschaftsgeschichte und historische Volkskunde. *Mitteilungen des Niederösterreichischen Landesarchivs, 3,* 16–28.
Sandgruber, R. (1982). Produktions- und Produktivitätsfortschritte der Niederösterreichischen Landwirtschaft im 18. und frühen 19. Jahrhundert. In H. Feigl (Ed.), *Die Auswirkungen der theresianisch-josephinischen Reformen auf die Landwirtschaft und die ländliche Sozialstruktur Niederösterreichs* (pp. 95–138). Wien: Niederösterreichisches Institut für Landeskunde.
Sandgruber, R. (2005). *Ökonomie und Politik: Österreichische Wirtschaftsgeschichte vom Mittelalter bis zur Gegenwart.* Wien: Ueberreuter.
Seufert, V., Ramankutty, N., & Foley, J. A. (2012). Comparing the yields of organic and conventional agriculture. *Nature, 485*(7397), 229–232.
Seyfang, G. (2006). Ecological citizenship and sustainable consumption: Examining local organic food networks. *Journal of Rural Studies, 22*(4), 383–395.
Sieferle, R. P. (1997). *Rückblick auf die Natur: Eine Geschichte des Menschen und seiner Umwelt.* München: Luchterhand.
Sieferle, R. P., Krausmann, F., Schandl, H., & Winiwarter, V. (2006). *Das Ende der Fläche: Zum gesellschaftlichen Stoffwechsel der Industrialisierung.* Wien, Köln, Weimar: Böhlau.
Singh, S. J., Haberl, H., Chertow, M., Mirtl, M., & Schmid, M. (Eds.). (2013). *Long term Socio-Ecological research: Studies in society-nature interactions across spatial and temporal scales.* Dordrecht: Springer.
Smil, V. (2001). *Feeding the world: A Challenge for the twenty-first century.* Cambridge: MIT Press.
Soto, D. (2014). Community, institutions and environment in conflicts over commons in Galicia, Northwest Spain (18th–20th centuries). *Workers of the World, 1*(5), 58–74.
Tello, E., Garrabou, R., Cusso, X., Olarieta, J. R., & Galan, E. (2012). Fertilizing methods and nutrient balance at the end of traditional organic agriculture in the Mediterranean bioregion: Catalonia (Spain) in the 1860s. *Human Ecology, 40,* 369–383.
Tello, E., Galan, E., Cunfer, G., Guzman, G., Gonzalez de Molina, M., Krausmann, F., et al. (2015). *A proposal for a workable analysis of Energy Return On Investment (EROI) in agroecosystems. Part I: Analytical approach.* Working Papers in Social Ecology 156. Institute of Social Ecology, Vienna.
Tello, E., Galan, E., Sacristan, V., Cunfer, G., Guzman, G., Gonzalez de Molina, M., et al. (2016). Opening the black box of energy throughputs in farm systems: A decomposition analysis between the energy returns to external inputs, internal biomass reuses and total inputs consumed (the Vallès County, Catalonia, c. 1860 and 1999). *Ecological Economics, 121*(1), 160–174.
Tilman, D. (1998). The greening of the green revolution. *Nature, 396*(6708), 211–212.
Winiwarter, V. (2001). Where did all the waters go? The introduction of sewage systems in urban settlements. In C. Bernhardt (Ed.), *Environmental problems in European cities in the 19th and 20th century* (pp. 105–119). Münster: Waxmann.
Winkelbauer, T. (1986). Bäuerliche Untertanen zwischen feudaler Herrschaft und absolutistischem Staat: Dargestellt am Beispiel der Waldviertler Grundherrschaften Gföhl und Altpölla vom 16. bis zum 18. Jahrhundert. Verein für Landeskunde von Niederösterreich, Wien.
Wolf, E. (1966). *Peasants.* New Jersey: Englewood Cliffs.

## Author Biographies

**Michael Gizicki-Neundlinger** holds BAs in Social Anthropology as well as History, a MSc in Social Ecology and has just completed his Ph.D. at the Institute of Social Ecology (SEC) Vienna. He has gained experience in Long-Term Socio-Ecological Research (LTSER) collaborating with the Center for Environmental History (ZUG) in Vienna, with the Rachel Carson Center for

Environment and Society (RCC) in Munich, with Potsdam Institute of Climate Impact Research (PIK) in Berlin and the Historical GIS Lab at the University of Saskatchewan, Canada. In the last years, he has worked and published within the international and interdisciplinary project "Sustainable Farm Systems". Therein, he was using a socio-metabolic approach to investigate the manifold biophysical interactions between social inequality and soil fertility.

**Simone Gingrich** holds a Ph.D. in Social Ecology and an M.Sc. in Ecology. She is senior researcher and lecturer at the Institute of Social Ecology Vienna, and member of the Center for Environmental History (ZUG), Vienna. She works in the field of Long Term Socio-Ecological Research (LTSER), analysing environmental change during industrialisation in Europe. Her work is inspired by the idea that today's sustainability problems can only be adequately tackled if understood in their historical context.

**Dino Güldner** holds an M.A. in History and is currently working on his Ph.D. dissertation at the Institute of Social Ecology, dealing with the industrialisation of agriculture. His research aims to draw lessons from the historical experiences and responses to universal challenges in agriculture to benefit its future sustainable intensification. His field of work comprises ecologically informed environmental history, Long Term Socio-Ecological Research (LTSER), soil science and the analysis of soil fertility management in pre-industrial agriculture.

# Chapter 7
# Food, Feed, Fuel, Fibre *and* Finance: Looking for Sustainability Halfway Between Traditional Organic and Industrialised Agriculture in the Czech Republic

**Eva Fraňková and Claudio Cattaneo**

**Abstract** In this chapter, we provide an in-depth analysis of a potentially sustainable local food system located in the Czech Republic, a small-scale organic family farm, involved in the Community Supported Agriculture scheme, with a traditional integrated farm structure combining cropland, grassland, and woodland, and a highly localised mode of both production, consumption and distribution. Both the biophysical and monetary profile of the farm is provided, and the biophysical characteristics benchmarked with pre-industrial era (1840') and current average data on organic and conventional Czech agriculture. The results show an interesting combination of traditional systems' characteristics (no artificial fertiliser inputs, significant human labour inputs, a significant level of closed internal material loops), and modern/industrialised features (input of fossil fuels related to mechanisation, prevalent market orientation and dependence on external, although mainly local markets). The concept of food localisation is employed to discuss the complex issues of sustainability on the farm level, and the nexus of Food-Feed-Fuel-Fibre production as discussed in the literature is extended to also include the aspect of Finance, too often neglected in current socio-metabolic studies.

E. Fraňková (✉)
Department of Environmental Studies, Faculty of Social Studies, Masaryk University, Joštova 10, 602 00 Brno, Czech Republic
e-mail: eva.slunicko@centrum.cz

C. Cattaneo
Barcelona Institute of Regional and Metropolitan Studies (IERMB), Edifici MRA, Autonomous University of Barcelona, 08193 Bellaterra, Barcelona, Spain
e-mail: claudio.cattaneo@uab.cat

C. Cattaneo
Facultat d'Economia i Empresa, Departament d'Economia i d'Història Economica, Universitat Autònoma de Barcelona, Edifici B, Campus UAB, 08193 Bellaterra, Spain

 
E. Fraňková et al. (eds.), *Socio-metabolic Perspectives on the Sustainability of Local Food Systems*, Human-Environment Interactions 7,
https://doi.org/10.1007/978-3-319-69236-4_7

**Keywords** Traditional organic agriculture · Industrialised agriculture
Organic farm · Localisation · Social metabolism · Local Multiplier (LM3)
Czech Republic

## 7.1 Introduction

During the last two centuries, agriculture in the West has gone through a major transition from traditional organic to predominantly industrialised agriculture (Fischer-Kowalski and Haberl 2007) and capitalist agri-food regime (McMichael 2009), with only a minority of land being managed in (modern, certified) organic mode of production. This has far-reaching implications for the sustainability of agri-food systems such as soil erosion and compacting, loss of biodiversity and landscape functionality, breaking up of local nutrient and energy cycles, water table contamination, etc. (Weiss 2010; Foley et al. 2011; Gomiero et al. 2011a; Tscharntke et al. 2012), and provokes an active quest for more sustainable models. Czech/Czechoslovak agriculture has gone through a transition resulting in a current metabolic profile comparable to the West, but via different pathways including a strong cooperative and agrarian movement in the 1920's, followed by collectivisation in the 50's, an extreme use of synthetic fertilisers in the 60's and 70's, and fast transformation, mainly towards capitalist large scale structures, but also including small-scale and organic regimes of production, in the 1990's; while still preserving partial imprints of these various regimes in its land use structure and management practices (Grešlová et al. 2015; Bičík et al. 2015), it provides an interesting context for looking for different, and possibly more sustainable agri-food systems.

For the last couple of decades, European agriculture has been expected to be "multifunctional" (Van der Ploeg and Roep 2003; Renting et al. 2009; Mouysset 2017). In reality, this means that besides producing significant amounts of Food, Feed, Fuel and Fibre (Fedoroff et al. 2010), which is already enough to create some uncomfortable trade-offs (West et al. 2010; Fernando et al. 2010; Giampietro and Mayumi 2015), EU agriculture is also expected to fulfil a wide range of other tasks—from generating significant economic income (Finance) and securing rural jobs, through ensuring sustainable landscape management and safeguarding biodiversity, to the capturing and sequestration of carbon (Rossing et al. 2007; West et al. 2010). These—often contradictory—objectives compete for limited resources (be it land, energy or finance) and create certain paradoxes in how society expects farmers to behave. In the context of recent debates on the unsustainability of current predominantly industrialised agriculture in the West (Hathaway 2016; Cramer et al. 2017), the question of how more sustainable agricultural systems—meeting all the above-mentioned expectations—could look like, is—still—on the table, requiring urgent attention.

To capture the profound transformation of agriculture that took place during the last two centuries, two types of agri-food systems can be contrasted: the pre-industrial, *traditional organic* farming systems (Tello et al. 2012), based exclusively on solar, animal and human energy inputs, with a significant share of subsistence production and typically heterogeneous land uses, integrated on a local or regional scale; and *industrial*, "high-external-input" (Giampietro et al. 2013) forms of agriculture which are capital-intensive, based mainly on fossil inputs (fuel, fertilisers, biocides) and advanced (bio)technologies, characterised by high specialisation and distant supply chains. Although traditional forms of agriculture still feed about half of the world's population (Fischer-Kowalski et al. 2011), the industrialised forms of agriculture are changing the biosphere at an unprecedented level (Foley et al. 2011; Erb et al. 2016), heavily influencing all the world's regions through complex globalised agri-food chains, and associated uneven economic and power relations (Mayer et al. 2015).

Despite the great advances in yields and labour productivity, the absolute growth of industrial food production was at a high cost. High intensification and specialisation, and the related spatial division between animal and crop production lead to major disruptions in nutrient cycles (Bouwman et al. 2013; Cordell and White 2014); intensified livestock breeding is one of the main drivers of climate change, biodiversity loss (Steinfeld et al. 2006), as well as intensification of land use, deforestation, increasing water use and water pollution (Gerber et al. 2011). Regarding productivity, both yield and labour productivity have grown within industrial agriculture, however, its energy efficiency has dropped significantly, for specific crops, regions, and management practices actually often below 1, i.e. more inputs than outputs in energy terms (Pimentel and Pimentel 2008; Smil 2013; Gliessman 2015). Not only the production of food, but also its distribution has become globally interconnected, inducing environmental impacts associated to long-distance transportation (Davis et al. 2011; Weber and Matthews 2008). Besides environmental struggles, many social, economic and political ones also arise in relation to the industrial agri-food regime, e.g. issues around GMO crops, patenting seeds and related intellectual property rights, land grabbing, indebtedness of farmers, especially in the Global South, the loss of local traditional knowledge, and struggles of peasants for decent livelihoods both in the North and South (Schrivastava and Kothari 2012; Holt-Giménez and Altieri 2013; Galt 2013; Edelman et al. 2014).

From this perspective, the traditional organic systems might seem more sustainable as they are typically (1) based on local renewable resources (solar, human and animal power), (2) more energy efficient, (3) based on integrated farm management in terms of land use (cropland, grassland, woodland), and integration of animal and crop production resulting in (4) highly closed nutrient cycles, and (5) a high level of self-sufficiency on both production and consumption sides (subsistence agriculture and local markets).[1] Consequently, many advocates of more sustainable agriculture call for re-adoption of traditional organic practices (Feenstra 1997; Pimentel et al. 2005; Altieri 2009). However, as discussed more thoroughly by Gizicki-Neundlinger et al. (Chap. 6 in this volume) traditional organic systems

[1]There are exceptions to this localness and close-loopness of traditional organic systems related to early processes of industrialisation and increased urbanisation. The imports of guano from Chile are one example, as well as the rural-urban relationships in pre-industrial Austrian agriculture. (Gizicki-Neundlinger and Guldner 2017).

—as practised historically till the mid-19th century in Europe and North America, and till nowadays in many regions of the Global South—have their own significant drawbacks such as social inequality (serfdom, slavery or less severe forms of strong social control), inability to provide stable provision of food (famines or chronical malnutrition) and potential long-term soil nutrient depletion (e.g. case of Kansas Great Plain agriculture described by Cunfer and Krausmann (2009)).

Thus, it seems that more sustainable agri-food systems should be somewhere between the traditional organic and current industrialised forms of agriculture, combining the better of the two. In many important aspects, such as soil quality (soil organic matter, soil biota), field biodiversity, genetic diversity, limited use of agrochemicals (and vast energy input embodied in these agrochemicals as well as industrial fertilisers), *current organic* agriculture—in which mechanisation is present but synthetic fertilisers, biocides and GMOs are not—seems to perform better than conventional systems (Pacini et al. 2003; Pimentel et al. 2005; Gomiero et al. 2011b; Guzmán and González de Molina 2016).[2] Also, in terms of energy efficiency, organic farming systems perform better in most documented case studies (for overviews see Gomiero et al. 2011b; Lynch et al. 2011). However, as only certain aspects of production are specified within the certification process, organic agriculture in the context of industrialised countries is often still based on large-scale, highly mechanised practices; organic products are traded on global markets, and closing cycles of nutrients and other materials, although part of the original organic concept, is not required in practice (Gomiero et al. 2011b).

For these reasons, the potential benefits of not only organic, but also small-scale mixed crop-livestock systems have been highlighted (Herrero et al. 2010), ideally combined with local, preferably direct marketing of the produce through systems such as Community Supported Agriculture (CSA), facilitating personal consumer-producer relationships[3] (Seyfang 2007). These characteristics are strongly articulated within food localisation, and more recently food sovereignty concepts. Organic (even if not certified) local food systems (LFS) are seen as

[2]Although neither being sustainable by definition (Gomiero 2011b), nor being homologous to the pre-industrial forms as characterised above, it shares a few important characteristics with it. The use of artificial fertilisers and biocides is not allowed and although using mostly fossil-fuel based machinery and electricity, a significant portion of external fossil energy inputs is spared from its metabolism. Leguminous plants, crop residues (re-ploughed or composted) and manure are used as alternatives, along with labour-intensive activities of weed and pest control. Finally, albeit not explicitly checked with the certification process, the achievement of a closer system of nutrient and energy flows inspires organic farming (Gomiero 2011b). Biodynamic agriculture and permaculture can also be expected to prove advantageous in these aspects, however there is not much scientific evidence available (Gomiero et al. 2011b).

[3]Or even merging the two roles through active participation of consumers in the production process, creating the role of a prosumer (PROducer+conSUMER). According to the European CSA Research Group (Urgenci 2016), CSA is based on the following principles: mutual assistance and solidarity (direct connections and shared risks between the farmers and those eating their food), agroecological farming methods, biodiversity and no GMOs, high quality food at prices fair both to producers and consumers, education on the realities of farming, and continual improvement (Urgenci 2016: 5).

promising, potentially more sustainable, alternatives to the globalised agri-food systems by many localisation proponents (Douthwaite 1996; Norberg-Hodge et al. 2002). However, although being discussed extensively, mostly it is the social and community aspects of LFS that have been researched (Seyfang 2007; Field Bennet 2009), while material, quantitative data on their impacts are notably missing (Martinez et al. 2010).

This chapter aims at filling this gap, explicitly assessing the sustainability of a local food system, constituted by a small-scale organic family farm which embodies characteristics of both the traditional and current organic local food systems, integrating animal and crop production and local distribution, partly via the Community Supported Agriculture (CSA) scheme. To capture the complexity of sustainability of this local food system (LFS), a combination of a social metabolism approach looking at the biophysical characteristics of the farm, and a monetary flow analysis using a specific indicator called Local Multiplier (see Sect. 7.3 for details) is applied.

The case study farm is located in the village of Holubí Zhoř, Czech Republic, a country that provides a very interesting context as it has shifted between East and West, throughout the Austro-Hungarian, then socialist and now EU regime, which all have influenced its land use and agricultural practices significantly (Grešlová-Kušková 2013; Grešlová et al. 2015). To benchmark the current data, we compare the farm's metabolism with the metabolism of the village of Holubí Zhoř in the pre-industrial era of c. 1840, focusing on biophysical characteristics of both the current and the historical small-scale farming systems, and with national average data for both periods (c. 1840 and 2000/12); moreover, we analyse the current functioning of the farm, using the concept of food localisation to place it within the current debate on more sustainable agri-food systems.

To summarise, the aims of the chapter are as follows:

1. to provide an in-depth study of a concrete, potentially sustainable LFS—a small-scale organic family farm—in both biophysical and monetary terms;
2. to benchmark the current farm's biophysical characteristics with the historical (national and village-level) and current (national) data on CZ agriculture to define its position between the traditional organic and industrialised agri-food systems;
3. to analyse the functioning of the farm in terms of compliance with food localisation characteristics; and
4. to discuss the main trade-offs and option space for balancing the biophysical, social and financial aspects of sustainability in agriculture (the Food-Feed-Fuel-Fibre *and* Finance nexus).

In the following Sect. 7.2, we introduce and operationalise the concept of food localisation and we further explain its relation to the sustainability of agri-food systems. Also, the concept of social metabolism (SM) is introduced as the main methodological framework for the study. Section 7.3 explains the details of the SM analysis, along with the so-called Local Multiplier (LM3) concept used for analysing monetary flows. The results are structured according to the main points of the study: Sect. 7.4.1 describes the process of metabolic transition of Czech agriculture to show that the case study farm combines biophysical characteristics of both traditional organic and current agriculture; Sect. 7.4.2 then provides further insights into the functioning of the farm (production, distribution, labour) according to the characteristics of food localisation—first in biophysical terms, followed by a detailed analysis of the related monetary flows (Sect. 7.4.3). The discussion (Sect. 7.5) then opens up the issues of complexity of the localisation agenda, the prevailing dependence of the current case study farm on external fossil fuel inputs, and the financial aspect of sustainability of the farm's operations in the context of the broader debate on sustainability of LFS. Section 7.6 summarises the main conclusions of the chapter.

## 7.2 Food Localisation and Social Metabolism: Conceptual and Analytical Framework

While sustainable food production is a global challenge and needs to be discussed in the global context, it has an inevitable local dimension. It is at the local level where people live and work, where environmental, economic, social, cultural and institutional issues are interlocked, and where the food is produced, processed, transported, traded, and consumed or wasted. Hence, understanding of the particular food production practices and their impacts at a local level (while keeping in mind their global implications) is crucial for any understanding of past and future perspectives.

Rising academic attention has been devoted, among other things, to so-called local food systems (LFS), which are rooted in the broader concept of economic localisation. According to the proponents of economic localisation (see Fraňková and Johanisová 2012 for a systemic analysis of the concept), LFS are characterised, among other things, by the following environmental, economic and social aspects:

1. the most feasible closed circulation of matter and energy, including management of waste as a resource;
2. preference for locally sourced factors of production (natural resources, labour and capital);
3. emphasis on sustainability of production and consumption;
4. attempts to shorten distances between production and consumption;
5. preference for local consumers and satisfaction of their needs;
6. preference for local ownership of factors of production;
7. emphasis on local circulation of money and local financial capital.

By putting these aspects into practice, localised food production is supposed to bring social, environmental and economic benefits such as: lower transport dependence resulting in less consumption of fossil fuels, lower $CO_2$ emissions, less waste from packaging, higher levels of local recycling, more closed cycles of matter and energy within the production system, and also stronger local economies showing a higher level of local circulation of money, lower dependence on foreign investments, and less dependency on, and more resilience towards, fluctuations of the global economy (Douthwaite 1996; Norberg-Hodge et al. 2002; Desai and Riddlestone 2002).

However, with the increasing academic interest in local food systems, there is a significant critique and opposition rising against the localisation argumentation, especially regarding the oft-presented simplistic connections between economic localisation and sustainability or its particular aspects (see also Sect. 3.2.2 in Chap. 3 in this volume). And even many proponents of localisation openly admit significant gaps in our knowledge of the social, environmental and economic impacts of food system localisation indicated above, especially the quantitative, material data used for critical investigation of the potential benefits and trade-offs of food localisation (Martinez et al. 2010).

This study aims to provide some material data for such a critical debate. Methodologically, the issue of the sustainability of food systems calls for complex, systemic insights, including the biophysical side of the problem (Gomiero et al. 2006). With the explicit aim of capturing this complexity, the methodological framework of social metabolism has been developed, focusing on the nexus between energy, material, land, and time use within agri-food systems (Giampietro 2003). It is already quite common to apply SM methods at a global and national level (Pelletier et al. 2011; Grešlová-Kušková 2013). However, at a local level, and especially in the Global North, the application of SM and other biophysically oriented approaches is still rather exceptional and focuses mainly on historical comparison, describing the metabolic transition in agriculture (Krausmann et al. 2003; Haberl and Krausmann 2007; Haas and Krausmann 2015, see Haberl et al. 2016 for an overview).

In this study, we also use historical comparison, but we go one step further and focus explicitly on the functioning of the current LFS; hence, we discuss it not only through the lens of agricultural transition, but also through the current concept of food localisation, relating the case study to literature engaged in the quest for more sustainable agri-food systems.

## 7.3 Social Metabolism and Local Multiplier: Materials and Methods

This paper presents a case study based on the social metabolism (SM) approach, applied on a local level, at a small-scale organic family farm based in the village of Holubí Zhoř, Czech Republic (see Fig. 7.1 for the exact location). It is focused on relevant material and energy flows, time use, land use, human and animal labour and derived indicators. These current biophysical characteristics of the farm are compared to the situation around 1840, just before the beginning of the industrial revolution, in Holubí Zhoř. Because of the different size of the two systems, mostly

**Fig. 7.1** Location of the case study farm within the Czech Republic

intensive variables, i.e. variables related to funds (e.g. hectare of land, hour of work) are presented.[4]

The present farm data refer to the period between 2000 and 2014, but mainly 2012 (primarily, full data were collected for the season of 2012, accompanied and cross-checked with many data from the previous and following years, 2011, 2013 and 2014). Several methods were combined to obtain the complex picture of the case study farm. Mostly, the data are based on direct field research: 8 farm visits in various seasons took place between May 2013 and October 2016; in total approx. 30 days were spent at the farm and approx. 10 other, shorter, personal interviews with the farming couple took place in Brno (on vegetable delivery days). During the farm visits, continuous short interviews took place, along with ongoing participant observation, including several days of physical work at the farm at different seasons of the year (spring sowing, summer weeding, autumn harvesting and winter post-harvest handling of the produce). Field notes were taken during all these periods. For the time use, a shadowing methodology (McDonald 2005) was applied for two weeks (one in August and one in November 2013). Beyond that, accounting data (all the individual receipts from single purchases of consumption goods, petrol, feeding additives etc. for the years 2012 and 2013), hand-written field farm notes on human labour inputs, on feeding amounts, and on part of the farm's sales from a sales diary) were digitalised into Excel sheets. Also, data from farm-management software (*Farmer's Portal*) was extracted, complemented by the publically available data, especially on land use. As typical for case studies, the triangulation

[4]The use of the intensive, or flow/fund indicators (such as GJ/ha or MJ/worked hour), enables one to maintain the reference to a fund, and thus the intensity of the flow is defined. Intensive indicators are not only useful for comparing flows related to agroecosystems with different areas, but also for the historical comparison of the same agroecosystem, particularly when the composition of a fund has changed over time (i.e. changes in the composition of cropland, pastureland and woodland with respect to total farmland). For further argumentation on the issue please see Fraňková and Cattaneo (2017) and Chap. 3 in this volume.

approach was applied, i.e. more data sources were combined to provide a multiple crosscheck for the data obtained.

In contrast to similar studies focused more on the history of LFS (Krausmann 2004; Tello et al. 2012; Gingrich et al. 2015, 2017, Chap. 5 in this volume), the current data are mostly based on direct field data collection, thus providing more detailed insight into the functioning of the farm system. Such a data set is quite exceptional, although some missing data still had to be modelled or estimated, such as straw/grain ratios, the Gross Calorific Value of agricultural products (Guzmán et al. 2014), and embodied energy in agricultural inputs (Aguilera et al. 2015). On the other hand, seasonal variations are taken into account only to a limited extent.

For the historical data, the *Franciscan (stable) cadastre* was the main source of information referring to the period between 1824 and 1852, for our case study village this being mainly 1843 (MPAB 2016). The cadastre provides a complex of historical maps and detailed descriptions (both textual and table) of agricultural and other production primarily on the village level. Typically, for each village a basic descriptive (textual) part of the document is available, providing information on population, livestock, and land uses, as well as the productivity of each land use type and a rough indication of human and animal feeding typologies (for a full account of the available information see Krčmářová and Arnold 2016: 219) Further, a Yield Value Protocol (*Schätzungselaborat*) is provided summarising the average net yields, amount of manure applied, and the monetary value for each quality class of each land use type present in the village. Further, partial data e.g. on yields from selected individual plots in several consequent years, the amount of seeds used for crops on certain plots, the amount of straw as a by-product, and also the amount of human and animal labour executed per unit of agricultural produce. are included in the records (MPAB 2016). For details on processing and analysing the data please see Fraňková and Cattaneo (2017).

As argued more thoroughly in Fraňková and Johanisová (2012), one of the important features of the localisation concept is its complexity and the interconnectedness of its particular aspects. For this reason, we combine an assessment of environmental impacts studied via the social metabolism concept with economic analysis analysing sources of monetary income of the farm on the one hand, and the structure of its expenditures on the other. Specifically, we apply the so-called local multiplier (LM3) methodology. LM3 looks at three rounds of spending of a particular system (a farm, in this case) within a year (2012): 1st Round: the original inflow of money (e.g. overall company turnover), 2nd Round: how that inflow is spent, i.e. how much of the money was spent locally, and 3rd Round: how the first-round locally-spent money is re-spent, i.e. again, how much of the second-round local money was again spent locally. LM3 concludes after three rounds because 80–99% of the total multiplier effect is contained in those first three rounds, and measuring further is unnecessarily burdensome. The overall process for LM3 is to calculate how much income is generated for the local economy and then turn that figure into a ratio. By summing up the local incomes generated from Rounds 1, 2, and 3, we get a ratio that shows how much local income was generated relative to what money entered the local economy initially. The final step therefore is to divide the total (Rounds 1–3) by

Round 1 to understand the relationship between the original inflow and the total local income generated, i.e. the size of the multiplication effect within the local economy (see the formula below) (NEF 2002).[5]

$$LM3 = (R1 + R2 + R3)/R1$$

The definition of "the local" area is established according to the combination of local conditions (administrative, historical, cultural, economic or other ones); usually it is an area encompassing a radius of approx. 20–30 km around the studied subject (NEF 2002). In our case, we established two areas as "local", one defined by a radius of 30 km (i.e. area of 2826 $km^2$) including the nearest bigger settlement of about 10,000 inhabitants, and another defined by a radius of 50 km (7850 $km^2$) including the regional capital city of Brno (with about 380,000 inhabitants); thus two indicators ($LM3_{30}$ and $LM3_{50}$) were calculated, and through their difference the influence of the regional economic centre of Brno was determined. To indicate the first and second round of spending, the accounting data of the case study farm from 2012 were used. These expenditures go basically either to employees of the farm or to its suppliers; thus, to indicate the proportion of local expenditures of these subjects in the third round of spending, questionnaires for both the employees and the suppliers were used.[6]

## 7.4 Results

### *7.4.1 The Metabolic Transition of Czech Agriculture (1830/1850–2000/2012)*

In the first half of the 19th century, Czech lands belonged to the Austrian Empire; both its agriculture, and the infant industrial production were advanced with respect

[5]Mathematically, the LM3 value is always between 1 and 3, LM3 = 1 indicating a complete immediate "leakage" of the money from the local economy, and LM3 = 3 the other extreme—a fully closed system where all the money keeps circulating within the defined area. Neither of the two extremes occur in real systems in the context of the Global North, as there are always some suppliers (typically of energy and fossil fuels, machinery, and services) who are distant.

[6]Following the recommended methodology (NEF 2002), five of the most important suppliers in the 30 km radius (representing 48% of total supplier expenditures) out of more than 40 were selected and asked to fill in the questionnaire, data from two of them (representing 34% of all supplier expenditures) were obtained; for the 50 km radius, six of the most important suppliers (representing 49% of expenditures) out of more than 60 were selected, data from three of them were obtained (27% of expenditures). Regarding employees, all three permanent ones (representing 94% of all employee expenditures) filled in the questionnaire (both for the radius of 30 and 50 km), seasonal workers were not included. We calculated average ratios of local expenditures for both the suppliers and the employees (for the 30 and 50 km radius separately), and these values were then extrapolated to all relevant employee and supplier expenditures.

to other regions of the Empire (Grešlová et al. 2015: 25). Agricultural practices, recorded in historical annals (see Sect. 7.3 for details), marked a turning point in the evolution of central European agriculture: basic three-year rotation systems, simple hand-tools for working the land, and the dominance of human and animal work were slowly coming to an end. In 1827, an improved wheeled plough was invented by the Veverka cousins, adumbrating more profound technological changes to come. New species were gradually introduced for cultivation, above all leguminous plants to improve soil fertility, but also potatoes, new root and fodder crops (Svobodová 2010: 5–10). More available feed allowed for higher cattle densities resulting in turn in more manure for further growth in yields (Krausmann 2004);[7] still, surplus crops for trading were scarce and limited to local markets (Fischer-Kowalski and Haberl 2007: 244–245). Feudal relations between country people and noble elites were still in place, though changing gradually to more autonomous ones based on material and monetary, instead of personal bondage, commitments (Svobodová 2010: 5–10). Serfdom was abolished in 1848 and the ownership structure started to change accordingly, from large manorial estates to smallholdings of less than 20 ha[8] (Grešlová et al. 2015: 25).

At that time, the village of Holubí Zhoř had 351 inhabitants in 59 houses and 87 households, all dedicated to agriculture, with 40 small householders with no land. The ordinary diet consisted of potatoes, milk, floury meals and bread, rarely meat or wine. Animals were fed with hay and straw in winter, otherwise from pastures, pigs on potatoes and residues, and horses on oats. Horses and oxen were used for draft power and put aside or sold to brandy makers when too old, cows were for both milk and meat, sheep for wool and meat, pigs and goats, as well as a considerable amount of poultry, for household needs. Most land was of class 2 and 3 (i.e. less productive), cultivated with a rotation of rye, barley and oats, potatoes and fallow land; marginally with wheat peas and beet. Pastures and forests were very fertile. Soils were poor due to the poor manuring conditions and the disproportion between cropland and grassland. Animals pastured on pastureland, on some crop sides, and on fallow land. Good yields allowed for brandy production or the sale at market of rye, oats and potatoes. The closest market was a weekly one in Třebíč, 3 h away.[9]

However, this traditional organic (Tello et al. 2012), solar-based agriculture (Sieferle 1997) was coming to an end as coal-based industry was booming, providing new machinery both for agricultural production and trade; after WWII, industrial fertilisers began to be used (González de Molina and Toledo 2014: 235). Around 1840, the majority of the Czech population (58.9%) still worked in

[7]Though a more significant rise in livestock numbers did not occur before the half of the 19th century, see Grešlová et al. (2015: 28).

[8]Small farms under 20 ha of size represented almost 85% of the total number of farms, though only 38% of agricultural land; farms above 100 ha represented only 0.3% in numbers, but the same share of land as small farms in the second half of the 19th century (Grešlová et al. 2015: 25).

[9]It is not possible to quantify the volume of traded goods based on the historical data available, however, it seems that only in years with above-average yields was a tradable surplus generated. In any case, the distribution was probably very much local (Třebíč is 27 km away from Holubí Zhoř).

agriculture, livestock was used primarily for labour and milk and only a little for meat (Grešlová et al. 2015: 28). Occupying 40–50% of agricultural land, cereal crops dominated land uses in a proportion that has been maintained until today, but average yields were only about 0.5 t/ha, just three times higher than seed inputs (Beranová and Kubačák 2010).

During the 20th century, Czech agriculture followed an intensification path, growing its dependency on fossil fuels, like other European regions but with certain regional specifics.[10] Table 7.1 summarises the main socio-metabolic indicators characterising the c. 1840 to 2010 metabolic transition, both on national, historical village and current farm levels. A biomass-based system shifted through a coal-based phase (Kušková et al. 2008: 400) into the current oil-based one: fossil energy inputs amount to 2.78–5.5 GJ/ha (Picková and Vilhelm 2009; MoE 2005: 110)[11] and 0.22 GJ/ha are in electricity (mainly produced from coal) (Picková and Vilhelm 2009). For this intensification, the number of agricultural workers has dropped from 1,474,000 in 1920 to 181,000 in 2000, and draft animal power from 0.025 GJ/ha in 1960 to zero from 1990 (Grešlová-Kušková 2013: 594, 596). The excessive use of fertilisers peaked between 1975 and 1985, and dropped significantly after the Velvet Revolution in 1989, from 389 kg/ha of agricultural land in 1985 to 82 kg/ha in 2000 (Kušková et al. 2008: 402; Grešlová et al. 2015: 30), but has grown again to the current level (2012) of 118 kg of artificial fertilizers per hectare (MoE 2014: 13). The use of pesticides has also lessened since 1989, though not so significantly, with current levels (2012) of approx. 1.62 kg of active pesticide component per ha of agricultural land (MoE 2014: 14).

Also in political and socio-economic terms, after joining the EU in 2004, Czech agriculture became part of the European Common Agricultural Policy (CAP) adopting (most of) its objectives and its system of subsidies, strongly shaping the management practices in Czech agriculture (Lokoč and Ulčák 2009). Thus, despite the differences in political regimes during parts of the 20th century, the current Czech metabolic profile is comparable to that of other Western European economies, with all its socio-economic and environmental implications (Kušková et al. 2008: 405; Ščasný et al. 2003; Krausmann et al. 2008).

After 1989, organic farming in its modern form (certified and defined by the law on organic agriculture) started growing, occupying 11.56% of agricultural Czech land in 2012. Czech organic agriculture is, however, characterised by medium-large farms of more than 100 ha (61.6% of organic area) while farms smaller than 10 ha occupy less than 1% of land—moreover, an amount that has so far been decreasing

---

[10]Most notably the socialist period of 1948–1989, which (re-)created the structure of large farm units (co-called, but not authentic, cooperatives) as a result of massive collectivisation; this structure prevails to a large extent even after the fall of the communist regime (Doucha and Foltýn 2006 in Grešlová et al. 2015: 24). However, the period of independent Czechoslovakia between 1918 and 1937 is also very interesting, with a strong agrarian political representation and advanced structure of both production, trading and consumption cooperatives (Feyerabend 1952/2007).

[11]Reliable data on energy consumption are still missing. The lower estimate is calculated from financial costs of the fuel, while the Ministry of Environment offers an estimate of twice as much.

**Table 7.1** Metabolic and land use transition indicators, comparison between c. 1840 and 2000/2012 for case study village (1843), case study farm (2012) and the national average (Czech lands/Czech Republic) for both time periods

| Time period | | c. 1840 | | 2012 | |
|---|---|---|---|---|---|
| Indicator | Unit | Case study village | National average | Case study farm | National average |
| Population density | cap/km$^2$ | 57 | 92 | 29 | 123 |
| Agricultural land per capita | ha/cap | 1.69 | 0.78 | 3,5 | 0.42 |
| Share of population in agriculture | % | almost 100% | 58.9 | n. a. | 8.5 |
| Farmland labour | h/ha | 491 | n.a. | 314 | n.a. |
| | h/cap/yr | 1.661 | n.a. | 1.597 | n.a. |
| Livestock density (LSU = Livestock Unit 500 kg) | LSU/km$^2$ | 26 | 24 | 12 | 33.5 |
| Draught animal power | GJ/ha | 0.76 | 0.025 | 0 | 0 |
| Energy input of human and animal labour in agriculture | MJ/ha | 1349 | 544[d] | 377 | 7.95[a] |
| Exo/Endo work input ratio | | 1.29 | n.a. | 21.6 | 116 |
| Share of biomass in total Domestic Energy Consumption (DEC) | % | n.a. | 94 | n.a. | 18 |
| Share of fallow in cropland | % | 34 | 14 | 0 | 2 |
| Application of artificial fertilisers on agricultural land | kg/ha | 0 | 0 | 0 | 117.6[b] |
| Application of pesticides and herbicides (in kg of active pesticide component) | kg/ha | 0 | 0 | 0 | 1.62[b] |
| Average yields of cereals | kg/ha | 781 | 805[c] | 2674 | 4530[b] |

*Source* National data compiled from Kušková et al. (2008), Grešlová et al. (2015), ČÚZK (2016), ÚZEI (2015), MoA (2013); village and farm data own research; n.a. = not applicable

[a]data from 2001

[b]data from 2012

[c]average from 1830–1848

[d]Own calculation based on data from Krausmann (2004) on human labour inputs for Upland Land Use System in Austria in 1830/1850, see Table VIII. For animal labour input, Krausmann provides only information on livestock units and area; the livestock weight, workload and energy input needed for calculation are based on average estimations from Hitschmann (1891: 66) in Gingrich et al. (2015) and Schaschl (2007: 127) in Singh et al. (2013)

[e]The yield for 2012 is taken from the Ministry of Agriculture (MoA 2013: 18). According to another resource (CSO 2015: 415), the average yield for cereals between 2010 and 2014 was 5270 kg/ha. Grešlová et al. (2015: 26) provide an average yield of cereals between 1989–2010 of only 1420 kg/ha. The difference is partly due to a different categorisation of land (agricultural land vs. utilised agricultural area), partly the yields have grown during the last decade compared to the previous period (1989–2010). However, still, the 1420 kg/ha seems very low in comparison to other data sources

year by year (MoA 2015: 10). Grassland represents 82.9% of organic area and fodder weights 93% of total Czech organic production (MoA 2013: 5–6). Although organic agriculture is generally more labour intensive than conventional (Gomiero et al. 2011b), such a large share of grassland implies an average of only 1.8 full employees/100 ha within organic agriculture versus a national agricultural average of 3.12. While small organic farms struggle to comply with complex bureaucracy, large ones are often profit driven, so they keep the smallest possible livestock densities in order to fulfill the EU requirements for subsidies, by maximising land and minimising labour (Lokoč and Ulčák 2009). Consequently, they provide only a very limited share of the food produced in the Czech Republic. However, there are organic and biodynamic (Demeter 2016) farms which explicitly try to manage the land and produce food with an explicit sustainability focus. To discuss to what extent one of these examples can be seen as a "successful" example, both in biophysical and financial terms, is the aim of this study.

### 7.4.2 *The Case Study Farm: Sustainability Halfway Through*

The present case study farm, based in the village of Holubí Zhoř (see Fig. 7.1), has a farming history reaching back to the period of c. 1840 (the farming family's surname already being present on historical maps from 1824, for both the farmhouse and the fields, see Fig. 7.2a). In 2012, the farming family comprises 4 adults (1 full-time and 3 part-time involved in agriculture) and 4 school-age children. The farm manages about 28 ha of land (see Fig. 7.2b), and produces vegetables, cereals, herbs, meat, eggs and occasionally cheese for the local market (see Sects. 7.4.2 and 7.4.3 for details), and more cereals, leguminous plants and grass/hay for feeding their own animals. An orchard has been planted, but the trees are still too young to provide any harvest. The woodland is used mainly for firewood to heat the farmhouse, timber only occasionally being harvested for the market. Manure is used only for fertilising the vegetable and potatoes plantations, the nitrogen input being otherwise assured by rotating crops of the leguminous plants, and by burying a significant amount of residual biomass back in the ground (for details on internal biomass reuse within the farm, including analysis of its energy balance, please see Fraňková and Cattaneo 2017). Although potentially self-sufficient in the majority of their diets, in reality the family sell all possible (certified organic) produce to the market and buys the majority of food for their own consumption at the market (see Sect. 7.4.3 for details).

On the present farm, a relatively high amount of working hours per hectare are employed. This is mainly due to the quite large fraction of land devoted to vegetable production which is a labour- intensive farming activity. However, being an organic farm, there are also lots of activities related to fertilisation (from manure management to cropland rotation management) that are more labour intensive than

**Fig. 7.2** **a** Historical map of the village of Holubí Zhoř (*Zhorz Holuby* in German) as created during the Franciscan cadastre mapping around 1824. Plots designated on the map as belonging to the ancestors (*Pospischil*) of the current farming family are highlighted in orange *Source* MPAB (2016). **b** Current map of the village of Holubí Zhoř, plots managed by the farming family in 2012 are highlighted in orange *Source* Farmer's portal

simply applying of synthetic fertilisers or biocides. The resulting 314 labour hours per hectare is a not very dissimilar value from the 491 h/ha estimated for the past. With an estimated energy consumption of 1.2 MJ/hour of labour, 377 MJ/ha are spent—while the national average is just below 8 MJ/ha. Hence, from the labour perspective the present farm is more similar to the past than to the present. However, when looking at all the mechanical energy applied (the exo/endo work input ratio), the metabolism of the present farm is very dissimilar to the past. In the past, the only source of exosomatic energy was animal power and it accounted for just a bit more than the human (endosomatic) energy input. 451 GJ of animal power and 350 of human labour were applied in the village overall, while in the present 227 GJ of mechanical power and 10.5 GJ of human labour are applied over an area which is 21 times smaller than in the past. This implies a present exo/endo input ratio of 21.6 compared to 1.29 in the past. As well, there is no more draught animal power, machines have entirely substituted work animals. Thus for the amount of endosomatic energy (labour) the present farm resembles the past while for the amount of exosomatic energy (mechanical power) it does not. But it also does not resemble the national average in which the exosomatic metabolic rate of the primary sector is 134 MJ/h, that is an exo/endo ratio of 116, much larger than the 21.6 estimated at the farm level.

Another indicator that shows how the present farm stands right between traditional organic and industrialised agriculture is the average cereal yield (barley): 2.674 kg/ha versus a national average of 4.530 and nearly 800 in the past: close to 1900 kg more than in the past and 1850 kg less than the present average.

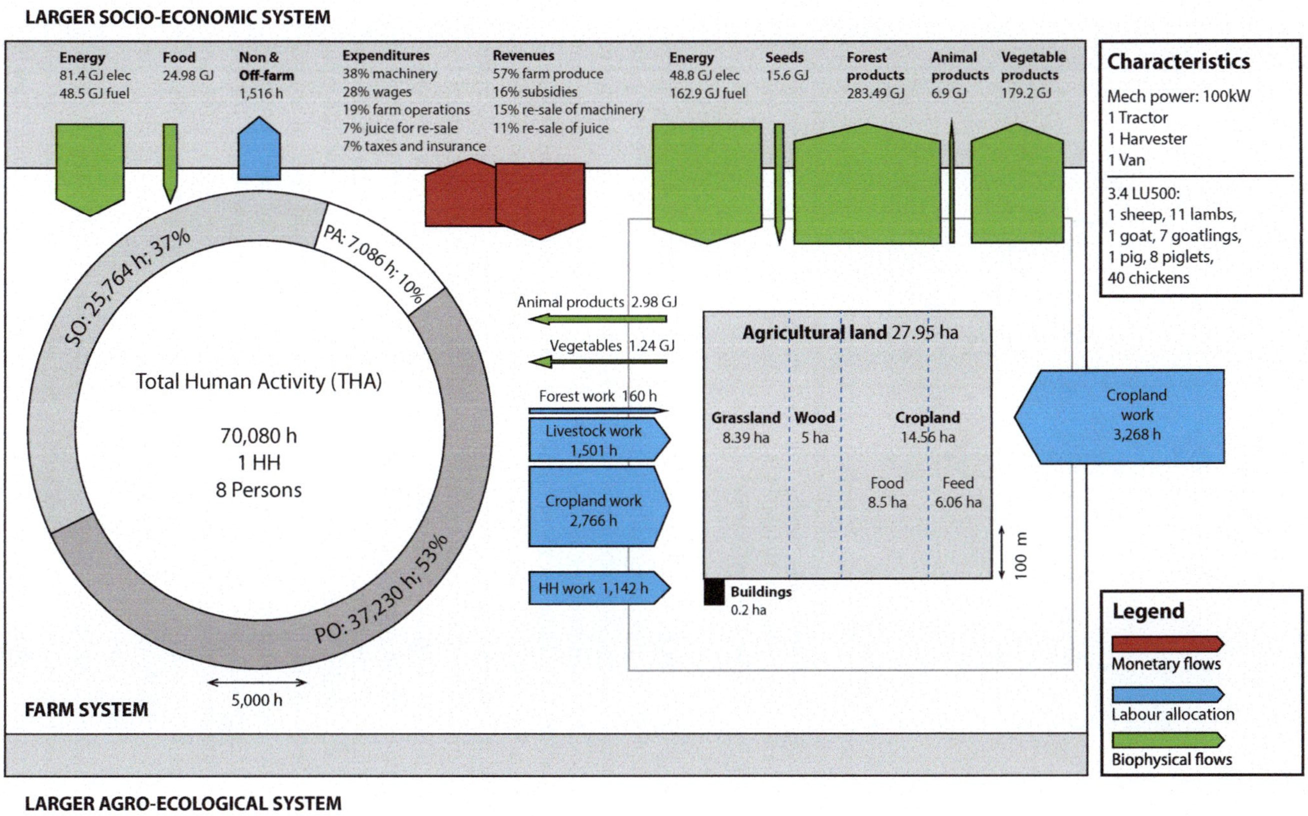


**Fig. 7.3** The main biophysical characteristics of the case study farm in 2012. Source: own calculations, the layout of the scheme based on the work of Scheidel et al. (2014) and Serrano-Tovar and Giampietro (2014), for a comparable picture representing a Cambodian village agricultural system see also Chap. 8 in this volume. *Note* THA = Total Human Activity; HH = Household; PO = Physiological Overhead; SO = Social Overhead; PA = Productive Activities; t = tons; l = litres. Arrows represent labour allocation, biophysical and monetary flows, they are scaled

From a spatial perspective, the farm's fertility depends on local recirculation of biomass flows: animals are fed with locally-produced feed and pasture which in turn contribute—along with an efficient crop rotation system—to maintaining soil fertility and relatively high yields. There are also farm expenditures in external inputs—such as barley seeds, potatoes or fruit trees—which are also bought from local suppliers (see Table 7.3 for details). Only non-renewable energy carriers originate from far away, making its metabolism not entirely local but in any case clearly different from conventional farming, which relies more heavily on synthetic fertilizers, biocides and biomass import from non-local sources. The main biophysical characteristics of the farm are summarised in Fig. 7.3.

### *7.4.3 Compliance with Localisation Characteristics*

Regarding biophysical aspects of localisation (see points 1–4 in Sect. 7.2), as already shown in the previous Sect. (7.4.2) and summarised in the biophysical profile of the case study farm (Fig. 7.3), there is a significant internal turnover of biomass (in terms of own production of animal feeding, both manure and leguminous plants used as natural fertilisers, and partly own production of seeds and seedlings—mainly part of grains and potatoes). External inputs are in the form of seed imported to the farm (of non-local origin), energy carriers (fossil fuels and electricity, also of non-local origin) and labour.

Regarding labour, as also apparent from the monetary flows (see Table 7.3), the vast majority of the farming work was performed locally. From the overall amount of 8500 h spent in the farm-related activities (see Table 7.4 for details), 61.6% was performed by the farming family itself, 26.3% by local waged labourers (living either directly in Holubí Zhoř, or in neighbouring villages, the most distant one being 25 km away from the farm), and only 12.1% by a (formally)[12] non-local waged labourer.

If we look at the most of significant man-made capital, the machinery, along with most of the construction materials (although these were not analysed directly as they were not part of the farm metabolic flows in 2012) were non-local. Where possible, the maintenance work (i.e. services) related to the man-made capital was, however, performed locally (see Table 7.3 for more details).

As for the market distribution of the farm produce, the main distribution channels are summarised in Table 7.2 (in terms of kilogrammes, revenues in Euros and average distance travelled within particular categories). Out of approx. 32 tons of products delivered to the market (21,500 kg of vegetables and fruits, 52 kg of

[12]According to the LM3 methodology (see Sect. 7.3 for details), the official registered residential place (both for individuals and companies) is decisive for defining if they are local or not. In the case of the non-local worker—officially his place of residence was 170 km from the farm, however, for most of the season he was accommodated, and also fed, at the farmhouse, and thus practically can be considered local, even though not accounted as such within the LM3 analysis.

**Table 7.2** Summary of all the distribution channels used for selling the farm products (about 32 tons in total) and their importance in terms of weight (kg), revenue (EUR) and average distance (km) travelled within particular categories. For "unknown" consumers, the overall average distance of 44 km is used. Source: own calculations

| Distribution channels | Proportion in weight [kg, %] | Proportion in revenues [EUR, %] | Average distance travelled in category [km] |
|---|---|---|---|
| CSA | 26.2 | 19.0 | 43 |
| Other farmers | 23.7 | 5.5 | 34 |
| Organic shops | 20.5 | 22.8 | 45 |
| Direct farm sale | 17.4 | 33.2 | 0 |
| Farmers' markets | 5.7 | 7.0 | 35 |
| Restaurants | 5.4 | 8.1 | 46 |
| Schools | 0.6 | 0.9 | 47 |
| Individual consumers | 0.3 | 0.4 | 29 |
| Wholesaler | 0.2 | 2.9 | 95 |
| Unknown | 0.1 | 0.2 | 44 |

dried herbs, 8800 kg of grains and 1750 kg of animal products—meat, milk products and eggs), about ¼ of the overall farm production is distributed through the CSA scheme (26.2% in weight), vegetables and fruit mostly going to Brno (43 km), and another ¼ to other farmers (23.7%), mostly grain used as feeding for animals at one organic pig farm (53 km) and one organic goat farm (20 km). Another 20.5% goes to organic shops, and 17.4% is sold directly from the farm (in both cases, besides vegetables, also a significant part of the animal products). The rest goes to farmers' markets, restaurants, schools, and individual consumers, the wholesaler is solely in the case of dried herbs.

The average distance travelled by one kg of product is 44.5 km (with Brno, the most important destination, 43 km away). The average distance travelled by one euro of product is 51 km; this is because dried herbs, the most valuable item, are sold in the more distant place (95 km). Accounting for products sold at the farm gate, the average distance decreases to 36 km (although we do not know how far the food travels with direct farm customers). In general, the system is very local and food miles are very low (Table 7.2).[13]

[13]As Schnell (2013) argues, it is very hard to provide some "average" comparable data on food miles that would be methodologically sound, only very partial data for specific cases, e.g. average food miles per ton of fresh produce delivered to one specific (Chicago Terminal Produce) market in the US exists—there, the average value was 1500 food miles (Pirog et al. 2001). For an "average food item" the value is supposedly significantly higher, however, sound data e.g. on the national level are not available (Schnell 2013).

### 7.4.4 The Energy Cost of Distribution

An important issue in local food systems refers to the food miles saved in the distribution of agricultural production. With 23% of global emissions embodied in traded goods (Davis et al. 2011), with food transportation representing 15% of the carbon footprint of food production and distribution (Weber and Matthews 2008), and with the energy cost of trans-oceanic transportation and feed handling and processing representing 17% of the gross calorific value of the feed traded (see Chap. 5 in this volume, referring to imports for feedlots in Vallés county, Spain), the case at hand shows that even on the level of the individual farm, (local) distribution activities can play a significant role within the farm system. It applies both in terms of time demand (8.3% of total farming time, see Table 7.4 for details), and in energy consumption: a van is used for distribution of farm products, and consumes an estimated amount of 636 litres of petrol per year. Assuming that 50% of the vehicle farm use is connected to distribution, and accounting for the embodied energy costs of the van and fuel production, a total of 15.913 MJ are consumed in distribution which, in turn, refer to 7.5% of all the non-renewable energies employed in the farm and to 8.6% of the total Gross Calorific Value of the farm products sold to the market.

These figures show that the farm's local distribution system can be considered more energy efficient than the global one (twice as much if we compare it to the data from Weber and Mathews 2008, see above). However, it should be noted that not all the distribution costs of the farm produce (both in time and energy terms) are accounted for.[14] Thus, the possibility of such a comparison (between a local farm system and a global food trading system) is limited because of the very different scale and characteristics of the two systems; for the debate on the complexity of the issue of system boundaries, and the limited comparability of different production and distribution systems see also Chap. 10 in this volume (for the case of organic tomatoes in Spain).

### 7.4.5 Financial Metabolism and the Local Multiplier (LM3)

Regarding the economic and financial aspects of localisation, above all the local ownership of factors of production, local financial capital and local circulation of money are emphasised.

[14]Whereas for most of the farm distribution channels (the CSA, other farmers, organic shops and the rest of the small ones) we can assume that no additional significant distribution is involved (i.e. the farm ensures the distribution to the final consumer, and e.g. the CSA members do pick up their vegetables predominantly by foot, bicycle or public transport), the 17.4% of the farm produce sold at the farm gate and its distribution costs (in energy, time and financial terms) are not included in the farm system metabolism. However, we can assume that all the indirect costs and embodied energy related to the complex distribution chains of the global distribution system are similarly not accounted for.

In terms of ownership, all physical man-made capital (e.g. buildings, machinery, etc.) is owned by the farming family. Out of the 28 ha land that is managed by the farming family, 93% is owned, and 7% is rented from another local owner. Financial capital is not very relevant as there were no external loans, or other forms of external capital (e.g. shares) involved in the financial turnover of the farm in 2012. Thus, the main localisation issue in the economic respect is the local circulation of money as analysed further in this section.

The overall financial turnover of the case study farm in 2012 was about 60,000 EUR (1.5 mil. CZK). Table 7.3 summarises both the main sources of income and main types of expenditures of the farm, along with their proportion of the total financial turnover and their localness. Following the methodology of Local Multiplier (see Sect. 7.3 for details), two areas of the "local" were defined, a radius of 30 km (including the nearest five market towns) and of 50 km around the farm (including the regional capital, the city of Brno).

Looking at the structure of incomes, it is clear that the main source of income (56.9%) comes from selling the farm's own produce to various subjects (the CSA group, other farmers, small shops, individuals etc., for their full overview see Table 7.2). Further, the farm income is composed of subsidies (16.3%), the sale of an old car and residuals of old machinery (15.3%), re-sale of juice (11.4%) and negligible investment revenues (0.2%). About a quarter of these incomes do come from a radius of 30 km, and almost 60% from a distance of up to 50 km from the farm. The vast majority of the farm produce is sold within the 50 km area, only grain and dried herbs travel to more distant places (non-local grain as feeding to 53 km distant organic pig farmer, dried herbs to a 95-km distant producer of organic teas and herbs). Another main source of non-local incomes are the agricultural subsidies (that can be considered as coming either from the Czech capital of Prague, 169 km away, or from the EU "capital" of Brussels, approx. 1100 km), and the revenue for selling an old car (69 km).[15]

Regarding expenditures, we applied fully the specific methodology of the Local Multiplier indicator (LM3, see Sect. 7.3 for details). The volume of local monetary flows related to the case study farm in the three rounds of spending, both within the 30 and 50 km local area, are visualised in Fig. 7.4. The first round of LM3 (i.e. the overall turnover of the case study farm) in 2012 was about 6,000 EUR (1.5 mil. CZK). Regarding the second round of spending (i.e. direct expenditures of the farm), if we define the 30 km radius as "local", about 33% of the total turnover was spent locally; 79% of this amount went to local employees and 21% to local suppliers. As all the

[15]If we further extend the "local radius", the farm incomes become even more significantly localised, 82.2% of incomes are local within a 100 km radius, 83.5% within 150 km, and 99.9% within 200 km (83.8% if we consider agricultural subsidies as coming from Brussels and not Prague). The Czech Republic is 493 km long and 278 km wide, i.e. the 200 km radius can be still seen as a sub-national scale; if defined by national borders, all incomes are national (or 84% if we consider agricultural subsidies as coming from Brussels and not Prague).

**Table 7.3** Main sources of income (a), and main types of expenditures (b) of the case study farm, along with their proportion of the total financial turnover of the farm and their localness (i.e. proportion of local monetary flows from the total incomes/expenditures within the respective detail category)

| Broad income category (proportion of all incomes) | Detail income category (proportion of all incomes) | Proportion of local incomes [%] | |
|---|---|---|---|
| | | 30 km radius | 50 km radius |
| *(a) Income analysis* | | | |
| Sale of farm produce (56.9%) | Sale CSA—vegetables and fruits (9.6%) | 0.0 | 100.0 |
| | Sale of other vegetables, fruits and animal products (41.9%) | 50.7 | 87.6 |
| | Sale of dried herbs (1.5%) | 0.0 | 0,0 |
| | Sale of grain (3.3%) | 14.4 | 14.4 |
| | Sale of pest control (0.5%) | 100.0 | 100.0 |
| Subsidies (16.3%) | Subsidies (16.3%) | 0.0 | 1.6 |
| Machinery (15.3%) | Sale of car (13.5%) | 0.0 | 0,0 |
| | Iron scrap (1.8%) | 0.0 | 100.0 |
| Re-sale (11.4%) | Re-sale juice (11.4%) | 18.1 | 92.7 |
| Investment revenue (0.2%) | Investment revenue (0.2%) | 0.0 | 0.0 |
| ALL INCOMES | | 24.3 | 59.9 |
| *(b) Expenditure analysis* | | | |
| Machinery (38%) | Van purchase (12.9%) | 0.0 | 0.0 |
| | Harvester purchase (10.2%) | 0.0 | 0.0 |
| | Fuel (9.3%) | 0.0 | 0.0 |
| | Machinery maintenance (5.6%) | 82.8 | 99.8 |
| Wage labour (28.3%) | Wage: farm and field work (27.7%) | 92.6 | 92.6 |
| | Wage: farmhouse maintenance (0.5%) | 100.0 | 100.0 |
| Farm production. operation and maintenance (19%) | Animal production (2.5%) | 17.2 | 78.2 |
| | Plant production (0.8%) | 40.7 | 40.7 |
| | Seeds and seedlings (4.9%) | 2.1 | 22.3 |
| | Other farming expenditures (4.7%) | 17.1 | 54.1 |
| | Farmhouse maintenance (0.4%) | 16.8 | 66.5 |
| | Electricity (4%) | 0.0 | 0.0 |
| | Water fees (0.4%) | 0.0 | 100.0 |
| | Mobile services (0.3%) | 0.0 | 0.0 |
| | Accountant services (0.8%) | 100.0 | 100.0 |
| Re-sale (7.8%) | Juice for re-sale (7.1%) | 0.0 | 0.0 |
| | Fruit and vegetables for re-sale (0.7%) | 12.8 | 35.3 |
| Taxes and insurance (7%) | Insurance (5.8%) | 0.0 | 0.0 |
| | Taxes (1.3%) | 0.0 | 0.0 |
| ALL EXPENDITURES | | 33.5 | 39.5 |

*Source* own calculations

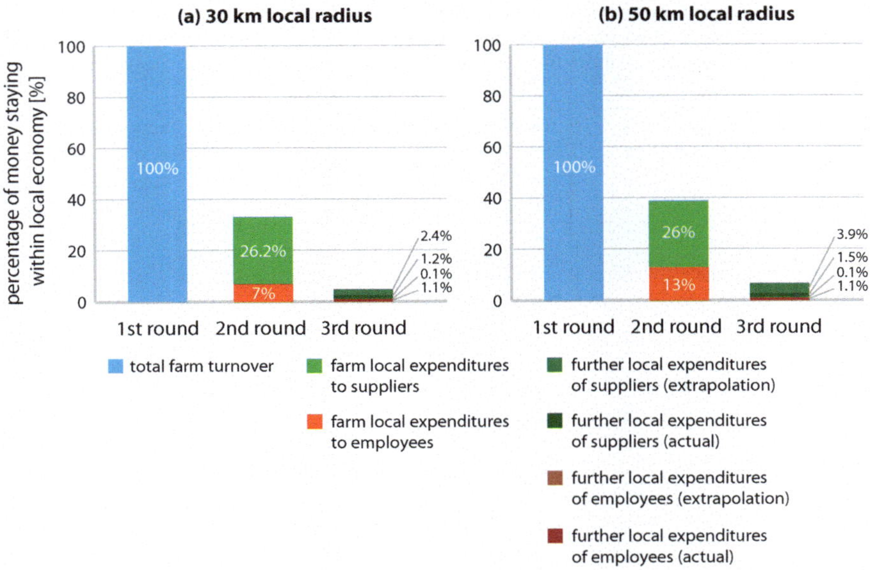


**Fig. 7.4** Local monetary flows in the three rounds of spending as captured by the Local Multiplier methodology (see Sect. 7.3 for details) for the radius of (**a**) 30 km and (**b**) 50 km. *Source* own calculations

employees are local, 100% of the employee expenditures were local, whereas only about 9.5% of all supplier expenditures were. Within the 50 km "local" radius, the amount of local expenditures is logically higher, but the difference (mostly reflecting the economic importance of the city of Brno) is not very significant: the proportion of overall local expenditures rises to 39% (composed of 67% local payments to employees and 33% to local suppliers). As still applies, all the employees (and thus also employee expenditures) are local, the proportion of payments to local suppliers from all the supplier expenditures rises to 18% if we extend the radius to 50 km (and include Brno).[16]

If we look at the third round of spending, i.e. the proportion of local expenditures exercised by the employees and the suppliers of the case study farm, we can see the following pattern: whereas the farm itself spends about 1/3 of its overall expenditures locally (33% in the 30 km radius and 39% in the 50 km one, see above), its employees spend on average only about 4% (4.3%, 4.4% resp.) of their wages locally—despite the fact that all the permanent workers are living in the local area; its suppliers, on the other hand, spend much more locally—51% on average for

[16]If we extend the "local" radius again, 45% of expenditures are made within a 100 km area, 60.2% within 150 km, and 77.4% within 200 km. Regarding the national scale, 89.8% of expenditures are made within the Czech Republic, the only foreign transaction being the second-hand purchase of the harvester from Schorndorf, Germany (622 km).

those within 30 km, and 41% on average for those within the 50 km radius (i.e. the more distant suppliers spend less money in the local area of the farm).

Counting the overall local multiplier effect using the LM3, for the 30 km radius we get the value of $LM3_{30} = 1.38$, and for the bigger local area the $LM3_{50} = 1.46$. This means that for every Czech crown spent by the case study farm, another 38 cents[17] (46 cents respectively) is generated as additional income for other subjects within the local economy.[18]

The most significant "leakages", i.e. biggest farm expenditures through which the money is leaving the local economy, are made within the second round of spending, where almost 2/3 of the farm turnover leaves the locality (67% for the 30 km radius and 61% for the 50 km one). Expenditures related to machinery (its purchase and fuel to run it) constitute the most significant outflow (48.8% of all non-local expenditures within the 30 km area), although its maintenance takes place mostly locally, followed by the purchase of juice for re-sale, insurance and taxes (10.6% of all non-local expenditures each), seeds and seedlings (7.2%), and electricity payments (6%). All other items—mostly related to farming activities and farm maintenance—do include some non-local expenditures, but to a limited extent.[19]

From the proportion of the farm produce that is sold via market, we can claim that the case study farm is strongly market oriented. Of its food produce expressed in energy terms, only 2.3% is for own consumption (which is estimated to cover only about 17% of dietary requirements of the farming family) while the rest is sold to markets, albeit local and partially "alternative" (e.g. the CSA, see Sect. 7.5 for further discussion on the issue). This raises the point that, at the present time, even if the agroecosystem resembles a past one, where forest, grassland and cropland uses are well integrated with each other so that it allows to be an almost closed system from the material perspective (see Sect. 7.4.1 for details), when we turn to the satisfaction of needs (traditionally the 4 Fs "Food, Feed, Fiber and Fuel"), present day market dynamics assume an almost hegemonic role so that the most important need satisfied by the agroecosystem turns out to be Finance.

The reason why this happens is due to the cost opportunity of not selling certified organic products at a premium price. A paradoxical situation thus occurs

[17]35 cents ("hellers") means 35/100 of one Czech crown.

[18]As explained in footnote 6, this value is based on extrapolation of the average proportions of local expenditures by employees and suppliers as obtained from a limited number of respondents (esp. regarding the suppliers). If we include only the third round of spending of the suppliers that actually filled in the questionnaire and provided their data (designated as "actual" expenditures in Fig. 8.4), i.e. without the extrapolation to all suppliers and to seasonal workers, the size of the multiplication effect would be a bit smaller, but comparable: for $LM3_{30actual} = 1.35$, and the $LM3_{50actual} = 1.42$.

[19]In the broad category of "Farm production, operation and maintenance" we see the big importance of Brno—whereas within the 30 km radius only e.g. 17.1 and 16.8% of expenditures in categories "Other farming expenditures" and "Farmhouse maintenance" are local, within 50 km distance (including Brno), significantly more (54.1 and 66.5%) expenditures qualify as local. With the "Animal production" and "Seeds and seedlings" categories we see the same trend.

when the farming family spends vast amounts of their time[20] producing high quality local organic food, of which 97.7% is then sold on the market. Only the income earned is then used to fulfill some of the basic needs, including food, that is purchased at a much lower price (and quality) mostly via the mainstream distribution channels: from the overall family expenditures, including food, cloth, leisure activities, and services etc., almost half (48.1% in monetary terms) is spent in supermarkets, 21.4% in local small shops, 14.2% at wholesalers and only 6.3% at organic shops, restaurants (2.3%) and other unspecified places (7.7%).

### *7.4.6 The Hourly Cost of Production*

Table 7.4 summarises the main farming activities and their proportion of the overall farming workload, from the revenue obtained from selling the products of the farm itself, and from the overall farm income in 2012. The total farm revenue is obtained with 8500 h of work at the farm (all family work, and paid work done by seasonal workers in the vegetable production). This would mean a gross revenue of about 7 Euros per hour. However, this also includes income from a one-off sale of machinery (that is outweighed by the purchase of another one, see Table 7.3), from the re-sale of juice that is not produced within the farm, and income from subsidies. Although these incomes also required a certain time allocation, it was negligible within the overall farm working time (if anything, income from subsidies might be matched by the "farm management" category as a significant proportion of this time is devoted to fulfilling the formal organic certification requirements). The income obtained solely from selling the direct produce of the case study farm adds up to about 29,000 Euros, and this amount results in a gross revenue of 3.4 Euro per hour.

As apparent, however, the gross revenues do not include all the farm operation costs. Salaries of the hired farm labourers (referring to 38% of the total farm working time) are in reality about 2.8 Euro/h (70 CZK/h). As for the work of the farming family (62%), if the same wages per hour as for the external workers applied, only about 33% of their working time is paid. The rest of the income is used to cover other operational costs of the farm. This fact, together with the paradox of the family mostly eating conventional food instead of their own produce (see above), raises serious questions about the financial aspect of the sustainability of the farm, and the hypothesis that the farm is not a profitable enough enterprise.

[20]As apparent from the time use analysis, the farm lady who works at the farm "full time" actually works about 170% of the "normal" full time (40 h per week in the Czech Republic) throughout the year (i.e. almost double the normal full time workload within the high farming season). Her husband works "only" about 100% of the normal fulltime at the farm, however, he has a parallel part-time job as a carpenter, mostly during the winter (when there are less farming activities). Moreover, the work performed by the farming family is only partially paid (see details further in the section), as its amount is enormous and if fully accounted for in the price of the products, they won´t be salable, even in the context of the "premium prices" of the local and/or organic products.

**Table 7.4** Overview of main farming activities and their proportion in terms of time allocation, share in revenues related to direct farm produce, and in overall farm incomes

| Farming activity | Vegetable production | Grain production | Meat and egg production | Milk production | Forest management | Distribution | Farm management |
|---|---|---|---|---|---|---|---|
| Proportion in farming time allocation [%] | 65.4 | 3.5 | 11.2 | 5.5 | 1.9 | 8.3 | 4.2 |
| Proportion in direct farm produce revenues [%]* | 59.8 | 5.6 | 18.0 | 2.9 | 13.6 | n.a. | n.a. |
| Proportion in overall farm incomes [%]** | 34.3 | 3.2 | 10.3 | 1.7 | 7.8 | n.a. | n.a. |

*Source* own calculations
*This category includes revenues from sales of: vegetables and herbs, grain, meat and eggs, milk products and timber (i.e. forest produce). This category corresponds to the broad category "Sale of farm produce" in Table 7.3 (a), however, timber is not included in the farm accounting as calculated in Table 7.3
**This category includes all revenues as calculated in Table 7.3 (a) plus income from selling the timber

## 7.5 Discussion

Localisation of the global food system has been advocated since the heyday of globalisation (La Trobe and Acott 2000). The local food movement attempts to counteract trends of economic concentration, social disempowerment, and environmental degradation in the food and agricultural landscape. The choice for "local" is therefore not only related to consumers' benefits in personal health, but to a wider framework, that encompasses the social and the cultural: "local" manifests high levels of social capital and relations of care, a more moral or associative economy, that has nonetheless to take care of the ambivalence of a negative "patriotic" act of exclusive localism with the positive result of melting producers and consumers, and enhancing mutualism, commensal relationships and more self-reliance (Hinrichs 2003).

Analogous to the expectations of multi-functionality in the case of European agriculture as discussed in the Introduction, local food (and LFS) also thus embodies a wide spectrum of values, standards and expectations as diverse as: quality (local, fresh, seasonal, sustainable source); justice (animal welfare, fair trade, work conditions); environment (climate change, water, land use, biodiversity, organic, soil); health (safety, nutritional, cultural); and social concerns (affordability, access, socio-economic status) (Lang 2010). Similarly, in the specific case of a CSA that combines a local food system with organic agriculture, the focus of the consumers is not only on the premium quality of organic food, but also on the local dimension that the production-distribution-consumption system assumes. In this way, CSA corresponds to more than one value, with a plurality of ethical concerns, close to what Lang identifies as "omnistandards" (2010).

In this chapter, we have focused on only some of these multiple aspects of localisation, namely those related to biophysical and economic aspects of localisation, as specified in Sect. 7.2. In these respects, we can claim the case study farm to be quite close to the ideal of localisation: as shown in Sects. 7.4.1–7.4.3, internal biomass flows are significant, including "natural" fertilising techniques (manure, together with other biomass leftovers used as a valuable source of nutrients); factors of production such as human labour are also locally sourced; neither artificial fertilisers nor biocides are used within the production process; the majority of the farm's produce is distributed locally, contributing to fulfilling one of the basic human needs by providing food of a high standard; and also in economic terms, a significant proportion of the farm-related monetary flows circulate within the local economy.

At the same time, through many of its biophysical characteristics (including its composition of land use, integration of animal, crop and forest production, and the significance of the internal biomass re-use) the farm is close to the traditional organic agriculture as practiced in the area before the industrial revolution (c. 1840). However, there are at least two significant differences from the traditional organic agriculture: (1) the use of machinery (and related strong dependence on fossil fuels), and (2) the fact that although distributed locally, the vast majority of the farm produce is sold to the market and—although potentially possible—the farm no longer fulfills the subsistence role for the farming family (or the village community). In these two respects, the farm is closer to the current industrialised forms of agriculture.

We argue that by this specific combination of the traditional and the new, this case is an interesting one, not only in the context of localisation but also for a broader debate on how more sustainable agri-food systems might look like. Further we discuss some of the sustainability implications and trade-offs this case reveals, especially regarding their biophysical and economic aspects.

First, following Tello et al. (2016) we observe a path towards energy optimisation in which biomass reused and external inputs achieve a better equilibrium than in the past, when agricultural inputs consisted mainly of biomass internally reused. This in fact implied an exaggerated biomass recycling effort that, in the present case, is spared so that a better Final Energy Return On Investment (FEROI) figure is achieved than in the past.[21] We recall that inputs are both Biomass Reuses that recirculate from within the agroecosystem (also known as Internal Inputs) and External Inputs—notably industrial ones derived from fossil fuels such as machinery and synthetic fertilisers. The consideration of both external and internal inputs into production fits what Tello et al. (2016) consider as the farmer's standpoint. The farmer's standpoint implies that in performing agricultural activities a farmer would tend to minimise, if possible, their labour by choosing the most efficient combination of internal and external inputs in terms of labour productivity. While in traditional organic farming this was limited (external industrial inputs did not exist so that the inputs to production consisted mainly of the recirculation of biomass—seeds, feed for animal traction, manure for fertilisation etc.) in the present, when one can simply turn the tractor's engine on and plough the land, buy highly productive seeds, and use biocides and synthetic fertilisers, the possibilities to minimise the labour effort are significantly increased. The result, from an energy efficiency perspective, is that the FEROI indicator has often dropped while labour efficiency has increased.

The case at hand is interesting because the perspective of minimising the labour effort is not adopted. Quite the contrary from the farmer's standpoint, the farming family is not interested in minimising the labour effort if it implies adoption of agricultural practices that are not coherent within their values—notably, the use of biocides or synthetic fertilisers.

From a broader agroecological perspective, both Internal and External Inputs play an important role: the presence of machinery that is easily "fed" with fossil fuels minimises the effort to maintain work animals fed with agricultural re-uses. On the other hand, fertilisation is still organic at the current farm—which implies a higher labour effort in recycling biomass (such as animal manure and re-ploughed

[21]The FEROI is calculated as the ratio between Final Product (FP) (i.e. External Output of agricultural products), and the sum of External Inputs (EI) and Biomass Reused (BR): FEROI = FP/(EI + BR). For further explanation of the importance and implications of this variable see Tello et al. (2016) and Chap. 2 in this volume.

harvest residues). From a labour perspective, it might be easier to maintain fertility with the simple input of synthetic fertilisers (in which case labour would be spared), however, the overall energy efficiency would drop. For this we can also claim that the farm stands half-way towards environmental sustainability: on the one hand it does use machinery implying certain External Inputs and labour savings, on the other, the potential for saving labour is not fully exploited as significant amounts of biomass are still reused, implying significant labour inputs along with the use of machinery, and a better balance between External and Internal Inputs, as mentioned earlier.

Second, although the farmer's recycling effort is less than in the past, the energy return per unit of labour is still lower than in cases of industrial agriculture where higher use of industrial inputs allows a larger output per unit of labour.[22] The combination of points 1 and 2 shows that, from the local food system perspective, agriculture in the present can be made more sustainable by the substitution of industrial inputs—of non-local origin—with a higher recirculation of biomass, crop rotation, natural (and locally sourced) methods of pest control etc.

Third, from the economic perspective, we can claim the current farm to be *market*-oriented (with 97.7% of its energy-equivalent production being sold in the market), however, it is not *profit*-oriented, i.e. its management decisions are not lead by the logic of maximising profits; for example, the way the case study farm participates in the CSA scheme can serve as an iconic example of this. In theory, the CSA should be beneficial for the farmer in several ways: the pre-payment of the farm produce at the beginning of the season, the fact that the CSA part of the produce is not (even partially) wasted, but distributed to the consumers at times and in amounts that are determined by the actual situation at the farm, not by the consumers' free choice, and also the price that should be fair both to the producer and to the consumers (Urgenci 2016). While the first two conditions apply in our case, the fairness of the price is questionable: first, the working time of the farmers is not fully accounted for, as shown in Sect. 7.4.3; and second, in monetary terms, the revenue from the vegetables sold via the CSA scheme is much lower that the revenue from any other distribution channel used by the farm (see also Table 7.2): whereas the real revenue from the CSA payments in 2012 was 4770 Euro, the potential revenue was more than double that, about 10,700 Euros if the same amount of vegetables, fruits and herbs had been distributed via other channels used by the farm, with an average price received from these channels.[23]

Also, in a broader sense, a significant proportion of the vegetable production within the farm land uses (see Sect. 7.4.1) is not driven by profit logic, actually

---

[22]See Fraňková and Cattaneo (2017) for a detailed comparison of the energy balances of the present case study farm and of the past village.

[23]It should be noted though that 2012 was the first CSA season for the case study farm, hence the conditions for both the producers and consumers were a trial. In the following season, the price went up a bit and the amount of delivered vegetables went down, partly because of lower overall vegetable production due to climatic conditions (a very dry season), and partly as a result of optimisation by the farmers.

quite the opposite. As already noted (Sects. 7.4.1 and 7.4.2) this production is very labour intensive, and thus brings significant costs (waged labour necessary to complement the family labour in the vegetable production constitute almost 30% of the overall expenditures of the farm, see Table 7.3). Although making much sense when considering the local food argumentation, this focus does not turn out profitable, and is actually quite exceptional in the context of Czech organic farming: as already noted (Sect. 7.4.1), both in terms of land use and produce, most of Czech organic farming is based on permanent grasslands and production of hay, driven by minimising labour and land management costs, and maximising subsidies.

Fourth, the farm is neither focused on maximising profit (in terms of financial planning towards maximising return on capital), nor interested in maximising growth. Instead, a sense of being enough (Dietz and O'Neil 2013) is perceived. Rather than specialising in a few products (be they crops or animal products) in which revenue per unit of land would be maximised, the farm is still well integrated between livestock and agricultural activities: animals are fed with crop residues, pasture and feed oriented cropland, and cropland is fertilised with animal manure so that this integration of activities contributes to a better closing of material cycles. Or else, instead of looking for more markets (be they restaurants or consumer cooperatives), cultivating more land or hiring more labourers, the farm size and land use composition has been kept similar from 1993, when the family got their fields back in restitution, until now (2017).[24]

From these and other examples,[25] we can claim that the farm management decisions are kind of "intuitive", not driven by one single rationale. Although it is not profit, a certain motivation to generate income is present, of course, as the farming family has to make their living. Also, by practising organic agricultureOrganic agriculture,[26] a strong environmental focus is present; however, at the same time some of the activities can be interpreted as energy inefficient (e.g. delivering small amounts of produce to individuals to meet their specific needs and wishes), hence neither is a strictly environmental-efficiency rationale applied. Also, the family farming tradition does play a strong role in the year-to-year decision of

---

[24]Although biocides and industrial fertilisers had not been applied in significant amounts since 1993, the organic certification was only gained in 2006.

[25]E.g. the farm lady was growing one specific variety of cabbage for a single customer that wanted it, and the farm lady did not want to "disappoint" her; another example being the limited consumption of own produce (see Sect. 7.4.3 for details): although generally motivated by getting the most possible income from the organic produce, and purchasing much cheaper food from conventional sources, for specific favourite articles of food this does not apply, own consumption thus being a combination of own lower-quality vegetables (but mostly in terms of visual or mechanical imperfection), and own "luxurious" products such as honey, pig fat or poppy seeds of excellent quality that could be marketed with a high-added value, but are consumed by the farming family in significant amounts as favourites for which the food quality cannot be compromised.

[26]And also by other activities related to ecosystem and landscape management, e.g. planting an orchard to protect soil from erosion, besides its production role.

whether to continue farming even though barely covering the operational costs and bringing neither profit nor sufficient income for bigger long-term investments. From a personal perspective, a strong sense of autonomy adds to the mixture of motivations as explicitly reflected by the farm lady—the freedom to decide for herself, to spend her time mostly outside in the field, to choose her own way of doing things.

Thus, we can argue that not only the consumers (as discussed extensively in the literature), but also the farmers apply a plurality of values and ethical concerns, the "omnistandards" as quoted earlier, although clearly these are not necessarily identical with the consumers' ones.[27] Also, we can interpret these omnistandards as an example of what could be called a "genuine multifunctionality" in agriculture, one that does not follow any form of single instrumental logic, but one that can hopefully contribute to cultivating more sustainable forms of agriculture in living practice. This is also very much in line with the notion of multi-purpose character and integrated use of natural resources as discussed in the context of "traditional peasant knowledge" literature (see Chap. 2 in this volume for further debate). In summary, we claim that the business model adopted in this case opens the space to categorise local food systems as an enterprise embedded in the wider context of social and nature-cultural relationships as Polanyi has previously observed (1944).

Our final remark regards the limitations of this study. Surely, the issue of the sustainability of agri-food systems is very complex and includes more aspects and indicators than we covered in our analysis; to be more comprehensive, nutrient balance, farm-associated biodiversity, and also the various social aspects of LFS should be analysed. By the selected aspects of the case study, we wanted to stress the importance of biophysical analysis in the context of food localisation argumentation, and the importance of addressing the issue of finance in the context of socio-metabolic studies. Our case study wanted to provide methodological tools, and a concrete example of such an analysis. Also, as already noted, our case is quite specific, and cannot be—even in the Czech context—simply generalised. The next steps for further work are thus to include more case studies to build up certain farm typologies (data for another two farms have already been collected to begin this), and to scale up the analysis to build scenarios on higher (regional and national) levels to see to what extent various LFS models are applicable on a larger scale, and with what implications for fulfilling all the needs in current agriculture, only beginning with the food, feed, fuel, fibre *and* finance.

[27]In many aspects, the farmers′ approach is close to ideals of food sovereignty as expressed e.g. in the Nyéléni declaration (Nyéléni 2007). For argumentation on the importance of including multiple criteria in discussing sustainability of LFS, but also multiple viewpoints (of consumers, farmers, policy makers, academics etc.) see also Chap. 3 in this volume.

## 7.6 Conclusions

This chapter provides an in-depth analysis of a potentially (more) sustainable local food system based in the Czech Republic, a small-scale organic family farm, involved in the Community Supported Agriculture scheme, with a traditional integrated farm structure combining cropland, grassland, and woodland, and highly localised mode of both production, consumption and distribution. In many biophysical aspects, such as composition of land use, integration of animal, crop and forest production, and significance of the internal biomass re-use, the farm is close to the traditional organic agriculture as practiced in the area before the industrial revolution (c. 1840). However, by the current market orientation of the farm it is far from the traditional, primarily subsistence forms of agriculture. Within the current industrialised system, it is hard to avoid a certain level of dependence on fossil fuel external inputs, both for financial and biophysical reasons (the so called land cost of sustainability, see Guzmán & González de Molina 2009) but we observe and discuss that there are several factors that determine why the current farm is not entirely adopting the apparent benefits of industrial agriculture: there are choices made at the management level that determine a higher labour effort and forego productivity, income and business opportunities. This apparent lack of economic rationality is explained by a system of production and distribution which is embedded in the context of the farming agents which give value to the local and the social dimension of the agri-food system under analysis. We conclude that the characteristics of the case study farm constitute a hybrid of socio-metabolic characteristics of the past and socio-economic ones of the present but that, nonetheless, they are far from being the same.

## References

Altieri, M.A. (2009). Agroecology, small farms, and food sovereignty. *Monthly Review, 61* (3).

Aguilera, E., Guzmán, G. I., Infante-Amate, J., Soto, D., García-Ruiz, R., Herrera, A., et al. (2015). Embodied energy in agricultural inputs. Incorporating a historical perspective. DT-SEHA n.1507. http://repositori.uji.es/xmlui/bitstream/handle/10234/141278/DT-SEHA%201507.pdf?sequence=1. Accessed May 20, 2016.

Bennet, C. F. (2009). *Reevaluating the community-building potential of community supported agriculture (CSA): A case study of the Washington State University CSA Program.* Master thesis of science in environmental science. Washington State University: School of Earth and Environmental Science.

Beranová, M., & Kubačák, A. (2010). *Dějiny zemědělství v Čechách, na Moravě a ve Slezsku.* Praha: Libri.

Bičík, I., et al. (2015). *Land use changes in the Czech Republic 1845–2010.* Cham: Springer International Publishing AG.

Bouwman, L., Goldewijk, K. K., Van Der Hoek, K. W., Beusen, A. H. W., Van Vuuren, D. P., Willems, J., et al. (2013). Exploring global changes in nitrogen and phosphorus cycles in agriculture induced by livestock production over the 1900–2050 period. *PNAS, 110*(52), 20882–20887. doi:10.1073/pnas.1012878108.

Cordell, D., & White, S. (2014). Life's bottleneck: Sustaining the world's phosphorus for a food secure future. *Annual Review of Environment and Resources, 39,* 161–188. doi:10.1146/annurev-environ-010213-113300.

Cramer, W., et al. (2017). Biodiversity and food security: from trade-offs to synergies. *Regional Environmental Change, 17,* 1257–1259. doi:10.1007/s10113-017-1147-z.

CSO. (2015). Statistical Yearbook of the Czech Republic 2015. Czech Statistical Office. https://www.czso.cz/csu/czso/statistical-yearbook-of-the-czech-republic-2015. Accessed 26 April 2016.

Cunfer, G., & Krausmann, F. (2009). Sustaining soil fertility: Agricultural practice in the old and new worlds. *Global Environment, 2*(4), 8–47. doi:10.3197/ge.2009.020402.

ČÚZK. (2016). Souhrnné přehledy o půdním fondu z údajů katastru nemovitostí ČR [Summary data on land use from the Land Register of the Czech Republic, data from 31.12.2015]. Český úřad zeměměřičský a katastrální, Praha. http://www.cuzk.cz/Periodika-a-publikace/Statisticke-udaje/Souhrne-prehledy-pudniho-fondu/Rocenka_pudniho_fondu_2016.aspx. Accessed 4 October 2017.

Davis, S. J., Peters, G. P., & Caldeira, K. (2011). The supply chain of $CO_2$ emissions. *PNAS, 108* (45), 18554–18559. doi:10.1073/pnas.1107409108.

Demeter. (2016). Production standards: For the use of Demeter, biodynamic and related trademarks. Demeter-International e.V. http://www.demeter.net/sites/default/files/di_production_stds_demeter_biodynamic_16-e.pdf.

Desai, P., & Riddlestone, S. (2002). Bioregional solutions for living on one planet. Schumacher Briefing No. 8, Bristol: Schumacher Society. Dartington, Devon: Green Books.

Dietz, R., & O'Neil, D. (2013). *Enough is enough. Building a sustainable economy in a world of finite resources.* Routledge, London.

Douthwaite, R. (1996). *Short circuit: Strengthening local economies for security in an unstable world.* Dublin: Lilliput Press. doi:10.1604/9781874675600.

Edelman, M., Weis, T., Baviskar, A., Borras, S. M., Jr., Holt-Giménez, E., Kandiyoti, D., et al. (2014). Introduction: Critical perspectives on food sovereignty. *Journal of Peasant Studies, 41,* 911–931. doi:10.1080/03066150.2014.963568.

Erb, K. H., Lauk, Ch., Kastner, T., Mayer, A., Theurl, M. C., & Haberl, H. (2016). Exploring the biophysical option space for feeding the world without deforestation. *Nature Communications, 7,* 11382. doi:10.1038/ncomms11382.

Fedoroff, N. V., et al. (2010). Radically rethinking agriculture for the 21st century. *Science, 327,* 833–834. doi:10.1126/science.1186834.

Feenstra, G. W. (1997). Local food systems and sustainable communities. *American Journal of Alternative Agriculture, 12*(1), 28–36. doi:10.1017/S0889189300007165.

Fernando, A. L., Duarte, M. P., Almeida, J., Boléo, S., & Mendes, B. (2010). Environmental impact assessment of energy crops cultivation in Europe. *Biofuels, Bioproducts and Biorefining, 4*(6), 594–604.

Fischer-Kowalski, M., & Haberl, H. (Eds.). (2007) *Socioecological transitions and global change: Trajectories of social metabolism and land use.* Cheltenham: Edward Elgar Publishing. doi:10.4337/9781847209436.

Fischer-Kowalski, M., Singh, S. J., Lauk, C., Remesch, A., Ringhofer, L., & Grünbühel, C. M. (2011). Sociometabolic transitions in subsistence communities: Boserup revisited in four comparative case studies. *Human Ecology Review, 18*(2), 147–158.

Foley, J. A., Ramankutty, N., Brauman, K. A., Cassidy, E. S., Gerber, J. S., Johnston, M., et al. (2011). Solutions for a cultivated planet. *Nature, 478,* 337–342. doi:10.1038/nature10452.

Fraňková, E., & Cattaneo, C. (2017). Organic farming in the past and today: Sociometabolic perspective on a Central European case study. *Regional Environmental Change.* doi:10.1007/s10113-016-1099-8.

Fraňková, E., & Johanisová, N. (2012). Economic localization revisited. *Environmental Policy and Governance, 22*(5), 307–321. doi:10.1002/eet.1593. John Wiley & Sons and ERP Environment.

Galt, R. E. (2013). Placing Food Systems in first world political ecology: A review and research agenda. *Geography Compass, 7*(9), 637–658. doi:10.1111/gec3.12070.

Gerber, P., Vellinga, T., Opio, C., & Steinfeld, H. (2011). Productivity gains and greenhouse gas emissions intensity in dairy systems. *Livestock Science, 139,* 100–108. doi:10.1016/j.livsci.2011.03.012.

Giampietro, M. (2003). *Multi-scale integrated analysis of agroecosystems.* Boca Raton: CRC Press.

Giampietro, M., & Mayumi, K. (2015). *The biofuels delusion: The fallacy of large-scale agro-biofuel production.* London: Routledge.

Giampietro, M., Mayumi, K., & Sorman, A. H. (2013). *Energy analysis for a sustainable future: Multi-scale integrated analysis of societal and ecosystem metabolism.* London: Routledge.

Gingrich, S., Haidvogl, G., Krausmann, F., Preis, S., & Garcia-Ruiz, R. (2015). Providing food while sustaining soil fertility in two pre-industrial Alpine agroecosystems. *Hum Ecol, 43,* 395–410. doi:10.1007/s10745-015-9754-0.

Gingrich, S., Theurl, M. C., Erb, K., & Krausmann, F. (2017). Regional specialization and market integration: Agroecosystem energy transitions in Upper Austria. *Regional Environmental Change, 46,* 1–14. doi:10.1007/s10113-017-1145-1.

Gizicki-Neundlinger and Guldner. (2017). Surplus, scarcity and soil fertility in pre-industrial austrian agriculture—The sustainability costs of inequality. *Sustainability, 9,* 265. doi:10.3390/su9020265.

Gliessman, S. R. (2015). *Agroecology. The ecology of sustainable food systems* (3rd ed.). Boca Raton: CRC Press.

Gomiero, T., Giampietro, M., & Mayumi, K. (2006). Facing complexity on agro-ecosystems: A new approach to farming system analysis. *International Journal of Agricultural Resources, Governance and Ecology, 5,* 116–144. doi:10.1504/IJARGE.2006.009160.

Gomiero, T., Pimentel, D., & Paoletti, M. G. (2011a). Is there a need for a more sustainable agriculture? *Critical Reviews in Plant Sciences, 30,* 6–23. doi:10.1080/07352689.2011.553515.

Gomiero, T., Pimentel, D., & Paoletti, M. G. (2011b). Environmental impact of different agricultural management practices: Conventional vs. Organic agriculture. *Critical Reviews in Plant Sciences, 30,* 95–124. doi:10.1080/07352689.2011.554355.

González De Molina, M., & Toledo, V.M. (2014). *The social metabolism: A socio-ecological theory of historical change.* Switzerland: Springer International Publishing. doi:10.1007/978-3-319-06358-4.

Grešlová-Kušková, P. (2013). A case study of the Czech agriculture since 1918 in a socio-metabolic perspective. From land reform through nationalisation to privatisation. *Land Use Policy, 30,* 592–603. doi:10.1016/j.landusepol.2012.05.009.

Grešlová, P., Gingrich, S., Krausmann, F., Chromý, P., & Jančák, V. (2015). Social metabolism of Czech agriculture in the period 1830–2010. *AUC GEOGRAPHICA, 50,* 23–35. doi:10.14712/23361980.2015.84.

Guzmán, G. I., & González de Molina, M. (2009). Preindustrial agriculture versus organic agriculture: The land cost of sustainability. *Land Use Policy, 26,* 502–510. doi:10.1016/j.landusepol.2008.07.004.

Guzmán, G. I., & González de Molina, M. (2016). *Energy in agroecosystems: A tool for assessing sustainability.* Boca Raton: CRC Press.

Guzmán, G. I., Aguilera, E., Soto, D., Cid, A., Infante, J., García Ruiz, R., et al. (2014). Methodology and conversion factors to estimate the net primary productivity of historical and contemporary 4 agroecosystems. DT-SEHA n.1407. http://repositori.uji.es/xmlui/bitstream/handle/10234/91670/DT-SEHA%201407.pdf?sequence=3. Accessed 26 May 2016.

Haas, W., & Krausmann, F. (2015). Transition-related changes in the metabolic profile of an Austrian rural village. *Working paper Social Ecology, 153* (Vienna: IFF Social Ecology).

Haberl, H., Fischer-Kowalski, M., Krausmann, F., & Winiwarter, V. (Eds.). (2016). *Social ecology: Society-nature relations across time and space.* Springer. doi:10.1007/978-3-319-33326-7.

Haberl, H., & Krausmann, F. (2007). The local base of the historical agrarian-industrial transition, and the interaction between scales. In M. Fischer-Kowalski & H. Haberl (Eds.), *Socio-ecological transitions and global change: Trajectories of social metabolism and land use* (pp. 116–138). Cheltenham, UK, Northampton, USA: Edward Elgar. doi:10.4337/9781847209436.00012.

Hathaway, M. D. (2016). Agroecology and permaculture: addressing key ecological problems by rethinking and redesigning agricultural systems. *Journal of Environmental Studies and Sciences, 6*(2), 239–250. doi:10.1007/s13412-015-0254-8.

Herrero, M., Thornton, P. K., Notenbaert, A. M., Wood, S., Msangi, S., Freeman, H. A., et al. (2010). Smart investments in sustainable food production: revisiting mixed crop-livestock systems. *Science, 327*(5967), 822–825. doi:10.1126/science.1183725.

Hinrichs, C. C. (2003). The practice and politics of food system localization. *Journal of Rural Studies, 19*(1), 33–45.

Hitschmann, H. H. (1891). *Vademecum für den Landwirth*, M. Perles.

Holt-Giménez, E., & Altieri, M. A. (2013). Agroecology, food sovereignty, and the new Green revolution. *Agroecology and Sustainable Food Systems, 37,* 90–102. doi:10.1080/10440046.2012.716388.

Krausmann, F. (2004). Milk, manure, and muscle power. Livestock and the transformation of preindustrial agriculture in Central Europe. *Human Ecology, 32,* 735–772. doi:10.1007/s10745-004-6834-y.

Krausmann, F., Fischer-Kowalski, M., Schandl, H., & Eisenmenger, N. (2008). The global sociometabolic transition: Past and present metabolic profiles and their future trajectories. *Journal of Industrial Ecology, 12,* 637–656. doi:10.1111/j.1530-9290.2008.00065.x.

Krausmann, F., Heberl, H., Schulz, N. B., Erb, K.-H., Darge, E., & Gaube, V. (2003). Land-use change and socio-economic metabolism in Austria—Part I: Driving forces of land-use change: 1950–1995. *Land Use Policy, 20,* 1–20. doi:10.1016/S0264-8377(02)00048-0.

Krčmářová, J., & Arnold, M. (2016). Traditional Agriculture as Cultural Heritage. Forgotten Agroforestry Practices Recorded in Textual Part of Nineteenth Century Tax Records. In M. Agnoletti, F. Emanueli (Eds.), *Biocultural Diversity in Europe*, Environmental History Series (pp. 211–231). Switzerland: Springer International Publishing.

Kušková, P., Gingrich, S., & Krausmann, F. (2008). Long term changes in social metabolism and land use in Czechoslovakia, 1830–2000: An energy transition under changing political regimes. *Ecological Economics, 68,* 394–407. doi:10.1016/j.ecolecon.2008.04.006.

La Trobe, H. L., & Acott, T. G. (2000). Localising the global food system. *International Journal of Sustainable Development and World Ecology, 7*(4), 309–320.

Lang, T. (2010). From value-for-money to values-for-money: Ethical food and policy in Europe. *Environment and Planning A, 42*(8), 1814–1832. doi:10.1068/a4258.

Lokoč, R., & Ulčák, Z. (2009). *Are the present agricultural policy instruments contradictory to their goals? The case of the Czech countryside* (Vol. 4, No. 1). Review on Agriculture and Rural Development, Hódmezövásárhely: University of Szeged.

Lynch, D. H., MacRae, R., & Martin, R. C. (2011). The carbon and global warming potential impacts of organic farming: Does it have a significant role in an energy constrained world? *Sustainability, 3*(2), 322–362. doi:10.3390/su3020322.

Martinez, S., Hand, M. S., Da Pra, M., Pollack, S., Ralston, K., Smith, T. et al. (2010). *Local food systems: Concepts, impacts, and issues*. ERR-97, U.S. Department of Agriculture, Economic Research Service. http://www.ers.usda.gov/Publications/ERR97/ERR97.pdf. Accessed 12 June 2012.

Mayer, A., Schaffartzik, A., Haas, W., & Sepulveda, A.R. (2015). Patterns of global biomass trade and the implications for food sovereignty and socio-environmental conflict. *EJOLT Report,* No. 20. doi:10.13140/2.1.1442.5128.

McDonald, S. (2005). Studying actions in context: a qualitative shadowing method for organizational research. *Qualitative research, 5*(4), 455–473.

McMichael, P. (2009). A food regime genealogy. *Journal of Peasant Studies, 36*(1), 139–170. doi:10.1080/03066150902820354.

MoA. (2013). Organic Farming in the Czech Republic. Yearbook 2012. [in Czech with English summary] Ministry of Agriculture, Prague. http://eagri.cz/public/web/file/533348/rocenka_EZ_2013_web.pdf. Accessed 4 October 2017.

MoA. (2015). Organic Farming in the Czech Republic. Yearbook 2014. [in Czech with English summary] Ministry of Agriculture, Prague. http://eagri.cz/public/web/file/533356/Roc_enka_EZ_2015_www_komplet.pdf. Accessed 4 October 2017.

MoE. (2005). Report on the Environment of the Czech Republic 2005. Ministry of Environment, Prague. https://www.mzp.cz/en/state_of_the_environment_reports_documents. Accessed 4 October 2017.

MoE. (2014). Report on the Environment of the Czech Republic 2014. Ministry of Environment, Prague. https://www.mzp.cz/C125750E003B698B/en/state_of_the_environment_reports_documents/$FILE/SOPSZP-environment_report-20160502.pdf. Accessed 4 October 2017.

Mouysset, L. (2017). Reconciling agriculture and biodiversity in European public policies: a bio-economic perspective. *Regional Environmental Change, 17,* 1421–1428. doi:10.1007/s10113-016-1023-2.

MPAB. (2016). *Zhorz Holuby. Catastral Schätzungs Operat*, signature 715, filing ("Karton") 280, file ("značka") D8, fund "Stable cadastre—Schätzungoperaten". [Archive material] Moravian Provincial Archives in Brno, Czech Republic.

NEF. (2002). *The money trail. Measuring your impact on the local economy using LM3*. London: New Economics Foundation.

Norberg-Hodge, H., Merrifield, T., & Gorelick, S. (2002). *Bringing the food economy home: Local alternatives to global agribusiness*. London: ZED Books.

Nyéléni, (2007). Declaration of Nyéléni. *Declaration of the Forum for Food Sovereignty*. Available at https://nyeleni.org/IMG/pdf/DeclNyeleni-en.pdf Accessed June 7, 2017.

Pacini, C., Wossink, A., Giesen, G., Vazzana, C., & Huirne, R. (2003). Evaluation of sustainability of organic, integrated and conventional farming systems: A farm and field-scale analysis. *Agriculture, Ecosystems & Environment, 95,* 273–288. doi:10.1016/S0167-8809(02)00091-9.

Pelletier, N., Audsley, E., Brodt, S., Garnett, T., Henriksson, P., Kendall, A., et al. (2011). Energy intensity of agriculture and food systems. *Annual Review of Environment and Resources, 36,* 223–246. doi:10.1146/annurev-environ-081710-161014.

Picková, A., & Vilhelm, V. (2009). Aspekty energetické náročnosti v zemědělství. *Ekonomika a management, 4,* 239–254.

Pimentel, D., & Pimentel, M. (2008). *Food, energy, and society* (3d ed.). Boca Raton: CRC Press.

Pimentel, D., Hepperly, P., Hanson, J., Douds, D., & Seidel, R. (2005). Environmental energetic and economic comparisons of organic and conventional farming systems. *BioScience, 55*(7), 573–582.

Pirog, R. S., Timothy, V. P., Kamyar, E., & Ellen, C. (2001). Food, Fuel, and Freeways: An Iowa perspective on how far food travels, fuel usage, and greenhouse gas emissions. Leopold Center Pubs and Papers 3.http://lib.dr.iastate.edu/cgi/viewcontent.cgi?article=1002&context=leopold_pubspapers. Accessed 4 October 2017.

Renting, H., Rossing, W. A. H., Groot, J. C. J., Van der Ploeg, J. D., Laurent, C., Perraud, D., et al. (2009). Exploring multifunctional agriculture. A review of conceptual approaches and prospects for an integrative transitional Framework. *Journal of Environmental Management, 90,* 112–123.

Rossing, W. A. H., Zander, P., Josien, E., Groot, J. C. J., Meyer, B. C., & Knierim, A. (2007). Integrative modelling approaches for analysis of impact of multifunctional agriculture: A review for France, Germany and The Netherlands. *Agriculture, Ecosystems & Environment, 120,* 41–57. doi:10.1016/j.agee.2006.05.031.

Ščasný, M., Kovanda, J., & Hák, T. (2003). Material flow accounts, balances and derived indicators for the Czech Republic during the 1990s: Results and recommendations for methodological improvements. *Ecological Economics, 45*(1), 41–57.

Schaschl, E. (2007). Rekonstruktion der Arbeitszeit in der Landwirtschaft im 19. Jahrhundert am Beispiel von Theyern in Niederösterreich. *Social Ecology Working Paper 96* (pp. 1–174).

Vienna: Institute of Social Ecology, IFF—Faculty for Interdisciplinary Studies, Klagenfurt University.

Scheidel, A., Farrell, K. N., Ramos-Marin, J., Giampietro, M., & Mayumi, K. (2014). Land poverty and emerging ruralities in Cambodia: Insights from Kampot province. *Environment, Development and Sustainability, 16*(4), 823–840. doi:10.1007/s10668-014-9529-6.

Schnell, S. M. (2013). Food miles, local eating, and community supported agriculture: Putting local food in its place. *Agriculture and Human Values, 30*(4), 615–628.

Schrivastava, A., & Kothari, A. (2012). *Churning the earth: The making of global India*. New Delhi: Viking/Penguin Global.

Serrano-Tovar, T., & Giampietro, M. (2014). Multi-scale integrated analysis of rural Laos: Studying metabolic patterns of land uses across different levels and scales. *Land Use Policy, 36,* 155–170. doi:10.1016/j.landusepol.2013.08.003.

Seyfang, G. (2007). Cultivating carrots and community: Local organic food and sustainable consumption. *Environmental Values, 16,* 105–123. doi:10.3197/096327107780160346.

Sieferle, R. P. (1997). *Rückblick auf die Natur. Eine Geschichte des Menschen und seiner Umwelt*. München: Luchterhand.

Singh, S. J., Haberl, H., Chertow, M., Mirtl, M., & Schmid, M. (Eds.). (2013). *Long term socio-ecological research: Studies in society-nature interactions across spatial and temporal scales*. Netherlands: Springer. doi:10.1007/978-94-007-1177-8.

Smil, V. (2013). *Harvesting the biosphere. What we have taken from nature*. Cambridge: Massachusetts Institute of Technology. doi:10.1111/j.1728-4457.2013.00617.x.

Steinfeld, H., Gerber, P., Wassenaar, T., Castel, V., Rosales, M., & De Haan, C. (2006). *Livestock's long shadow: Environmental issues and options*. Rome: FAO.

Svobodová, K. (2010). *Stav pěstování zemědělských plodin a chovu hospodářských zvířat na jižní a jihovýchodní Moravě v polovině 19. století ve světle stabilního katastru* [The state of the growing of the agricultural plants and of the farm animals breeding on the south and south-east Moravia in the half of the 19th century in the light of the stable land-registry]. [Dissertation (in Czech)]. Department of History, Faculty of Arts, Masaryk University in Brno, Czech Republic. https://is.muni.cz/th/8274/ff_d?furl=%2Fth%2F8274%2Fff_d;so=nx;lang=en. Accessed April 18, 2016.

Tello, E., Garrabou, R., Cussó, X., Ramón Olarieta, J., & Galán, E. (2012). Fertilizing methods and nutrient balance at the end of traditional organic agriculture in the Mediterranean bioregion: Catalonia (Spain) in the 1860s. *Human Ecology, 40,* 369–383. doi:10.1007/s10745-012-9485-4.

Tello, E., Galán, E., Sacristán, V., Cunfer, G., Guzmán, G. I., González de Molina, M., et al. (2016). Opening the black box of energy throughputs in farm systems: A decomposition analysis between the energy returns to external inputs, internal biomass reuses and total inputs consumed (the VallÃ¨s County, Catalonia, c.1860 and 1999). *Ecological Economics, 121,* 160–174.

Tscharntke, T., Clough, Y., Wanger, T. C., Jackson, L., Motzke, I., Perfecto, I., et al. (2012). Global food security, biodiversity conservation and the future of agricultural intensification. *Biological Conservation, 151*(1), 53–59. doi:10.1016/j.biocon.2012.01.068.

Urgenci. (2016). Overview of Community supported agriculture in Europe. *European CSA Research Group*. http://urgenci.net/wp-content/uploads/2016/05/Overview-of-Community-Supported-Agriculture-in-Europe.pdf. Accessed June 1, 2017.

ÚZEI. (2015). Statistická šetření ekologického zemědělství. Základní statistické údaje (2014) [Statistical analysis of organic agriculture. Basic statistics (2014)]. Ústav zemědělské ekonomiky a informací, Brno. http://eagri.cz/public/web/file/533363/Statisticka_setreni_ekologickeho_zemedelstvi_2014_finalverze.pdf. Accessed 4 October 2017.

Van der Ploeg, J. D., & Roep, D. (2003). Multifunctionality and rural development: the actual situation in Europe. In G. van Huylenbroeck & G. Durand (Eds.), *Multifunctional agriculture; A new paradigm for European agriculture and rural development* (pp. 37–53). Hampshire, England: Ashgate.

Weber, C. L., & Matthews, H. S. (2008). Quantifying the global and distributional aspects of American household carbon footprint. *Ecological Economics, 66,* 379–391. doi:10.1016/j.ecolecon.2007.09.021.

Weis, T. (2010). The accelerating biophysical contradictions of industrial capitalist agriculture. *Journal of Agrarian Change, 10,* 315–341. doi:10.1111/j.1471-0366.2010.00273.x.

West, P. C., Gibbs, H. K., Monfresa, C., Wagner, J., Barford, C. C., Carpenter, S. R., et al. (2010). Trading carbon for food. *PNAS, 107*(46), 19645–19648.

## Author Biographies

**Eva Fraňková** works as an Assistant Professor at the Department of Environmental Studies, Faculty of Social Studies, Masaryk University in Brno, Czech Republic. Her long-term research interests and passions include the concept of eco-localisation, sustainable degrowth and various grass-root alternative economic practices including eco-social enterprises, local food initiatives etc. Recently she has been involved in the mapping of heterodox economic initiatives in the Czech Republic, and in research on the social metabolism of local food systems. She is also involved in several NGOs—the Association of Local Food Initiatives, the Society and Economy Trust, and NaZemi (OnEarth), a global education and Fair Trade organisation in the Czech Republic.

**Claudio Cattaneo** works as a postdoctoral researcher at the Barcelona Institute of Regional and Metropolitan Studies and as associate professor of ecological economics at the Department of Economics and Economic History at the Autonomous University of Barcelona, Catalonia, and at the School of Economics of LIUC University, Castellanza, Italy. He also teaches MSc Ecological Economics at the School of Geosciences, University of Edinburgh, Scotland. He is senior researcher at the Research and Degrowth association in Barcelona and a member of the informal Squatting Europe Kollective network. His research interests and passions include the socio-metabolic analysis of agro-ecosystems and landscape ecology, degrowth, the squatters' movement, and alternative communities and grass-root practices.

# Chapter 8
# Leapfrogging Agricultural Development: Cooperative Initiatives Among Cambodian Small Farmers to Handle Sustainability Constraints

**Arnim Scheidel, Bunchhorn Lim, Kimchhin Sok and Piseth Duk**

**Abstract** Many small farmers across Cambodia are currently facing multidimensional sustainability challenges, such as the need to produce sufficient food for home consumption and income generation, while keeping pressures on land, labour and the environment at bay. This chapter illustrates these challenges through the socio-metabolic analysis of a non-industrialized rice farming village in Kampot Province. Apart from these challenges, the chapter also describes how some villagers have adopted a series of 'low-capital' and cooperative innovations and initiatives to handle some of these issues. At the same time, they have partly bypassed more conventional pathways such as green revolution techniques and the transition to fossil LP gas fuels. The adopted initiatives include agroecological techniques such as the System of Rice Intensification (SRI), to increase yields while reducing farming inputs; a small-scale biogas system for cooking and lighting; a community bank to address villagers' financial needs; a community-operated paddy rice bank to manage transitory food shortages; and a rice mill association to increase farmers' market performance. These developments can enhance the sustainability of resource use patterns, understood to be strongly embedded in local socio-economic dynamics. Diffusion of such cooperative, knowledge-based initiatives in the small-scale farming economy therefore bears the potential to leapfrog more conventional

A. Scheidel (✉)
International Institute of Social Studies (ISS), Erasmus University Rotterdam (EUR), Kortenaerkade 12, 2518 AX The Hague, The Netherlands
e-mail: arnim.scheidel@gmail.com; scheidel@iss.nl

B. Lim
Department of Commerce, Kampong Thom Province, Cambodia
e-mail: limbunchhorn@gmail.com

K. Sok · P. Duk
Ecosystem Services and Land Use Research Centre (ECOLAND), Royal University of Agriculture (RUA), Phnom Penh, Cambodia
e-mail: sokkimchhin@gmail.com; sokkimchhin@rua.edu.kh

P. Duk
e-mail: duk.piseth@gmail.com



E. Fraňková et al. (eds.), *Socio-metabolic Perspectives on the Sustainability of Local Food Systems*, Human-Environment Interactions 7,
https://doi.org/10.1007/978-3-319-69236-4_8

agricultural development pathways. Simultaneously, they can foment the creation of local agroecological knowledge, cascading resource uses and the closing of nutrient cycles, as well as economic democratization and a fairer participation of farmers in the food trade chain. Cooperative agricultural development may thus be vital for local sustainable food systems.

**Keywords** Cooperative economy · Leapfrogging · System of rice intensification · Community finance · Biogas digester · Paddy rice bank · Cambodia

## 8.1 Introduction

Cambodia's small-scale farming sector is a central pillar of the largely rural economy. It not only represents a source of livelihoods for around 75% of the population, but is also central to food security, contributing with agricultural production to both subsistence household consumption and to the increasing demands of a growing urban population (NIS 2010). As agents that actively shape land use patterns, which are at the core of many environmental challenges (Erb 2012), small farmers play a crucial role for sustainability, not only in Cambodia, but also globally, where they account for the vast majority of farms today (Mayer et al. 2015; Fraňková et al., Chap. 1 in this volume).

Yet, nowadays many small farmers across Cambodia face multidimensional challenges in creating and maintaining sustainable rural livelihoods. Hit by a wave of land grabbing, the availability of agricultural land for small farmers and access to forests that have traditionally supplied them with important livelihood resources, has rapidly declined on the country level (Scheidel et al. 2013; Jiao et al. 2015; Scurrah and Hirsch 2015). In spite of small land entitlements, local farming systems are nevertheless required to produce sufficient food for household consumption as well as enough agricultural goods for the market for income generation. While commonly taken development paths, such as green revolution techniques or rural-urban migration to increase incomes through non-farm work, offer some solutions to the challenges of land shortage, they can also produce new problems across other dimensions. For example, an increasing use of fertilizers would enable the boosting of paddy rice yields, which are below the Southeast Asian average. However, it also comes with increasing expenditure on agricultural inputs, which many smallholder in Cambodia cannot afford (Theng et al. 2014). Moreover, increases in well-known environmental pressures on soils and water bodies follow (Tilman 1999), as well as rising greenhouse gas (GHG) emissions through fertilizer production and application (Snyder et al. 2009). Such 'simple' solutions in one dimension may thus directly produce new challenges across other sustainability dimensions (Giampietro 2003).

This chapter aims to illustrate and discuss these challenges, based on a socio-metabolic analysis of a non-industrialized rice farming village in Kampot province, Cambodia. Apart from addressing the challenges faced, we also present a series of innovative alternative developments, adopted by villagers to deal with a series of sustainability issues. Fundamental to the sustainability question is whether alternative pathways exist that would allow for leapfrogging, i.e. bypassing of the well-known environmental and social problems of conventional development pathways. Within this context, the adequate diffusion of cleaner production technologies has received much attention (Perkins 2003). In this chapter, we illustrate that not only technological, but also institutional and organizational, changes can support leapfrogging, particularly through the diffusion of knowledge- and cooperative-based resource management models and practices. Such changes may enhance sustainable resource use patterns, for example through enabling local nutrient cycling, or the creation of place-based agroecological knowledge. Furthermore, they can also facilitate democratic control over resources as well as supporting fairer participation in food trade chains (Tello and González de Molina, Chap. 2 in this volume).

After providing background information on Cambodia and the farming village presently discussed (Sects. 8.2 and 8.3), a detailed analysis of multidimensional challenges through the lens of societal metabolism follows (Sect. 8.4). The chapter then goes on (Sect. 8.5) to specifically discuss how some of the challenges are addressed through innovative alternative developments: (i) agroecological techniques (System of Rice Intensification) to increase yields while reducing water, seeds and fertilizer needs; (ii) small-scale biogas systems to make cascading use of manure and to reduce firewood demand for cooking; (iii) a community bank to address villagers' financial needs to adopt new assets; (iv) a paddy rice bank to manage transitory food shortages; and (v) a rice mill association to increase small farmers' market performance while retaining valuable by-products. We close the chapter (Sect. 8.6) by arguing that resource use patterns are strongly embedded within socio-economic dynamics. Many of the presently described alternative developments therefore bear the important potential to enhance the livelihoods and the sustainability of resource use patterns of land constrained farmers through the diffusion of low-capital, cooperative and knowledge-based agricultural models and practices.

## 8.2 The Rural Economy of Cambodia

Cambodia is largely a rural economy, with paddy rice being the most important staple food and agricultural product. Although rice production is largely dedicated for household consumption and local market supply, the country has steadily increased surplus production for export. The rice distribution system has developed quickly since 1996; the first year after the civil war that a stable rice production surplus was achieved. While Cambodia is still a small player in the world rice market, even compared to neighbouring countries like Laos, Vietnam and Thailand, production and trade has been constantly growing (ADB 2012). During 2013, Cambodia produced 9.4 million tons of paddy rice, of which it exported 361,246 tons of milled rice, with a value of around $260 million (FAO 2015). Much of the rice production is based on small-scale farming, whereas currently 70% of the 3.3 million ha of household agricultural holdings are dedicated to rice crops (NIS 2015).

Cambodia's small-farmer sector has experienced turbulent transformations during the last decades, which have imposed profound changes and challenges on small farmers. During the rule of the Khmer Rouge (1975–1979), private ownership of land was abolished and large parts of urban as well as rural populations were uprooted and forced to collectively cultivate fields in those areas most suitable for rice production, i.e. mostly in Northern and Central Cambodia. It was estimated that up to 2.8 million people lost their lives during the Pol Pot regime (Heuveline 1998). This radical agrarian collectivization came to an end under the subsequent Vietnamese occupation and the People's Republic of Kampuchea (PRK) (1979–1989). Many farmers then returned to their original villages in search of their families and farming lands. Subsequently, 'solidarity groups' (*krom samakhi*) were established by the PRK, comprising of 10–15 families, which were recognized as central units of rural development. In 1989, private property was reintroduced under the transitional State of Cambodia (1989–1993), followed by processes of land registration and titling, and the family farm as the main unit of agricultural production increasingly returned. After the 1993 transition, in which Cambodia turned into a constitutional monarchy operated as market-oriented democracy, inequality in landholdings through land concentration increased notably. The government further started to set up a concession economy in order to develop the rural sector. In particular, since the turn of the millennium, vast tracks of rural areas have been transformed due to the granting of Economic Land Concessions (ELCs) for agro-industrial development (for details see Chandler 2008; Diepart 2015).

Within the scope of this chapter, we are unable to provide a detailed account of all the implications these changes have had for the current rural sector. Yet it is relevant to point out that this recent and tumultuous history of rural Cambodia has much contributed to the current situation of lack of agricultural infrastructure and weak land governance (Diepart 2015; Grimsditch and Schoenberger 2015; Scurrah and Hirsch 2015). Moreover, Cambodia has experienced a rapidly growing rural labour force. This is partly due to the Khmer Rouge genocide, which massively diminished population cohorts of people aged above 35–40 (that is, born before the genocide). In turn, cohorts below that age have also grown rapidly due to a post-war baby boom (Diepart 2015). Consequently, rural areas are increasingly unable to absorb the growing rural labour force based on family farming; this is also because more than 2 million hectares of available land has been granted as ELCs to domestic and international agro-business developers (Licadho 2015). This has not only caused a massive land grab crisis, in which more than 700,000 people have so far been adversely affected—by such land deals, through overlapping land concessions and loss of forest livelihood resources—but also that land is becoming an increasingly scarce resource at the country level (Leuprecht 2004; Licadho 2009; CCHR 2013; Scheidel 2016). ELCs are thus posing limits to expansion of smallholder agriculture on the country level and consequently, many Cambodian small farmers, including the villagers of the case study introduced below, find themselves in need of making a living based on very small land entitlements. Intensification strategies are becoming prominent options to increase production, as well as out-migration, which increases incomes from other livelihood activities. The vulnerability of farmers, already making a precarious living, may increase further with the effects of climate change—changes in precipitation patterns, floods and droughts. Cambodia is in fact among the countries most vulnerable to climate change (Yusuf and Francisco 2009).

In this chapter, we share the challenges of land constrained small farmers, as well as the creative ways to deal with them, by reporting from a village (of which the name is kept anonymous) located in Kampot Province. In Kampot, only a few ELCs were granted, yet it is among the more densely-populated regions in Cambodia, where demographic changes also play out. Most family farms depend on non-industrialized wet season rice farming, and access to common pool forest resources is limited. The whole commune in which the village is located had only 2 ha of forest land during the time of field research (2011). Yet it is comprised of five villages and has a total population of 3954 villagers.[1] Basic infrastructure, such as roads and wired electricity, were generally not developed much in the region (NCDD 2009). While on average Cambodian farming households hold 1.6 ha of agricultural land (excluding homelot), with 47% of them having less than 1 ha, in Kampot Province average holdings are at 1.03 ha/household, with 59% below 1 ha (NIS 2015). In the studied village, almost 70% of all households own less than 1 ha, and average land holdings are down at 0.91 ha/hh (own survey data). The studied village is thus an illustrative example from which to learn how smallholders have been able to deal with little and limited access to land (Scheidel et al. 2014).

## 8.3 Concepts, Methods, Data Sources

### *8.3.1 Concepts and Methods*

The methods employed in this chapter draw from the concept of societal metabolism and the related MuSIASEM approach (Multi-scale integrated analysis of societal and ecosystem metabolism) (Giampietro and Mayumi 2000a, b; Giampietro et al. 2009). The concept of societal/social metabolism focusses generally on the appropriation, transformation and disposal of materials and energy by a given society in order to create and reproduce itself. MuSIASEM offers an accounting framework that allows structuring a quantitative analysis of how selected flows (e.g., biophysical flows, monetary flows) are produced, traded and consumed by certain structural elements of a socio-economic system, such as through land uses,

[1]The administrative units in Cambodia are structured as follows: village, commune, district, province, country.

and people and their activities. Following the conceptual framework of Georgescu-Roegen (1971), these structural elements, which need to be maintained and which provide important transformative services within the economic process at a given rate, are termed *funds*. MuSIASEM is an 'open grammar' instead of a closed and standardized accounting framework, which means that the analyzed flows and funds need to be initially defined according to the research focus (see Tello and González de Molina, Chap. 2, of this volume for a discussion of different socio-metabolic approaches).

For the analysis of rural systems, MuSIASEM applications have commonly focussed on the production and consumption of biophysical and monetary flows through the funds 'land use' and 'human activity' (Gomiero and Giampietro 2001; Giampietro 2003; Ravera et al. 2014; Serrano-Tovar and Giampietro 2014). This chapter follows this approach; however, we note that there are also other funds worth analyzing, such as machinery or livestock, which we discuss here only in general terms (see for example Chap. 5 by Padró et al. of this volume, for a detailed analysis of the role of the livestock fund). Land use is commonly subdivided into categories of non-colonized land (i.e., largely natural forests, rivers) and colonized land (agriculture and livestock land, buildings and roads). Categories of human activity, although often overlapping, can be divided into physiologically relevant activities (sleeping, eating and personal hygiene), socially relevant activities (education, leisure time, and local cooperation activities), and productive activities (household work including family care work, farm and livestock work and other productive activities). Note that various options exist for grouping activities into meaningful categories. For example, the activities that are here, and generally within MuSIASEM, grouped under the categories of Physiological Overhead (i.e., sleeping, eating and personal hygiene) and Social Overhead (i.e., education, leisure time, meeting friends and community members), have been grouped by Singh et al. (2010) under the category Person System. This needs to be taken into account when comparing different time use studies of rural systems.

While both land uses and human activity (i.e. labour time) act as crucial livelihood resources, they are also constraints as their availability is limited and they need to be maintained and reproduced (Pastore et al. 1999; Grünbühel and Schandl 2005). In addition to the common focus on land and labour dynamics within agrarian studies, MuSIASEM and other time use studies allow us to widen the perspective by focussing not only on *productive* land and labour, but also on the *non-productive* and *reproductive* uses and how they are intrinsically related to each other. In this chapter, we quantify allocation of land use and human activity across the above-mentioned categories in order to understand the role they play as both producing and consuming elements of the local food system. The scale of analysis is the village economy. With regard to the studied monetary flows, we reconstructed all income and expenditure flows across defined categories, based on a randomized household survey (see Sect. 8.3.2 Data Sources). With regard to biophysical flows, we did not aim to address all possible flows but rather those that were most important to understanding the functioning of the local food system and its direct interaction with the local environment from a rural livelihoods perspective, i.e. production of agricultural goods, consumption of farming inputs, and estimation of firewood needs for cooking. Units were left as reported (e.g., kg of paddy rice production, litres of gasoline use, etc.) as they represent meaningful units in understanding the village economy and related challenges from a farmers' perspective.

In order to address the sustainability issues of local food systems, it is crucial to look at biophysical and environmental dimensions in direct relation to socio-economic dimensions; hence to employ a nexus perspective. To arrive at such a multidimensional reading of the farming system, graphical representations commonly used in rural MuSIASEM are employed here. Figure 8.1 will show the allocation of flows and funds within the village economy, following the scheme developed by Serrano-Tovar and Giampietro (2014) and Scheidel et al. (2014). While this enables a good graphical representation of the village economy in terms of land use, time use, and related production and consumption activities, Table 8.1 further provides a series of related numeric indicators. Finally, MuSIASEM's Impredicative Loop Analysis (ILA) has been adapted for a nexus analysis to illustrate how the 'simple' conventional solution of closing the yield gap through agricultural intensification in fact turns into a multidimensional challenge. An ILA typically quantifies the interactions between several flow/fund elements that belong to different scales of the systems, hence showing the forced quantitative relations between the whole and parts of the whole (Giampietro 2003). However, this chapter employs the same type of graphical visualization in a looser way, with its main purpose being to show the forced linkages between different flow and fund elements. The visualization shown in Fig. 8.2 consists of four axes where each axis represents a different flow/fund element: total agricultural land, total paddy rice production, fertilizer inputs, and selected monetary flows. The forced relationship between these extensive values becomes visible at their interface, i.e. through the generated intensive variables, such as for example yield (kg/ha), or pressures from agro-chemicals (kg/ha). Hence, this type of graph is used here to illustrate the linkages between environmental and socio-economic dimensions.

### *8.3.2 Data Sources*

Data employed are mainly primary data, collected by the authors between March and May 2011 with the help of three enumerators. Primary data collection methods included a random household survey to collect information on demography, livelihood activities, land uses, income and expenditure, time use and local cooperation activities at the household (hh) level. The total random sample covered 86 out of 195 hh, which is a representative sample at the village level with a confidence level of 8%, assuming normal distribution of selected characteristics across the surveyed population. Cambodian Riels were converted to US dollars, based on the exchange rate of \$1 = 4100 Riels, which was a common exchange rate in the village and nearby towns at the end of 2010 and beginning of 2011. Detailed survey design and data processing methods are reported elsewhere (Scheidel 2013).

Furthermore, 19 semi-structured interviews were conducted with key informants, including both local government officials, and representatives from grassroots cooperation groups, who explained and shared data on the new alternative developments discussed in this chapter. Participant observation was crucial to understand, validate and contextualize the collected information of the survey as well as the interviews. While survey data are limited to the agricultural year 2010–2011, the new initiatives adopted by the villagers were tracked through phone interviews until May 2016. These interviews focused mainly on the questions of how they have developed since 2011.

Firewood demand for cooking was estimated for the share of households (95%) that reported to fully depend on firewood as cooking fuel. During field research, no single household used LP gas or wired electricity as a cooking source. The remaining 5% mainly used the new small-scale biogas system as cooking fuel (Sect. 8.5.2). Average firewood consumption values for rural Cambodia were kindly provided by GERES,[2] who report between 32 and 40 kg of firewood/hh/week, depending on the cooking stove design (personal communication). Based on the more conservative value of 32 kg/week, an average monthly firewood demand of 139 kg/hh was estimated. This is in line with another study from a similarly resource-constrained area in Svay Rieng province (UNDP 2008), in which households consumed around 145 kg of collected firewood per month. No data could be estimated for the use of palm fronds and animal dung as a cooking source, however, they are assumed to play a role as cooking fuel in times of fuelwood shortages. No information was available for the use of cooking fuelwood for other purposes, such as palm sugar production. Hence, estimated fuelwood demand represents a conservative approximation. Humanure was roughly approximated using average data from agricultural villages on dry matter/capita/day from Gotaas (1956). Animal manure data could not be estimated due to a lack of data regarding livestock composition.

## 8.4 The Village Economy from a Socio-metabolic Perspective

### *8.4.1 Sustainability Chances and Challenges*

We now turn to a discussion of several sustainability challenges of the small farmer village from a socio-metabolic perspective. Figure 8.1 shows the village metabolism during the agricultural year 2010/2011 in terms of their funds (land use and human activity) and the associated production and consumption of biophysical and monetary flows. Note that these values represent aggregated average values for the village economy, whereas differences exist at the household level in terms of

[2]The Group for the Environment, Renewable Energy and Solidarity (GERES) is a non-governmental organization, specialized in the promotion of sustainable and renewable energy use. It has a strong presence in Cambodia and Southeast Asia. http://gsea.regions.geres.eu/.

household demography, distribution of productive assets and incomes. For instance, the top quintile holds 33% of agricultural land, while the lowest quintile holds only 7% of all agricultural land. While the implications of such differences are relevant to understand individual household livelihood strategies (Scheidel et al. 2014; Scheidel and Farrell 2015), in this chapter we focus centrally, on the sustainability challenges of the village economy as a whole.

Broad categories of human activity are represented on the left side of Fig. 8.1. 58% of the annual time budget is spent on activities relevant for physiological well-being, such as sleeping, eating and personal hygiene. 22% of the annual time budget is spent on activities with social relevance, particularly education (4%), meeting friends or fellow villagers, and leisure (18%). 20% of time is spend on productive activities, including household chores, farm work and non-farm work. Land use types are represented on the figure's right side. According to the village chief and survey data, the total village area amounts to 195 ha, of which 170 ha (87%) is agricultural land and 21 ha is residential land (11%). The remaining 2% is comprised of river canals, meadows and roads. Around 11 ha of land from outside the village has been rented by villagers for further paddy rice production, increasing access to agricultural land to about 181 ha.[3] Total amount of livestock, relevant for food production, income generation and agricultural labour were estimated to amount to 435 cattle, 211 pigs, 14 buffalos, and 5000 hens and ducks.[4] Livestock, particularly cattle, is fed in the dry season through grazing on villagers' agricultural land. In the wet season, when open grazing spaces fall short, agricultural by-products, i.e. rice straw from previous yields, play an important role as livestock fodder, which is maintained in good quality throughout the year thanks to traditional storage techniques.

The produced, traded and consumed flows that leave and enter the village are represented at the top and the bottom of Fig. 8.1, while those that stay within the village are represented in the area termed 'village system'. No data were available on the amount of money, time and physical assistance (i.e. paddy rice stocks) that entered the village through NGOs and governmental programmes. Nevertheless, Fig. 8.1 indicates them as relevant flows of our analysis, as they have had an important influence on how the village has developed (see Sect. 8.5). The figure shows that at the village level, gross revenues and expenditure flows are roughly able to cover each other; however there are no savings at all (red arrows). Further, much of the produced food (i.e. paddy rice) is used for household consumption, whereas livestock production plays a crucial role for sale and income generation (green arrows). Also, the allocation of productive activities remains largely within the village, and only a smaller share of them leaves the village for non-farm work (blue arrows). A detailed numerical characterization through indicators derived from this socio-metabolic analysis can be found in Table 8.1.

[3]No information on land rented out to neighbouring villagers was found and according to interviews it did not play an important role.

[4]No information on livestock composition was available that would allow the calculation of LU500 units.

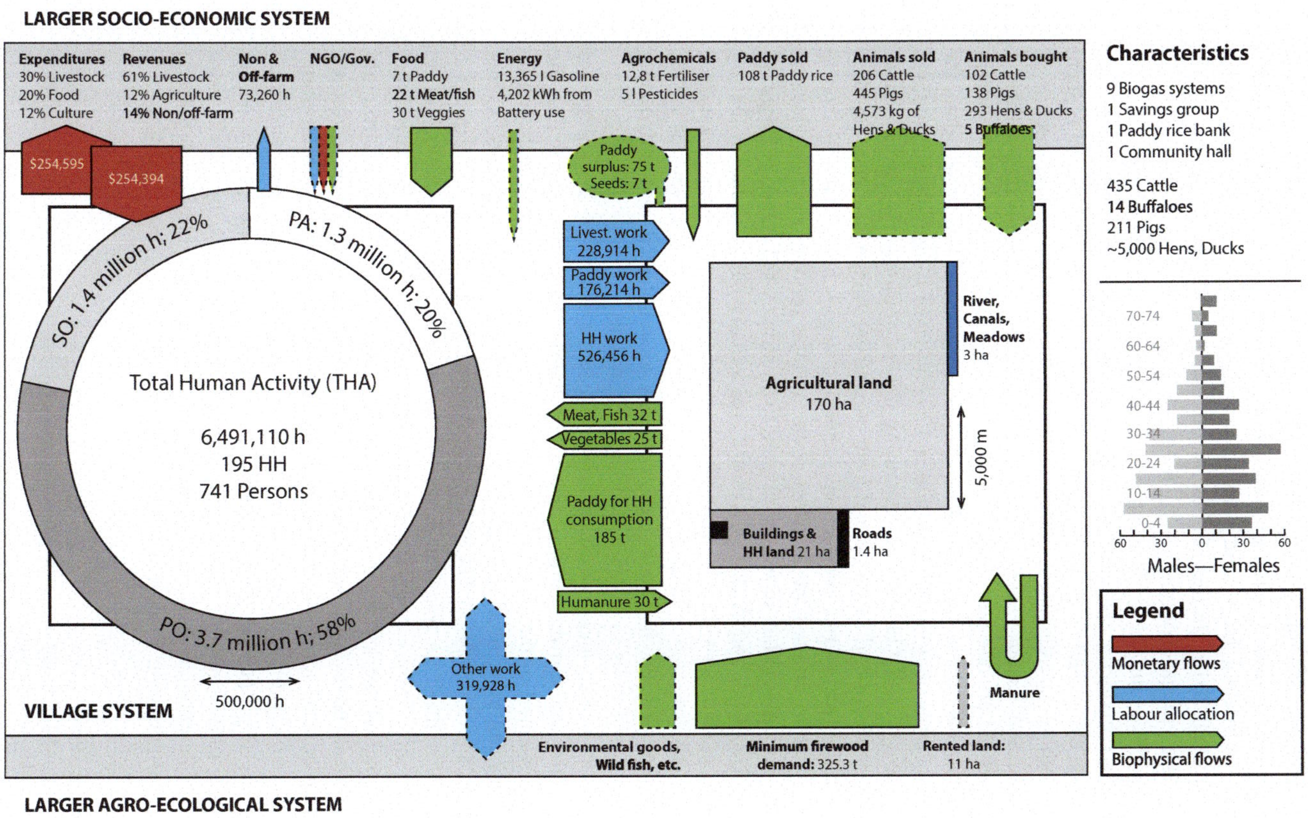


**Fig. 8.1** The societal metabolism of the rice farming village. Own elaboration based on survey data, adapted from Scheidel et al. (2014). *Note* THA = Total Human Activity; HH = Household; PO = Physiological Overhead; SO = Social Overhead; PA = Productive Activities; t = tons; l = litres. For a detailed description of how time use accounts were calculated, see Scheidel (2013). Arrows represent labour allocation, biophysical and monetary flows. They are generally scaled, but where no data have been available, they are represented in dashed lines

**Table 8.1** Multidimensional food system performance indicators, derived from the MuSIASEM analysis at the village level

| Relevance | Description | Performance indicator | Value | Unit | (First value) | (Second value) |
|---|---|---|---|---|---|---|
| Economic | Gross revenues rice farming (land use) | Gross monetary returns per land area[c] | 510.60 | $/ha | 27,158 | 53.2 |
| | Gross revenues rice farming (time use) | Gross monetary returns per hour of activity[c] | 0.52 | $/h | 27,158 | 52,715 |
| | Gross revenues livestock (land use) | Monetary returns per land area | 799.76 | $/ha | 155,115 | 194 |
| | Gross revenues livestock (time use) | Monetary returns per hour of activity | 0.68 | $/h | 155,115 | 228,944 |
| | Gross revenues off-farm work | Monetary returns per hour of activity | 0.25 | $/h | 3363 | 13,665 |
| | Gross revenues non-farm work | Monetary returns per hour activity | 0.56 | $/h | 33,126 | 59,595 |
| | Labour dependence on external labour market | Share of labour hours allocated outside the village | 4% | % (h/h) | 59,595 | 1,327,830 |
| | Financial dependence on external incomes[b] | Share of gross income from work outside village | 17% | % ($/$) | 43,567 | 254,394 |
| | Expenditure farming agro-inputs (fert./pest.) | Costs per land area[a] | 44.26 | $/ha | 8003 | 181 |
| | Expenditure farming non-family labour | Costs per land area[a] | 21.17 | $/ha | 3828 | 181 |
| | Expenditure livestock maintenance | Costs per land area[a] | 37.79 | $/ha | 6834 | 181 |
| | Expenditure livestock buying new animals | Costs per land area[a] | 392.09 | $/ha | 70,898 | 181 |

(continued)

**Table 8.1** (continued)

| Relevance | Description | Performance indicator | Value | Unit | (First value) | (Second value) |
|---|---|---|---|---|---|---|
| Agronomic | Average land holdings | Total agricultural land owned per household | 0.91 | ha/hh | 176.64 | 195 |
| | Food sovereignty (total production) | Self-sufficiency in years of consumption covered by total annual production | 1.96 | Years | 375 | 192 |
| | Food sovereignty (after market sale) | Self-sufficiency in years, after selling paddy rice | 1.39 | Years | 267 | 192 |
| | Paddy rice yields per agricultural land[a] | Paddy rice production per land area | 2076 | kg/ha | 375,000 | 180.8 |
| | Paddy rice yields per labour time invested | Paddy rice production per hour | 2.1 | kg/h | 375,000 | 179,214 |
| Social | Time use with physiological relevance | Share of sleeping, eating, personal hygiene out of THA | 58% | % (h/h) | 3,766,577 | 6,491,160 |
| | Time use for social activities | Share of education and leisure out of THA | 22% | % (h/h) | 1,396,781 | 6,491,160 |
| | Upon which formal community meetings | Share of formal community meetings out of social activities | 0.6% | % (h/h) | 7905 | 1,396,781 |
| Ecological | Fertilizer use | Amount of fertilizer per land area[a] | 70.67 | kg/ha | 12,778 | 181 |
| | Pesticide use | Amount of pesticides per land area[a] | 0.03 | l/ha | 5 | 181 |
| | Demographic pressure | Persons per ha of agricultural land | 4.37 | Persons/ha | 741 | 169.73 |
| | Land use pressure | Colonized land per total village land (%) | 98% | % (ha/ha) | 192.39 | 195.39 |
| | Fossil energy use (gasoline) per HH | Gasoline consumption | 68.54 | l/hh/year | 13,365.44 | 195 |
| | | | 21.55 | KwH/hh/year | 4202 | 195 |

(continued)

**Table 8.1** (continued)

| Relevance | Description | Performance indicator | Value | Unit | (First value) | (Second value) |
|---|---|---|---|---|---|---|
| | Electricity use per HH | KwH/hh from battery use | | | | |
| | Cooking fuel wood needs | Annual cooking fuelwood demand (low estimate) | 1669 | Kg/hh/year | 325,372 | 195 |

*Note* Dimensions of relevance are only separated for structuring purposes, yet in practice they overlap largely. [First value] and [second value] are presented to indicate the absolute numbers, on which basis the relative indicators were constructed. [a]Includes rented agricultural land; [b]includes remittances, pensions, non-farm work. [c]Gross monetary returns on rice farming were calculated by accounting only for the share of land and labour inputs required for the share of rice sold on the market (29%). *Source* own survey data

From a sustainability perspective, the village shows positive performances in some aspects as well as large challenges in other dimensions (Table 8.1). Among the positive social aspects is that villagers are 'rich in time', meaning that in comparison with other (agrarian) societies, they have a comparatively large share of time allocated to the physiological overhead, as well as a large share dedicated to social and leisure activities [compare e.g. with Grünbühel and Schandl (2005) for neighbouring Laos; and NIS (2007) for average Cambodia]. This, it can be argued, allows social well-being and resilience, as people have time for cultural and social activities. Villagers have also been able to build up a strong network of cooperation, which is largely maintained by only a small share of social activities (0.6%) dedicated to formal community meetings and activities (Scheidel and Farrell 2015). Finally, the strong allocation of productive activities within the village economy itself also enables flexible family care, as many activities can be combined (e.g. household work and provision of elderly or children). Increasing out-migration can challenge such flexible arrangements based on proximity.

From a food systems perspective, the village economy is highly localized and able to be completely self-sufficient in terms of paddy rice, meat and vegetable production. Hence food sovereignty is assured at the village level from a *production* perspective, although *distribution* issues also matter. Distribution is currently achieved through trade on local markets: while some villagers sell their vegetable and meat surplus, others buy it from the market. Finally, from an environmental perspective, positive aspects include the absence of pesticide use (on average, only 0.03 l/ha), as well as the low consumption of fossil energy. Gasoline use amounted on average to about 69 litres per household, and electricity consumption through use of car batteries was on average less than 22 kWh/hh/year. During the time of field research, no single household accessed wired electricity; hence sources were limited to batteries, which were recharged in neighbouring villages using diesel generators.

With regard to livelihood challenges, we can see that access to the main productive asset 'land' is quite low, with average agricultural holdings of 0.9 ha/hh, and also well under the low national average of 1.6 ha/hh (NIS 2015). Further, demographic pressure in 2011 was at 4.4 persons/ha, while population dynamics mirror the countrywide dynamics (see population pyramid, Fig. 8.1). The village has a young population with a growing labour force, whereas more young people are entering the economy than elderly people retiring. This drives the need for either further land for small farming, or non-farm jobs to enter other economic sectors. Yet the village has reached already its biophysical limits of agricultural expansion: 98% of all village land is already colonized land, i.e. under residential, agricultural, or infrastructure use (Table 8.1). Hence, rather than an expansion of agriculture, either intensification, or changes in livelihood strategies will be required to overcome biophysical limits of the current village economy. With average paddy rice yields at around 2.1 t/ha, there is technically space to increase yields through green revolution techniques to the national average of 2.9 tons/ha (MAFF 2014). However this also comes with trade-offs discussed later (Sect. 8.4.2).

Regarding economic challenges, Fig. 8.1 shows that the village economy is hardly able to cover its own expenditure without an annual deficit. In fact, agriculture only contributes about 12% to the gross monetary revenues. Table 8.1 shows that rice farming has on average a monetary return of 510.60$/ha and 0.52$ per labour hour; however, only 29% of produced rice is for market sale, while the remaining stays in the village for subsistence and food security purposes. Livestock production in turn accounts for 61% of gross incomes, with a gross revenue rate of around 800$/ha and 0.68$/h of time investment, but it also requires initial capital to invest in livestock that can be sold later on, when fully grown. Currently, the village economy is unable to meet expenditures solely through working in agriculture and livestock production. Non-farm work, usually involving seasonal migration to the new industries (garment sector, construction work, etc.), has a medium average return of 0.56$/h. Yet it depends on seasonal availability and, further, may provoke social challenges through out-migration, disrupting family life. At the time of field research, the external labour market absorbed only 4% of all productive activities, which, however, contributed to 17% of gross monetary revenues (including remittances and pensions).

Finally, there are also environmental challenges. In 2008, all households depended on firewood as cooking fuel (NIS 2008) and in 2011 only 10% of all households had adopted new sources by making use of biogas produced through animal and humanure. Conservative estimates indicate a total firewood demand of at least 325 tons per year (Fig. 8.1), but the lack of access to forests makes local firewood supply problematic. All households collect firewood and other cooking fuels, such as palm fronds from their own homelots and the few meadows available. Households further reported purchasing firewood from external traders or collecting it themselves from remote forests. Being constrained in local supply, it can be assumed that other fuel sources such as animal dung are also used (cf. UNDP 2008). Yet this practice further withdraws organic material and nutrients from the agricultural system and has been associated with declining soil fertility (OECD/IEA 2007). While demand for cooking fuel is generally not considered as a main driver of deforestation (ibid.), it does increase pressure on forest ecosystems and requires a relevant share of household time allocated to firewood collection. Furthermore, firewood use as primary cooking fuel is associated to detrimental health effects through exposure to indoor pollution (WHO 2002).

### *8.4.2 Conventional Solutions: New Tensions and Trade-Offs*

In relation to these sustainability challenges, a series of somewhat conventional solutions exists in terms of practices that are either commonly proposed or commonly employed to overcome challenges of the smallholder economy. These are, for example, green revolution techniques to close the yield gap and to fully exploit available agricultural land (Godfray et al. 2010). They also include rural-urban migration to seek new incomes with higher economic returns in other economic sectors (Hecht 2010; Kelly 2011), as well as the replacement of traditional biomass cooking fuels by modern energy forms such as Liquefied Petroleum Gas (LPG), stoves and cylinders (UNDP 2008). Yet such conventional development pathways produce a series of new tensions and trade-offs across other sustainability domains.

The most commonly proposed solution to enhance the agricultural enterprise is closing of the yield gap through green revolution techniques, i.e. through modernization of irrigation, fertilization and introduction of high yielding seed varieties. It is out of our scope to review the broader implications and limitations of such techniques for development and the environment here.[5] However, from a nexus perspective we briefly highlight the direct implications of increasing fertilizer use by creating new pressures across other sustainability dimensions (Fig. 8.2).

Figure 8.2 shows the paddy rice metabolism of the village economy in relation to actual and potential fertilizer use, embedded in a series of biophysical constraints, economic dynamics and decisions to be made by the farmers. Paddy rice production is limited by biophysical constraints in terms of the amount of agricultural land that can be accessed by the villagers (Fig. 8.2a). As of 2011, available agricultural land comprised 170 ha of paddy land located within the village, plus 11 ha of paddy land rented from surrounding villages. Given that currently no more village land can be colonized for agriculture, and that land has become scarce regionally and countrywide, land is assumed here as a biophysical constraint that cannot easily be overcome by colonizing new land or further 'importing' land from neighbouring areas. Hence, changing the quantity of land use is unlikely, but rather changes in the quality and management of land use can be expected in order to increase production.

However, decisions on increasing fertilizer use (Fig. 8.2c) to boost yields also depend on economic constraints, i.e. fertilizer costs, which during 2010–2011 were on average at 0.63$/kg of industrially produced fertilizer (Fig. 8.2b) (own survey data, see also Theng et al. 2014). The sale of larger shares of paddy produce during

[5]For a review and discussion see for example Pingali (2012) and Patel (2012).

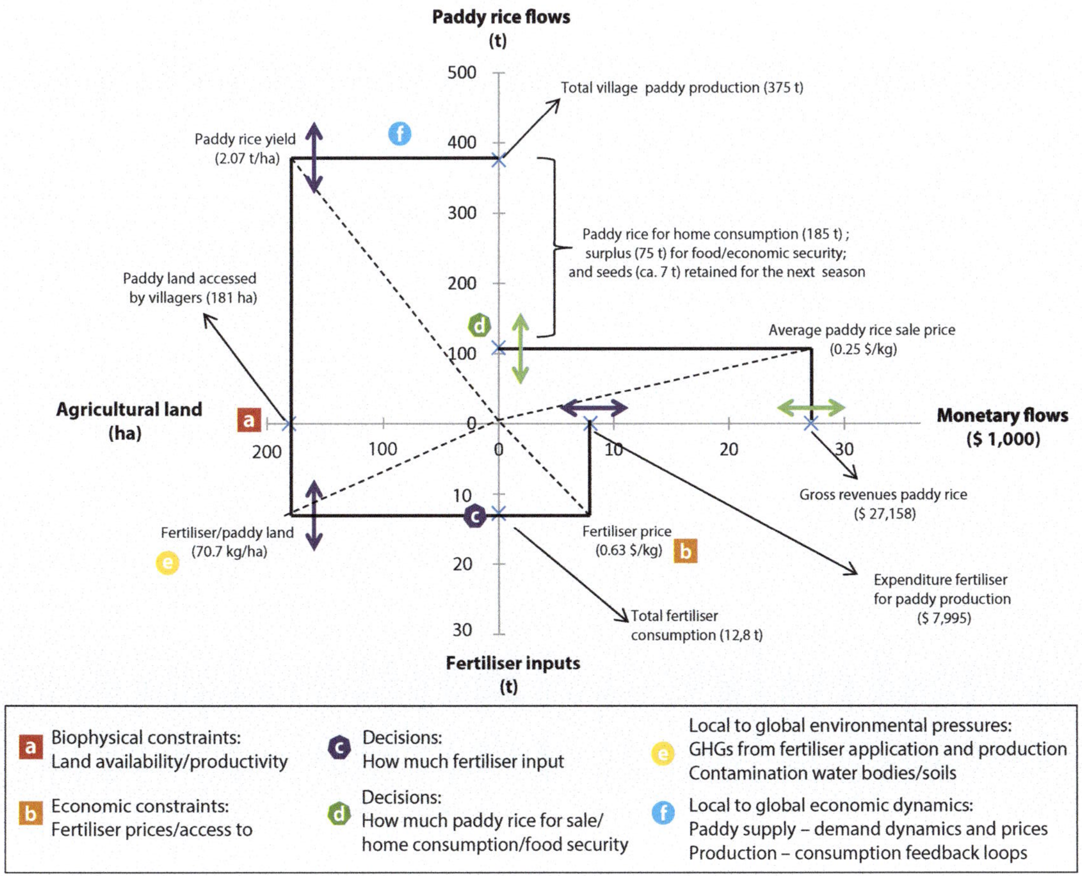


**Fig. 8.2** The paddy rice metabolism of the small-farmer village, embedded in biophysical and economic constraints, farmers' decisions and potential pressures on the environment (2010–2011 agricultural year). Right X-axis = selected revenues and expenditures (1000$); Left X-axis = Agricultural land accessed by villagers (own land plus rented land); Upper Y-axis = Paddy production and sale (tons); Lower Y-axis = Total fertilizer input (tons). Small differences are due to rounding issues. *Source* Own elaboration based on survey data. GHGs = greenhouse gases

the previous year, to increase agricultural returns in order to be able to invest in agro-inputs, would create new tensions on food security, particularly for those households who produce below the village average (Fig. 8.2d and green arrows). For many farmers, the extra costs that fertilizers entail in fact pose a barrier to further usage, particularly when they face problems of limited surplus production and monetary liquidity such as in this village. In such cases, increased returns (higher yields) on investments (additional fertilizer costs) come too late to make it a viable option. Use of credits for buying inputs is an option, but further increases the economic burden, whereas a part of the increased paddy rice production would then just be used to repay interest rates of around 2–3%/month (Theng et al. 2014 see also Sect. 8.5.3). Further, rising yields cannot be expected to increase economic revenues linearly, as a growing paddy rice supply can also be expected to lower paddy rice prices due to supply-demand dynamics (Fig. 8.2f). In fact, in the village and other areas, it is common that paddy rice prices rise before harvest, when local supply falls short, and fall directly after harvest, when supply increases.

Beyond these economic dynamics, the adoption of such apparently simple green revolution techniques is directly linked to well-known environmental issues (Fig. 8.2e). Increasing fertilizer use drives environmental pressure on water bodies and soils (Tilman 1999). It also has relevant climate 'rucksacks' associated with fertilizer production, trade and application. Snyder et al. (2009) estimate that production and transport of 1 kg of N-fertilizer is associated with a Global Warming Potential (GWP) of more than 4 kg of $CO_2$. Further, application of 100 kg for rice cropping produces on average around 1.25 kg of the highly climate damaging trace gas $N_2O$, which corresponds to a GWP of 370 kg of $CO_2$ (Leisz et al. 2007). In a scenario in which the village increases fertilizer use to 100 kg/ha, the additional amount of fertilizers consumed (ca. 5.3 tons) would imply a roughly estimated additional GWP, through fertilizer production, transport and application, of more than 40,000 kg of $CO_2$/year. To make this value palpable, this corresponds to the emissions generated through driving a passenger car in Europe for more than 300,000 km.[6]

To sum up, the implications of green revolution techniques to close the yield gap are not straightforward or linear, but are rather embedded in complex system dynamics comprising biophysical, economic and social dimensions. As illustrated in Fig. 8.2, they are far more than just a simple technological fix. The fix triggers feedback loops within production and consumption of the socio-economic system, which may create new pressures across different dimensions. Such feedback-loops might in fact just result in a vicious cycle, where the amount of production indeed increases, but also intensification is further pushed to increase. While in such a scenario, environmental pressures may increase over time, economic returns for farmers do not necessarily follow due to system changes and consumption-production

[6]Comparisons to $CO_2$ emission from car driving are solely for illustrative purposes. They are based on the current maximum allowed EU limit value of 130 g $CO_2$/km for new cars, referring to the generation of $CO_2$ during its use (not including during the production of the car). http://ec.europa.eu/clima/policies/transport/vehicles/cars/index_en.htm.

feedback loops. As Tello and González de Molina (Chap. 2 in this volume) have stated, increased land and labour productivity achieved by the Green Revolution was historically generally accompanied by a steady decline of the net-value of production retained by farmers.

Similarly, other common solutions to other challenges also come with new trade-offs. For generating sufficient incomes, livelihood diversification and migration is playing an ever-increasing role (Ellis 2000; Kelly 2011). When it occurs out of necessity, because of lacking access to land rather than as a desired change pursued by rural dwellers, then the implications for the household may be manifold. They may include acceptance of badly paid jobs, or long-term migration with significant impacts on family relations, including the issue of who will take care of kids and the elderly. The future will show, how these rural changes, currently underway in Southeast Asia, play out, and what kind of new questions they will pose with regard to the sustainability of local food systems.

With regard to environmental challenges, particularly the demand for cooking fuels as discussed above, conventional development paths suggest a change in source, by moving from biomass to more modern types of energy sources, such as LPG stoves (OECD/IEA 2007; UNDP 2008). While this resolves some problems (i.e. pressure on forest ecosystems, human health) they also create new ones through replacing genuinely renewable energy sources with fossil ones, thus increasing dependency on externally supplied energy. From a nexus perspective, such trade-offs across dimensions and scales need to be carefully considered when imagining local sustainable food systems, in order to avoid the repetition of unsustainable development pathways, through which many countries have already gone.

## 8.5 Cooperative Initiatives to Support Sustainability of Local Food Systems

Biophysical constraints such as land shortage, in combination with other drivers of change, are posing severe challenges on the small farmer economy. Escapes from this situation, based on conventional development pathways, are not straightforward. Yet there are many new developments under way, which may allow local food systems in the global South to take a different pathway from the conventional agricultural development path. While in this context, the concept of leapfrogging has received much attention (Perkins 2003), it has been narrowly focused on technological solutions. Since resource uses are strongly embedded in complex socio-economic dynamics, local institutions and new sets of practices in how the local economy is organized do also have impacts on resource use patterns. In this final section, we report on five alternative and low-capital intense developments that have been pursued by villagers. Some of these are new technologies and agricultural techniques, and others represent new small-scale cooperative economic models. In this section we will review their functioning and their immediate implications on sustainability, as well as their adoption in the village and elsewhere in Cambodia (see Table 8.2 for an overview of their key requirements and key benefits from a sustainable food systems perspective).

### *8.5.1 System of Rice Intensification: Agroecological Techniques to Enhance Yields and Reduce Inputs*

Some households have started to adopt the System of Rice Intensification (SRI), which is an agroecological rice cropping practice initially developed by poor farmers in Madagascar. It is comprised of a set of cropping techniques that help to save seeds, fertilizer and water, based on the use of traditional rice varieties (Uphoff 1999; Stoop et al. 2002). The set of practices can furthermore be adopted and combined according to farmers' possibilities. Practices include, for example, selection of the best seeds for sowing in a nursery bed; the transplanting of only good seedlings into the field; careful transplantation of single seedlings at the two-leaf stage; careful water management by keeping the soil moist; mulching the soil with organic matter; the use of compost; planting green manure; and selection of the best seeds for the next crop. Since its discovery, SRI has expanded globally and has been reported to have the potential to increase rice yields by up to 20–100%. At the same time, required seed inputs is argued to be reduced by up to 90% and water use by up to 50%.[7] SRI has received positive response in Cambodia since the turn of the millennium, when many Cambodian farmers started to adopt SRI techniques (Anthofer 2004).

SRI was initially promoted in the village through the NGO CEDAC (Cambodia Center for the Study and Development in Agriculture). Survey results suggested that during 2010–2011, around 10% of the households had used between three to eight SRI techniques on some parts of their land. In early 2016, the number of SRI farmers has grown to 44%. However, some of the more labour-intensive practices, such as transplantation of seedlings, were replaced by less labour-intensive practices.[8] While the reported cases are too few to provide significant and detailed evidence regarding differences in yields, labour inputs, and the like, the few SRI farmers that we interviewed did mention higher yields. This is in line with a countrywide study that showed a significant yield increase of 41% through SRI techniques, maintained over several years (Anthofer 2004).

---

[7]See SRI International Network and Resources Center, Cornell University: http://sri.cals.cornell.edu/.

[8]Transplantation of the seedlings is an agricultural technique through which farmers transplant seedlings from a nursery bed to the paddy field after finishing land preparation. It requires more time and effort than spreading the paddy seeds directly on the field. The latter method is mostly practiced on upland or less-watered rice fields. It is less time consuming, but associated with lower yields and higher seed inputs.

On the country level, the Ministry of Agriculture, Forestry and Fisheries (MAFF) started to support SRI by hosting the national SRI secretariat, which it has done since 2004. During 2012, at least 101,719 ha came under SRI management, corresponding to around 150,000–200,000 households (SRI-Rice 2015). The use of agroecological rice cropping techniques such as SRI is relevant for leapfrogging conventional green revolution techniques due to their reduced requirements on synthetic fertilizer, water and seeds. SRI further promotes the creation of place-based agroecological knowledge, whereas SRI farmers reported constantly trying out new methods or changing the set of techniques employed, according to their needs. However, in order to understand its sustainability potential, it is necessary to have a good understanding of farmers' reasons for adoption and non-adoption of SRI techniques (Moser and Barrett 2003; Ches and Yamaji 2016).

### 8.5.2 Small-Scale Biogas Production to Reduce Climate Gases, Firewood and Fertilizer Needs

The adoption of small-scale biogas systems had just started to spread in the village. Natural biodigesters represent a simple but innovative technology to produce biogas at the household level, based on human and animal manure. Such biogas systems consist of a small tank, constructed underground, with a capacity of 4–15 $m^3$ to process animal dung, humanure and other organic wastes through anaerobic digestion into methane gas ($CH_4$). The obtained biogas can be used for cooking and lighting. The fully digested waste can be further used as organic fertilizer; when compared to undigested manure, nutrients are generally better taken up by soils (KOICA/UNEP/CAPS 2011; NBP 2015).

During 2011, 5% of the village households reported having installed a natural biodigester thanks to a provincial support programme that provided $150 of subsidies for each installed digester. The most commonly adopted size was either 4 or 6 $m^3$, whereas construction costs amounted to $400 and $500 respectively. A 6 $m^3$ biodigester—the most commonly used in Cambodia (KOICA/UNEP/CAPS 2011)—can produce daily biogas up to 1.6–2.4 $m^3$, which corresponds to 4–6 h of biogas stove use, or 16–24 h of biogas lamp usage. Based on commonly used, low-tech wood stoves, 1 $m^3$ biogas can replace about 5 kg of firewood (NBP 2015). Based on the lower estimates of 1.6 $m^3$ daily biogas production (6 $m^3$ digester), monthly production would allow to save up to 240 kg of firewood, which covers household cooking fuel needs, while still allowing the use of biogas for lighting purposes. In early 2016, villagers reported that 15% of households had had biodigesters installed. While wired electricity has meanwhile reached the village, the adoption of small-scale biogas systems can be expected to further increase, as firewood continues to be a scarce resource, and households tend to use electricity for different purposes, such as lighting or watching TV.

On the country level, household biodigesters started to be promoted in 2006. Up to 2014, about 22,000 systems were installed across Cambodia, with 96% of the systems reported as still operating satisfactorily several years after installation. According to the NBP programme, a biodigester can save on average 50$/year of expenditure on chemical fertilizers, with 90% of all users reported to be applying the produced bio-slurry as organic fertilizer (NBP 2015). In summary, small-scale biodigesters bring benefits across multiple sustainability dimensions: reduction of firewood needs; decrease in climate aggressive methane emissions released during uncontrolled decomposition of livestock manure; enhanced applicability of manure as organic fertilizer; reduced time requirements for firewood collection, and improved health through reduction of indoor air pollution and increased sanitation. Economic benefits accompany lowered resource needs, i.e. through reducing expenditures on fertilizers and firewood. With regard to the sustainability of local food systems, biodigesters thus support a cascading resource use and closing of local nutrient cycles. They further show a strong potential for leapfrogging conventional development pathways through avoiding a transition to fossil-fuelled cooking and lighting fuels such as LP gas.

### *8.5.3 Community Banking to Acquire New Assets and Avoid Capital Outflow*

The adoption of new assets, such as a biodigester, also requires access to financial capital. In order to foster household saving as well as community access to cheaper credits, the villagers established a small-scale cooperative banking (SSCB) system, also known as community finance, savings groups, or credit unions (Evans and Ford 2003). Similar to those credit unions established by European farmers during the mid 19th century (Fairbairn 1991; Goglio and Leonardi 2010), such cooperative savings groups establish a community fund through pooling of individual savings, in order to provide credits at defined terms to members. Savings, credits and interest rates are managed under a well-defined set of institutional agreements which is established, maintained and modified by all-members meetings and by a SSCB committee, elected democratically every three years. After learning and participating in the group, members can access cheaper loans than from external micro-credit. They may serve not only to solve short-term needs for credit, but also to expand their income sources through investment in new livelihood activities, such as livestock raising or the growing of vegetables to diversify household production.

In early 2011, the SSCB group had 168 members. At the end of the 2010 banking year, the total capital fund, pooled from individual savings, amounted to about $35,000. Based on this capital fund, credits were provided to villagers that allowed them to diversify their livelihood activities through investment into additional livelihood assets, as well as to overcome transitory money shortages. The interest rate was democratically set around 1 percentage point lower per month than external micro-credit providers: during 2010 it was at 3%/month. In 2011 it was reduced to 2.5%/month, with a lowered interest rate of 2%/month for poor households.[9] While these rates still seem high, they are below global averages of micro-credit interest rates (Kneiding and Rosenberg 2008). Furthermore, the returns from interest payments stay in the village instead of flowing out to external micro-credit providers, thus supporting village capitalization (Ward and Lewis 2002). Since its establishment, the group has developed well and in early 2016, it reported 228 members and a capital of 680 million Riels (more than $166,000).

In rural Cambodia, where farmers have expressed high credit needs (ADB 2001; Ballard 2006), such a kind of savings-led microfinance has become increasingly widespread to provide farmers with access to financial services. Such groups were actively promoted through NGOs like CEDAC or Oxfam. Oxfam, for instance, started a promotion initiative in 2005 and since then has reached approximately 110,000 farmers that became members of 6000 groups, with the majority of the members being women (Oxfam 2014). Saving groups have become increasingly relevant to farmers across Cambodia and now there has been increased focus on the sustainability of these groups, particularly regarding the achievement of autonomous functioning after promoting agencies phase out (EMC 2015).

While such savings groups do not necessarily avoid the dynamics of debit and credit, which can produce or reinforce local inequalities, they also have environmental implications. As Gerber (2013) argues, credit and related interest rates may have detrimental effects on the environment as the need to repay high interest rates, in addition to the initial credit capital, further generates the need for increased surplus production. This may further lead to higher rates of biomass extraction based on (unsustainable) agricultural intensification, which may be associated with environmental degradation through the discharge of agro-chemicals into the environment. Having democratic control over financial resources and services may allow the lowering of interest rates, in comparison with conventional micro-credit providers, as well as the ability to make decisions about which types of credits are granted. Savings groups may thus allow dampening interest rate pressures on resource extraction and the environment (Scheidel and Farrell 2015).

### 8.5.4 The Cooperative Paddy Rice Bank: Overcoming Transitory Food and Seed Shortages

To deal explicitly with food and seed shortages, that may appear before the new harvest or when rice prices are high, the village community established a paddy rice bank. Initially supported by German's technical cooperation agency (GIZ), with

[9]Interest rates are as reported by community representatives and not inflation adjusted. Annual inflation in Cambodia was at 4.00 and 5.48% in 2010 and 2011, respectively (World Bank 2015).

shared knowledge, a storage place and a start-up paddy rice capital fund, the paddy rice bank provides rice credits to villagers that have become short on food. Rice credits need to be repaid with interest rates, which are also paid in paddy rice. Similar to the cooperative bank, it is democratically managed by the villagers themselves, based on a defined set of local institutional agreements.

In 2010, the bank had 190 members (rice borrowers) and a paddy rice stock of 28 tons, of which 7 tons were provided as rice credits to villagers. Based on average paddy rice consumption values of 223 kg/cap/year (ACI 2005), the total stockpile could ensure staple food security for around 126 persons during a full year. Interest rates amounted to half a basket of rice (ca. 12 kg) for every three baskets borrowed, equalling an interest rate of 16%. This was also the maximum amount a villager could borrow. Each borrower required up to five persons from other households who agree to share the liability of paying the rice back. A small part (6%) of the total benefits created from interest payments were used to compensate the efforts of the rice bank committee in maintaining the cooperative. The remaining share was largely used to support community activities such as the repairing of roads or dams. Although the number of members has declined over recent years, in 2016, the paddy rice bank was still active, with around 78 members and approx. 26 tons of paddy rice stock. 10 tons of paddy rice stock was previously sold by the community to finance repairs and the construction of a water dam benefitting the whole village.

Such small paddy rice banks have been commonly supported across Cambodia by NGOs, which provided a small paddy rice stock to assist with rice production. Initially, only a small number of families were allowed to borrow paddy rice, which they had to return after harvest season for paddy rice production. Interest rates were used to increase local paddy rice stocks, with between 5 and 10% of the interest generally being retained to develop their local areas with projects such as roads, canals and so on. In operation since 2003, with the support of the government and NGOs, more than 12,000 Cambodian families were reported to have joined such groups. The paddy rice stocks of small paddy rice banks across Cambodia have increased steadily, with several thousand tonnes of paddy rice being stockpiled in rural areas to support villagers.

Regarding the sustainability of local food systems, this cooperative bank helps to overcome problems of 'paddy rice liquidity' as well as to increase village food sovereignty, by facilitating at defined rules the allocation of paddy rice flows between harvests (time) as well as between households (space). In contrast to the role of local markets in the distribution of rice production, the paddy rice bank is under democratic control, avoiding price volatility, speculation, and the exit of rice flows in periods of scarcity. It is also an important alternative to external credit or agricultural intensification in overcoming transitory shortages. However, it also produces a situation in which the poorer households—in terms of those lacking access to land for sufficient food production—are financing village community activities through their paddy rice interest payments. This is a drawback that should be considered. The lowering of interest rates to a minimum may help to reduce such negative social aspects.

### 8.5.5 Rice Mill Associations to Increase Market Performance and Use of Valuable Organic Residuals

The establishment of a rice mill association in the neighbouring village represents another interesting development. It was set up to enhance and stabilize paddy rice prices through collective rice trade, and to enhance market performance by producing high-quality milled rice. The association was established in 2008 with support from NGOs, and is now managed through farmers who have invested to become shareholders. The Ministry of Agriculture, Forestry and Fisheries (MAFF) acknowledged the association in June 2009 and provided initial funds of 1,500,000 Riels (around $365) and a rice milling machine. Members initially came from 11 villages in 5 different districts.

The association buys rice from its members and other farmers at a stable price, this is then sold collectively, either as paddy or as milled rice. Hence it acts as a 'paddy rice cartel' to increase bargaining power, enabling farmers to increase their share in the food value chain. In addition, the rice mill association also provides credits to villagers and initially also conducted other activities such as seed conservation, which, however, was stopped due to a lack of resources. 70% of the benefits generated from trading rice and providing credits are split among the shareholders, whereas 17% is distributed among the democratically elected committee members as compensation for their management efforts. 10% are further retained for maintenance of the rice mill infrastructure and 3% are retained for capacity building activities. In 2016, the president reported that the association was still active with a total of 37 members and a shareholder capital of 30 million Riels (ca. $7300). Trading of paddy rice had to stop during the previous year because of a lack of transport options. The total number of members decreased because some members living outside the district had joined other nearby rice mill associations that had been established in their areas. Yet, due to its secondary function as credit provider, the rice mill association has continued to be an active cooperative. Compared to other producer associations studied by the authors (cashew, cassava, handicraft associations), financial involvement seems to play an important role in assuring long-term commitment to local producers' associations.

In Cambodia, the rice marketing system is largely dominated by small traders and processors even though Cambodian rice is increasingly valued on international markets and export rates are growing fast. The changes in traded quantity were also accompanied by quality improvements through enhanced milling capacities, accompanied by the introduction of food safety certification programmes. Within this move from trading basic paddy to high-quality milled rice, small-scale rice mill associations as described above play an important role countrywide. They reduce intermediaries between producers and consumers and support small farmers to increase their shares of benefits along the rice value chain. This also has direct implications for resource use and sustainability. Firstly, achieving stabilization or even increases of paddy rice prices may help to avoid a race to the bottom among competing sellers of such primary commodities. Consequently, for the same

**Table 8.2** Overview of cooperative and knowledge-based initiatives among Cambodian small farmers, their key requirements, and their key benefits from a sustainable food systems perspective

| Initiative | Key requirements | Key benefits from a sustainable food systems perspective |
|---|---|---|
| System of rice intensification (SRI) | • Detailed knowledge and skills<br>• May require additional labour, depending on adopted techniques | • Fosters place-based knowledge creation for agroecological fitness, enhanced yields and improved paddy quality<br>• Reduces natural resource inputs, i.e., seeds, water<br>• Reduces use of agro-chemicals and related GHG and fossil energy rucksacks attached to their production, trade and application |
| Small-scale biogas system (Biodigester) | • Basic knowledge<br>• Low-capital investment (Biodigester construction) | • Enables cascading resource use and closing of nutrient cycles through enhanced manure/humanure management<br>• Reduces GHG gases from enhanced manure/humanure management and reduced fossil fuel demand for cooking/lightening<br>• Reduces firewood demand and pressures on forests<br>• Frees labour time required for firewood collection<br>• Reduces indoor pollution and health risks |
| Small-scale cooperative banking (SSCB) | • Detailed knowledge and skills<br>• Cooperation between farmers | • Supports economic democratization of village capitalization and fairer participation of farmers in the financial economy<br>• Avoids capital outflow from local interest payments to large microfinance corporations<br>• Fosters saving and enhances access to (cheaper) credit<br>• Supports acquisition of household assets, such as a biogas system |
| Paddy rice bank | • Detailed knowledge and skills<br>• Cooperation between farmers<br>• Low-capital investment: physical storage place | • Supports democratic control over community food storage<br>• Increases food security at the village level<br>• Supports community activities through food supply |

(continued)

**Table 8.2** (continued)

| Initiative | Key requirements | Key benefits from a sustainable food systems perspective |
|---|---|---|
| Rice mill community | • Detailed knowledge and skills<br>• Cooperation between farmers<br>• Low to medium capital investment: storage place, rice milling facilities | • Supports economic democratization of paddy rice trade and fairer participation of farmers within the food trade chain through increasing farmers' share in the value chain, enhancement of their bargaining power, and reduction of intermediaries<br>• Supports closing of nutrient cycles by retaining by-products such as rice husk for local uses<br>• Supports conservation of local (traditional) seeds |

*Source* Own elaboration. For further elaboration of the key benefits from a sustainable food systems perspective, see Tello and González de Molina, Chap. 2 in this volume

amount of biomass extraction (and associated environmental impacts) farmers may obtain higher economic returns than without the rice mill association. Secondly, farmers can retain biomass by-products, such as husk separated after milling, which have many useful applications, ranging from soil aeration in horticulture, to animal bedding, or compost production (Badar and Qureshi 2014). Considering that around 20% of paddy weight is husk (IRRI 2013), the village could theoretically retain up to 75 tons of extracted biomass for local uses, while further reducing need for synthetic fertilizers. Hence, the establishment of such cooperatively-managed distribution and trade channels may therefore further support local sustainable food systems through local nutrient cycling, reduction of intermediaries between producers and consumers, and increasing the economic benefits that small farmers may obtain from food trade.

## 8.6 Conclusions

Many small farmers in Cambodia and elsewhere face a difficult situation, trying to maintain and enhance their farming-based livelihoods, while keeping pressures on land, labour and the environment at bay. Land shortage is a key issue to be dealt with, and against the backdrop of a countrywide and global land grab crisis, it can be expected to remain a central challenge for rural dwellers over the next years. However, there are many new developments underway, able to support local food systems through a series of cooperative, knowledge-based and low-capital intensive initiatives. They offer important potentials to leapfrog more conventional development pathways and to support the sustainability of local food systems.

This chapter has particularly discussed the role that the System of Rice Intensification (SRI), small-scale biogas systems, a paddy rice bank, community finance and a rice mill association can play for farmers who, although facing severe land shortage, are still in control of their means of production. As a set of initiatives that have diffused, thanks to cooperation between farmers and NGOs, they can contribute to sustainable food systems through a series of processes. Particularly, SRI fosters place-based knowledge creation that supports agroecological fitness, enhanced yields and improved paddy quality, while lowering natural and synthetic farming inputs. The small-scale biogas system is a clever but simple invention to enable cascading resource use and closing of nutrient cycles through enhanced manure/humanure management. Reduced demand for firewood and the labour required for its collection, a decreased dependence on fossil energy demand for cooking, and reduced air pollution are just some of its benefits. Community banking of both financial capital and rice stocks are examples of how to enhance economic democratization regarding the management, access to, and use of community resources, while supporting a fairer participation of farmers in the financial economy. Finally, a community-operated rice mill association also helps farmers to have a fairer share of the food trade chain, while enabling them to retain by-products such as rice husk that support the closing of local nutrient cycles (Table 8.2).

As many of these alternatives have only recently found diffusion, it is difficult to identify their long-term potential for leapfrogging conventional agricultural development. Moreover, rural systems in the Global South are currently undergoing rapid transformations (Kelly 2011; Ravera et al. 2014). Yet their study will be increasingly important for further imagining a path towards local sustainable food systems. While, with regard to the question of leapfrogging, much attention has been paid to the diffusion of new and clean technologies, we have illustrated that it is further relevant to look at the diffusion of social innovations and cooperative initiatives. They do not require much capital or technology, but rather new skills and an enhanced understanding of models of local cooperative economies—including their implications for resource uses. While some of them are spreading fast thanks to an increasingly connected information and actor network, the future will show us the long-term benefits as well as the new challenges they may pose to local food systems.

## References

ACI. (2005). Final Report for the Cambodian Agrarian Structure Study. Prepared for the Ministry of Agriculture, Forestry and Fisheries, Royal Government of Cambodia, the World Bank, the Canadian International Development Agency (CIDA) and the Government of Germany/G. Bethesda, Maryland.

ADB. (2001). *Participatory poverty assessment in Cambodia*. Manila: Asian Development Bank.

ADB. (2012). Technical Assistance Consultant' s Report: The Rice Situation in Cambodia.

Anthofer, J. (2004). *The potential of the System of Rice Intensification (SRI) for poverty reduction in Cambodia*, Berlin.

Badar, R., & Qureshi, S. A. (2014). Composted rice husk improves the growth and biochemical parameters of sunflower plants. *Journal of Botany, 2014,* 1–7. doi:10.1155/2014/427648.

Ballard, B. (2006). Land tenure database development in Cambodia. In M. Torhonen, P. Groppo (Eds.), *Land reform—Land settlement and cooperatives* (pp. 71–82). Food and Agriculture Organization (FAO).
CCHR. (2013). *Cambodia: Land in conflict. An overview of the land situation*. Phnom Penh.
Chandler, D. (2008). *A history of Cambodia* (4th ed.). Boulder: Westview Press.
Ches, S., & Yamaji, E. (2016). Labor requirements of system of rice intensification (SRI) in Cambodia. *Paddy and Water Environment, 14,* 335–342. doi:10.1007/s10333-015-0503-1.
Diepart, J.-C. (2015). *The fragmentation of land tenure systems in Cambodia: Peasants and the formalization of land rights*. Paris.
Ellis, F. (2000). The determinants of rural livelihood diversification in developing countries. *Journal of Agricultural Economics, 51,* 289–302. doi:10.1111/j.1477-9552.2000.tb01229.x.
EMC. (2015). *Sustainability study of savings group programs in Cambodia for CARE, Oxfam, and Pact*. Phnom Penh.
Erb, K. H. (2012). How a socio-ecological metabolism approach can help to advance our understanding of changes in land-use intensity. *Ecological Economics, 76,* 8–14. doi:10.1016/j.ecolecon.2012.02.005.
Evans, A., & Ford, C. (2003). *A technical guide to rural finance*. Madison: World Council of Credit Unions.
Fairbairn, B. (1991). *Farmers, capital, and the state in Germany, c. 1860–1914*. Saskatoon: University of Saskatchewan.
FAO. (2015). *FAO Statistical Databases*. Food and Agriculture Organization of the United Nations.
Georgescu-Roegen, N. (1971). *The entropy law and the economic process*. Cambridge, London: Harvard University Press.
Gerber, J.-F. (2013). The hidden consequences of credit: An illustration from rural Indonesia. *Development and Change, 44,* 839–860. doi:10.1111/dech.12045.
Giampietro, M. (2003). *Multi-scale integrated analysis of agroecosystems*. Florida: CRC Press LLC.
Giampietro, M., & Mayumi, K. (2000a). Multiple-scale integrated assessment of societal metabolism: Introducing the approach. *Population and Environment, 22,* 109–153.
Giampietro, M., & Mayumi, K. (2000b). Multiple-scale integrated assessments of societal metabolism: Integrating biophysical and economic representations across scales. *Population and Environment, 22,* 155–210.
Giampietro, M., Mayumi, K., & Ramos-Martin, J. (2009). Multi-scale integrated analysis of societal and ecosystem metabolism (MuSIASEM): Theoretical concepts and basic rationale. *Energy, 34,* 313–322.
Godfray, H. C. J., Beddington, J. R., Crute, I. R., et al. (2010). Food security: The challenge of feeding 9 billion people. *Science, 327*(80), 812–818.
Goglio, S., & Leonardi, A. (2010). *The roots of cooperative credit from a theoretical and historical perspective*. Trento: European Research Institute on Cooperative and Social Enterprises.
Gomiero, T., & Giampietro, M. (2001). Multiple-scale integrated analysis of farming systems: The Thuong Lo Commune (Vietnamese Uplands) case study. *Population and Environment, 22,* 315–352.
Gotaas, H. B. (1956). Composting: Sanitary disposal and reclamation of organic wastes. Geneva: World Health Organization (WHO).
Grimsditch, M., & Schoenberger, L. (2015). *New actions and existing policies: The implementation and impacts of Order 01*. Phnom Penh.
Grünbühel, C., & Schandl, H. (2005). Using land-time-budgets to analyse farming systems and poverty alleviation policies in the Lao PDR. *International Journal of Global Environmental Issues, 5,* 142–180.
Hecht, S. (2010). The new rurality: Globalization, peasants and the paradoxes of landscapes. *Land Use Policy, 27,* 161–169.

Heuveline, P. (1998). "Between One and Three Million": Towards the demographic reconstruction of a decade of Cambodian history (1970–79). *Population Studies (New York), 52,* 49–65. doi:10.1080/0032472031000150176.

IRRI. (2013). *A second life for rice husk.* Rice Today April–June.

Jiao, X., Smith-Hall, C., & Theilade, I. (2015). Rural household incomes and land grabbing in Cambodia. *Land Use Policy, 48,* 317–328. doi:10.1016/j.landusepol.2015.06.008.

Kelly, P. F. (2011). Migration, agrarian transition, and rural change in Southeast Asia. *Critical Asian Studies, 43,* 479–506. doi:10.1080/14672715.2011.623516.

Kneiding, C., & Rosenberg, R. (2008). *Variations in microcredit interest rates.* CGAP policy brief, July 2008.

KOICA/UNEP/CAPS. (2011). *Biogas digesters for Cambodians: A multi-partner national biodigester program in Cambodia.*

Leisz, S. J., Rasmussen, K., Olesen, J. E., et al. (2007). The impacts of local farming system development trajectories on greenhouse gas emissions in the northern mountains of Vietnam. *Regional Environmental Change, 7,* 187–208. doi:10.1007/s10113-007-0037-1.

Leuprecht, P. (2004). *Land concessions for economic purposes in Cambodia. A human rights perspective.* Special Representative of the Secretary General of Human Rights in Cambodia, Phnom Penh.

Licadho. (2009). *Land grabbing and poverty in Cambodia: The myth of development.* LICADHO —Cambodian League for the Promotion and Defense of Human Rights, Phnom Penh.

Licadho. (2015). Cambodia's concessions. http://www.licadho-cambodia.org/land_concessions/. Accessed 23 Nov 2015.

MAFF. (2014). *Ministry of Agriculture, Forestry and Fisheries (MAFF).* http://www.maff.gov.kh/index.php/overview.html. Accessed 12 June 2016.

Mayer, A., Schaffartzik, A., Haas, W., & Sepulveda, A. R. (2015). *Patterns of global biomass trade: Implications for food sovereignity and socio-environmental conflicts.*

Moser, C. M., & Barrett, C. B. (2003). The disappointing adoption dynamics of a yield-increasing, low external-input technology: The case of SRI in Madagascar. *Agricultural Systems, 76,* 1085–1100.

NBP. (2015). *National Biodigester Programme Cambodia.* http://nbp.org.kh/Default.aspx?lang=en. Accessed 19 May 2016.

NCDD. (2009). *Kampot Data Book.* Phnom Penh: National Committee for Sub-National Democratic Development (NCDD).

NIS. (2007). *Cambodia socio-economic survey 2004—Time use in Cambodia.* Phnom Penh, Cambodia: National Institute of Statistics (NIS), Ministry of Planning.

NIS. (2008). *Cambodia general population census 2008.* Phnom Penh, Cambodia: National Institute of Statistics (NIS), Ministry of Planning.

NIS. (2010). *Cambodia socio-economic survey 2009.* Phnom Penh, Cambodia: National Institute of Statistics (NIS), Ministry of Planning.

NIS. (2015). *Census of agriculture of the Kingdom of Cambodia 2013.* Phnom Penh.

OECD/IEA. (2007). Chapter 15 Energy for cooking in developing countries. In *World Energy Outlook 2006* (pp. 419–445). Paris: International Energy Agency (IEA) and Organisation for Economic Co-operation and Development (OECD).

Oxfam. (2014). *Study on women's empowerment and leadership in saving for change groups in Cambodia.*

Pastore, G., Giampietro, M., & Ji, L. (1999). Conventional and land-time budget analysis of rural villages in Hubei Province, China. *CRC Critical Reviews in Plant Sciences, 18,* 331–357.

Patel, R. (2012). The long green revolution. *The Journal of Peasant Studies, 6150,* 1–63. doi:10.1080/03066150.2012.719224.

Perkins, R. (2003). Environmental leapfrogging in developing countries: A critical assessment and reconstruction. *Natural Resources Forum, 27,* 177–188. doi:10.1111/1477-8947.00053.

Pingali, P. (2012). Green revolution: Impacts, limits, and the path ahead. *Proceedings of National Academy of Sciences, 109,* 12302–12308. doi:10.1073/pnas.0912953109.

Ravera, F., Scheidel, A., Dell'Angelo, J., et al. (2014). Pathways of rural change: An integrated assessment of metabolic patterns in emerging ruralities. *Environment, Development and Sustainability, 16,* 811–820. doi:10.1007/s10668-014-9534-9.

Scheidel, A. (2013). *From MuSIASEM theory to practice: Reflections and experiences from field research in Kampot Province, Cambodia.* Working Paper on Environmental Sciences. Institute of Environmental Science and Technology (ICTA) Autonomous University of Barcelona (UAB), Barcelona.

Scheidel, A. (2016). Tactics of land capture through claims of poverty reduction in Cambodia. *Geoforum, 75,* 110–114. doi:10.1016/j.geoforum.2016.06.022.

Scheidel, A., & Farrell, K. N. (2015). Small-scale cooperative banking and the production of capital: Reflecting on the role of institutional agreements in supporting rural livelihood in Kampot, Cambodia. *Ecological Economics, 119,* 230–240. doi:10.1016/j.ecolecon.2015.09.008.

Scheidel, A., Farrell, K. N., Ramos-Martin, J., et al. (2014). Land poverty and emerging ruralities in Cambodia: Insights from Kampot province. *Environment, Development and Sustainability, 16,* 823–840. doi:10.1007/s10668-014-9529-6.

Scheidel, A., Giampietro, M., & Ramos-Martin, J. (2013). Self-sufficiency or surplus: Conflicting local and national rural development goals in Cambodia. *Land Use Policy, 34,* 342–352. doi:10.1016/j.landusepol.2013.04.009.

Scurrah, N., & Hirsch, P. (2015). *The political economy of land governance in Cambodia*, 23.

Serrano-Tovar, T., & Giampietro, M. (2014). Multi-scale integrated analysis of rural Laos: Studying metabolic patterns of land uses across different levels and scales. *Land Use Policy, 36,* 155–170. doi:10.1016/j.landusepol.2013.08.003.

Singh, S. J., Ringhofer, L., Haas, W., et al. (2010). *Local studies manual—A researcher's guide for investigating the social metabolism of local rural systems.* Vienna: Institute for Social Ecology, IFF Vienna, Alpen-Adria University of Klagenfurt.

Snyder, C. S., Bruulsema, T. W., Jensen, T. L., & Fixen, P. E. (2009). Review of greenhouse gas emissions from crop production systems and fertilizer management effects. *Agriculture, Ecosystems & Environment, 133,* 247–266. doi:10.1016/j.agee.2009.04.021.

SRI-Rice. (2015). SRI International Network and Resources Center, Cambodia. In Cornell University—College of Agriculture and Life Science. http://sri.ciifad.cornell.edu/countries/cambodia/. Accessed 20 May 2016.

Stoop, W. A., Uphoff, N., & Kassam, A. (2002). A review of agricultural research issues raised by the system of rice intensification (SRI) from Madagascar: opportunities for improving farming systems for resource-poor farmers. *Agricultural Systems, 71*, 249–274.

Theng, V., Khiev, P., & Phon, D. (2014). *Development of the fertiliser industry in Cambodia: Structure of the market, challenges in the demand and supply sides, and the way forward.* Phnom Penh.

Tilman, D. (1999). Global environmental impacts of agricultural expansion: The need for sustainable and efficient practices. *Proceedings of the National Academy of Sciences of the United States of America, 96,* 5995–6000. doi:10.1073/pnas.96.11.5995.

UNDP. (2008). *Residential energy demand in rural Cambodia: An empirical study for Kampong Speu and Svay Rieng.*

Uphoff, N. (1999). Agroecological implications of the system of rice intensification (SRI) in Madagascar. *Environment, Development and Sustainability, 1,* 297–313.

Ward, B., & Lewis, J. (2002). *Making the most of every pound that enters your local economy.* London: The New Economics Foundation (NEF).

WHO. (2002). *The health effects of indoor air pollution exposure in developing countries.* Geneva.

World Bank. (2015). *Data Catalog.*

Yusuf, A. A., & Francisco, H. (2009). *Climate change vulnerability mapping for Southeast Asia.* Singapore.

## Author Biographies

**Arnim Scheidel** is currently a postdoctoral researcher at the International Institute of Social Studies (ISS), The Hague, the Netherlands. He maintains close ties as affiliated researcher with the Institute of Environmental Science and Technology (ICTA), the Autonomous University of Barcelona (UAB), Spain, where he worked as researcher from 2010 and received his Ph.D. in Environmental Sciences in 2013. His general research interests include rural change, food system research, societal metabolism and ecological economics, with a particular focus on better understanding the political ecology of development and the emergence of alternative grassroots practices. Geographically, Arnim's focus has been on Southeast Asia as well as Europe, where he has done extensive research stays and field work. He is also an active contributor to the Atlas of Environmental Justice (EJatlas).

**Bunchhorn Lim** received his Bachelor degree in 2011 from the Faculty of Agricultural Economics and Rural Development at the Royal University of Agriculture (RUA), Phnom Penh, Cambodia. Currently, he is continuing his training with a Master's in Business Administration (MBA) at Norton University, Phnom Penh. He was a research assistant at the Economic Institute of Cambodia from 2012 to 2013 and now works in the Trade Development office at the governmental Department of Commerce in Kampong Thom province. His professional work and research interests include rural livelihoods and household economics, where he has done extensive research in Kampot province. More recently, he has started to study the market development of agri-products in Kampong Thom Province and is interested in understanding how related value chains can be improved for local farmers.

**Kimchhin Sok** is a researcher and lecturer at the Ecosystem Services and Land Use research center (ECOLAND) and the Faculty of Agricultural Economics and Rural Development of the Royal University of Agriculture (RUA) in Cambodia. Currently, he is finishing his Ph.D. in Economics and Management Science at the University of SupAgro Montpellier in France. In the past, he graduated with a Bachelor's and a Master's degree in agro-socioeconomics. Before and during his Master's research, he was involved in ecosystem services and agrarian systems research in his country, Cambodia, as well as in Burkina Faso, Morocco, France, and Myanmar. His education and research interests are related to the topics of the transition of agrarian/agricultural systems, household economies and livelihoods, and ecosystem services in the context of developing societies, with the aim of supporting sustainable development and natural resources conservation. At ECOLAND of RUA, he is also the secretary and a researcher.

**Piseth Duk** has been a researcher and lecturer at the Faculty of Agricultural Economics and Rural Development, Royal University of Agriculture (RUA), Phnom Penh, Cambodia. His research and teaching interests include rural regional management and livelihoods, agricultural development and land use planning. He recently turned from theory to practice and opened the De Burlap Cafe in Phnom Penh, where delicious and locally grown food is served.

# Chapter 9
# A Socio-metabolic Transition of Diets on a Greek Island: Evidence of "Quiet Sustainability"

**Panos Petridis and Julia Huber**

**Abstract** In the search for sustainable food systems, the Mediterranean diet occupies a prominent place, from the point of view of health, by standards of ecological sustainability and as promoting a culture of moderation and conviviality. Focusing on the Greek island of Samothraki, this chapter tells the story of a community which finds itself in the middle of a dual transition, socio-metabolically, from a traditional agrarian lifestyle to a modern industrial society, and nutritionally, towards a westernization of diets. We aim at understanding current dynamics and identify potential leverage points for sustainability, from a socio-metabolic perspective. Despite an increasing dependence on imports, our findings highlight the significant role of agricultural self-provisioning and informal food networks, as an example of "quiet sustainability". We propose to reinforce these sustainable elements of local tradition by associating them with values that find resonance within the community, such as health, localness and quality. There is the potential to support a better utilisation of local produce and make adherence to the Mediterranean diet and culture more attractive and economically viable.

**Keywords** Sustainable diets · Socio-metabolic transition · Mediterranean diet · Quiet sustainability · Samothraki · Food self-provisioning

P. Petridis (✉) · J. Huber
Institute of Social Ecology, Alpen-Adria University, Schottenfeldgasse 29, 1070 Vienna, Austria
e-mail: panos.petridis@aau.at

J. Huber
e-mail: j.huber@respact.at

J. Huber
respACT—Austrian business council for sustainable development, Wiedner Hauptstraße 24/11, 1040 Vienna, Austria

 
E. Fraňková et al. (eds.), *Socio-metabolic Perspectives on the Sustainability of Local Food Systems*, Human-Environment Interactions 7,
https://doi.org/10.1007/978-3-319-69236-4_9

## 9.1 Introduction

### *9.1.1 The Mediterranean Diet, A "Sustainable Diet" in Transition*

Globally, industrialized food production has been the cause of a series of sustainability issues, and this trend is further reinforced not only by population growth, but also by shifting consumer preferences (notably the increase of meat heavy diets), and a 'supermarket culture'. Despite divergent views on how to tackle the issue—from bio-technological fixes to radical system restructuring—, there is a nevertheless general consensus on the need to look for alternative, more sustainable solutions (Gomiero et al. 2011). Even though local is not by definition more sustainable (since local food can also be produced in a manner that is environmentally harmful), some degree of localisation has been proposed to be beneficial for the sustainability of food production and consumption for a number of environmental, social and economic reasons, such as closing nutrient loops, reducing travel distances and waste generation, improving farm energy efficiency, maintaining land heterogeneity, conserving site-specific knowledge, improving sense of community and local employment, attaining local economic multipliers and achieving a democratic control of resources (Fraňková and Johanisová 2012; Tello and González de Molina, Chap. 2 in this volume).

The Mediterranean diet is a good example of a diet providing for concerns of health, biodiversity, local origin, traditional culture and sustainability (Burlingame and Dernini 2011). Mediterranean food consumption patterns have developed over the past 5000 years or more. Spreading from the Fertile Crescent, they were influenced by the successive conquests of different civilizations as well as by the dietary rules of three monotheistic religions (Judaism, Christianity and Islam). As a consequence, the Mediterranean diverse food consumption and production patterns represent a variable historical and environmental mosaic (Dernini and Berry 2015), rather than one single diet. The Mediterranean 'diet', derived from the Greek word *díaita* (δίαιτα), to mean *way of life*, rather indicates the existence of some common dietary characteristics in Mediterranean countries, such as the intake of high amounts of plant-based foods (cereals, fruits, vegetables, legumes, tree nuts, seeds and olives), with olive oil as the principal source of added fat, moderate to high intakes of fish and seafood, moderate consumption of eggs, poultry and dairy products (cheese and yoghurt), low consumption of red meat and a moderate intake of alcohol where it does not contradict religious norms altogether (Bach-Faig et al. 2011).

The term "Mediterranean diet" was popularised following the 'Seven Countries Study' of nutrition by researcher Ancel Keys in the 1950s and 1960s. Keys and his team looked at the dietary patterns of sixteen different populations in seven countries and investigated the link between lifestyles, diets and the occurrence of cardiovascular diseases. One of the major findings was that people living in Crete and other parts of Greece, as well as in southern Italy, had very high life expectancies and very low rates of heart disease and certain types of cancer, despite relatively limited medical services available (Keys et al. 1984; Keys 1995). The findings have been reconfirmed in the decades that followed (e.g. Nestle 1995; Trichopoulou et al. 2003; Wahlqvist et al. 1999).

In recent years, there has been a reappraisal of the Mediterranean diet, mainly by academia and international organisations, as an environmentally friendly plant-based dietary pattern (CIHEAM/FAO 2015), associated with sustainable production and consumption practices (Bach-Faig et al. 2011) and low environmental impact (e.g. Baroni et al. 2007; Burlingame and Dernini 2011; Dernini et al. 2013; Duchin 2005; Gussow 1995). This has led to a gradual shift in the notion of the Mediterranean diet, from a healthy pattern to a "sustainable diet" (e.g. CIHEAM 2012; Dernini and Berry 2015; FAO/Bioversity 2010).

The concept of sustainable diets was first introduced in the 1980s (Gussow and Clancy 1986), and has been revived recently using the Mediterranean diet as the prime example (FAO/Bioversity 2010, 2012). In an international symposium on "biodiversity and sustainable diets", organised in 2010 by FAO and Bioversity, sustainable diets were defined as "...those diets with low environmental impacts which contribute to food and nutrition security and to healthy life for present and future generations. Sustainable diets are protective and respectful of biodiversity and ecosystems, culturally acceptable, accessible, economically fair and affordable; nutritionally adequate, safe and healthy; while optimizing natural and human resources" (FAO/Bioversity 2010).

It is clear that the concept of "sustainable diets" goes well beyond concerns of efficiency, or technical fixes, as it combines issues of health and ecological sustainability together with promoting a culture of frugality and moderation. In the case of the Mediterranean, this food culture, sustained over the centuries in harmony with the environment, was additionally recognised by UNESCO as an "intangible cultural heritage of humanity" (UNESCO 2013):

> The Mediterranean diet involves a set of skills, knowledge, rituals, symbols and traditions concerning crops, harvesting, fishing, animal husbandry, conservation, processing, cooking, and particularly the sharing and consumption of food. Eating together is the foundation of the cultural identity and continuity of communities throughout the Mediterranean basin. It is a moment of social exchange and communication, an affirmation and renewal of family, group or community identity. The Mediterranean diet emphasizes values of hospitality, neighbourliness, intercultural dialogue and creativity, and a way of life guided by respect for diversity. It plays a vital role in cultural spaces, festivals and celebrations, bringing together people of all ages, conditions and social classes. It includes the craftsmanship and production of traditional receptacles for the transport, preservation and consumption of food, including ceramic plates and glasses. Women play an important role in transmitting knowledge of the Mediterranean diet: they safeguard its techniques, respect seasonal rhythms and festive events, and transmit the values of the element to new generations. Markets also play a key role as spaces for cultivating and transmitting the Mediterranean diet during the daily practice of exchange, agreement and mutual respect (UNESCO 2013).

Over the past decades, the Mediterranean agricultural and rural cultural landscapes forming the cornerstones of the Mediterranean identity have increasingly been influenced by globalization, industrial development, population growth, rapid urbanization, growing incomes and the integration of women into the labour market (CIHEAM/FAO 2015). As a result, food production and consumption patterns have also been changing in ways that profoundly affect human health as well as agroecosystems, such as the increasing trend of overexploitation of water resources (De Marco et al. 2014).

The Mediterranean region is therefore believed to be passing through a "nutritional transition" in which problems of undernutrition coexist with overweight, obesity and food-related chronic diseases (CIHEAM/FAO 2015; Popkin 2004; Rosenbloom et al. 2008). During the last decades, we have been observing a degradation of the Mediterranean food culture, and a 'westernization' of diets (e.g. Aounallah-Skhiri et al. 2011; León-Muñoz et al. 2012; Vareiro et al. 2009; Varela-Moreiras et al. 2010), i.e. a steep increase in the consumption of animal products, lipids other than olive oil (added lipids as well as those embodied in other foods) and sugar (Alexandratos 2006), enhanced among other things by technical change in the food industry and the rapid diffusion of fast-food outlets. This shift is even more pronounced in countries that traditionally used to display the highest adherence to the Mediterranean diet, such as Greece (Da Silva et al. 2009), and there are indications that this trend has further accelerated during the recent economic downturn (Bonaccio et al. 2014).

### *9.1.2 Food Consumption Patterns in Greece*

Up until the mid-1960s, the majority of the Greek population followed a more or less agrarian lifestyle, and a diet that was very close to the "ideal-type" Mediterranean diet as described in the previous section:

> a high intake of fruits, vegetables (particularly wild plants), nuts and cereals mostly in the form of sourdough bread rather than pasta; more olive oil and olives; less milk but more cheese; more fish; less meat; and moderate amounts of wine, more so than other Mediterranean countries (Simopoulos 2001).

Notwithstanding the still high consumption of fresh fruits, vegetables and olive oil, today's food consumption patterns in Greece have significantly moved away from this traditional (healthy) pattern, particularly among the young (Alexandratos 2006; Kontogianni et al. 2008; Papadaki et al. 2007; Tyrovolas et al. 2009; Yannakoulia et al. 2004). People now consume significantly more sweets, meat and dairy products, and although they keep the tradition of having lunch at home, they often skip other meals (KEPKA 2006; Scoullos and Malotidi 2007). Indicatively, from the mid-1960s until the early 2000s, on average, there has been an increase in total food intake from 2900 to 3700 kcal per day, in fat consumption from 92 to 152 g/day, in meat consumption from 33 to 88 kg per year and in sugar consumption (raw sugar equivalent) from 18 to 32 kg per year (Alexandratos 2006).

Causes of this shift can be traced back to a series of socio-economic changes, such as the integration of women into the labour market, rapid economic growth, and Greece's entry into the European Union in 1981. Increased purchasing power, together with an opening of borders, resulted in a rising demand for food diversity, while work schedules modified behavioural patterns (Vareiro et al. 2009). The prevalence of overweight and obesity in Greece has now become a major issue: In 2011 about half of the Greek population was overweight (53.7%) and one fifth (20.1%) obese (Dernini and Berry 2015). Moreover, these trends have also been associated with ecosystem mismanagement, as well as with socio-economic problems (CIHEAM/FAO 2015).

The recent Greek economic crisis has further increased pressure on the Greek population. An analysis of EU-27 household final consumption expenditures showed that Greece, together with the Baltic countries, suffered most from the economic and financial crisis, having registered a nearly 15% loss of actual individual consumption (in volume terms) between 2008 and 2011 (EUROSTAT 2013). Calculations of the Greek food basket indicating the monthly budget required for an adequate food intake by reference households show a discrepancy between available wage income and food prices. According to the food basket, the monthly budget required for a healthy diet in Greece is €295 for a single person and €915 for a family of two adults and two children, while the minimum wage is currently €683 before taxes (EC 2015). Hence, a full time employee earning the minimum wage needs to spend around half of her take-home income on food consumption alone.

### *9.1.3 Agriculture in Greek "Marginal" Rural Agrarian Landscapes*

Due to the strong links between food and rural development, rural landscapes in the Mediterranean have been described as food-based landscapes (CIHEAM/FAO 2015). In order to achieve a better understanding of food consumption patterns in rural areas in Greece, it is worthwhile highlighting a few historical notes on the evolution of agriculture and family farming. During Ottoman times, the style of production in Greece was characterised by integrated agriculture/animal husbandry and common management of resources, and, by following a strategy of diversification of production, storing and redistribution to markets (Horden and Purcell 2000) was oriented toward self-sufficiency. Mountainous and insular areas in particular, due to their "marginality", often enjoyed relative autonomy and were characterised by common management and decision making, leading to often flourishing diverse economies, further reinforced by maritime trade (Kizos 2008). However, following the creation of the Greek State, these areas lost the comparative advantage of having small protected markets, were considered as less favourable and were gradually abandoned (Kizos 2008).

The modernisation and mechanisation of agriculture were accomplished by the 1970s (Moisides 1986), but only for the fertile plains. As a result, a large gap emerged between modernized, mechanized, intensive agriculture in coastal areas and a few large plains, and traditional, non-mechanized, food-producing farms on the islands and in the mountains (Damianakos 1997). For the majority of small-scale Greek farmers that fall in the second category, "agriculture is not an occupation, but a social condition to which they submit because the economy provides no credible occupational alternative" (Damianakos 1997, 191–192).

Such family farms have been historically open and extrovert in character in relation to the surrounding economy and regional markets (Kasimis and Papadopoulos 2013). However, increasing dependency on the market did not affect the foundations and goals of such family farms, namely to support the families and reproduce their means of production, rather than actively pursuing profits. Despite the fact that production was aimed for the market, it was largely non-mechanized, and families sold only what they did not consume (Damianakos 1997).

### *9.1.4 Introduction and Historical Background of the Food System on Samothraki*

Beyond national surveys and statistics, looking deeper at the local level can help us identify some of the drivers behind ongoing trends in food consumption patterns and potential links with rural management practices. Our case study, the island of Samothraki, lies in the north eastern corner of the Aegean Archipelago, 42 km from the Greek mainland, and has a total surface area of about 178 $km^2$ (Fig. 9.1). It is a mountainous island with its highest peak rising to 1611 m, creating two distinct microclimates: in the north, a wet microclimate, with numerous freshwater streams, centuries-old oriental plane forests and rich vegetation, and in the southwest a more typical Mediterranean climate and vegetation, where wheat fields, olive groves, vineyards and shrublands dominate the landscape.[1] Most of the land area is covered by natural grasslands and shrublands (used for grazing), while about 20% constitutes arable land, olive groves and complex cultivation (CORINE database). The island has been inhabited since prehistoric times and its "Sanctuary of the Great Gods", a large temple complex, was celebrated as a spiritual centre devoted to the Kaveiria mysteries for several centuries up until 400 AD. During Byzantine and Ottoman times, the island came under Venetian, Genoan and Ottoman rule. In the 19th century, Samothraki was violently depopulated by the Ottoman army and in 1912 became part of the modern Greek state.

[1]In a typology of the Greek landscape produced by Tsilimigkas and Kizos (2014) Samothraki appears at the same time as a coastal area with high elevation and high slope gradient, covered mainly by forest and semi-natural vegetation, and therefore falls in the category of "marginalised" rural areas in contrast to the country's fertile plains (Damianakos 1997).

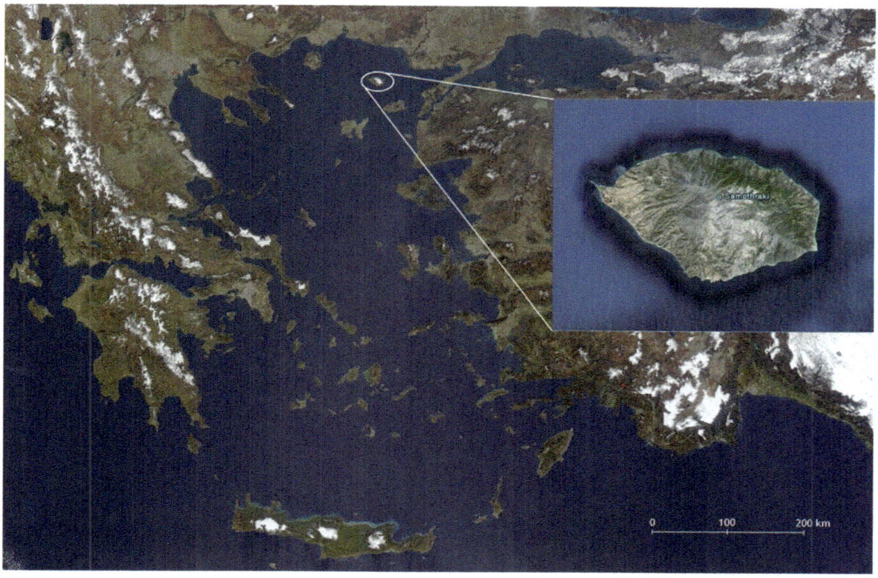


**Fig. 9.1** The location of Samothraki in the NE Aegean Sea. *Source* © NASA, nasaimages.org

The island's resident population has dropped from 4200 people in 1951 to 2859 inhabitants currently (ELSTAT 2012), mainly due to labour migration to the Greek mainland and Germany. The island thus has a low population density (15 persons/km$^2$). About 40% of the population have only completed primary school, 45% have a secondary school degree and 15% university education (ELSTAT 2012). The primary sector on the island accounts for 22% of the economically active population and includes animal husbandry, fishery, agriculture, olive cultivation and, to a small extent, wine cultivation. However this number excludes many households that own small gardens or a few animals and many pensioners who still assist in agricultural work. The secondary sector is relatively small and consists of one olive press, a municipal wheat mill, a small winery, some construction and mining activity, as well as some bakeries, a slaughterhouse and one dairy. The tertiary sector has increased profoundly over the past decades and today employs more than 60% of the professionally active population in trade, services and tourism (ELSTAT 2012). Samothraki attracts about 40,000 visitors per year, in a relatively short season (Fischer-Kowalski et al. 2011).

Up until the mid-1960s, Samothraki was almost entirely self-sufficient in terms of food production, and adhering to local, seasonal and—one can say—sustainable dietary patterns. Food consumption patterns complied with the traditional Greek dietary pattern (Simopoulos 2001). The population was "poor" compared to today's standards, and lived a fully agrarian lifestyle. Obesity was not an issue, on the contrary, undernutrition was experienced more often. Nevertheless, the self-sufficient lifestyle resulted in a relative high independence from wider socio-economic influences and had a limited impact on the local environment. From a socio-cultural perspective, food played a major role in peoples' lives and food practices were traditionally passed on from generation to generation. In addition, the Greek Orthodox Church also shaped people's food consumption patterns by promoting fasting rules throughout the year.

The main agricultural products were legumes, olives and grains, allegedly since the Stone Age and until recently—as evidenced by several mills dating some decades ago. Hence, bread and olive oil has traditionally played an important role in peoples' diet. Fishing was not pursued much during the Middle Ages due to widespread piracy, but ever since the Early Modern Age fish consumption has become increasingly popular. Meat has always been part of peoples' diet; the local population hunted animals, such as birds and rabbits, and kept goats and pigs. In addition, during the warmer periods of the year, a variety of fruits and vegetables were cultivated. During the colder periods, people added wild greens (chórta) to their diet. Snails were a popular source of protein following rain. However, without fridges and freezers, food storage was difficult. Meat was stored in pig fat and cheese was pickled in brine.

Goat and sheep herding in particular has a long tradition on Samothraki and is still the dominant agricultural activity, involving two thirds of all farmers at least partially. Until the 1970s, farmers on Samothraki followed an extensive animal husbandry model and did not depend on any external inputs. The goats were utilized efficiently and not just kept for their meat, but also for milk, yoghurt and cheese production. Thereafter, the population of goats and sheep increased to a peak of about 60–80,000 animals, mainly sustained by EU CAP subsidies, but remains largely underutilised. It is currently estimated at 50,000. This number still significantly exceeds the island's carrying capacity, estimated at around 20,000 animals (Evros 2004) which leads to overgrazing and extreme soil erosion.

Socio-metabolically speaking, the island society finds itself in the middle of a transition from a traditional agrarian society to modern industrialisation, maintaining elements of both. Labour migration to Germany brought higher incomes and modified mentalities. Agricultural subsidies further complemented incomes, but also led to severe overgrazing. In addition, the effects of globalization and the spread of modern communication technologies resulted in a steep increase of imported goods (also packaging waste) to the island, and a diffusion of more westernized diets. This has led to visible effects on health, but also to greater dependence on international markets, and vulnerability to food provision shortages, either caused by natural events, such as storms, or strikes on the mainland. Currently, a modern supermarket culture coexists with traditional dietary habits and practices, related to local food cultures and the utilization of local opportunities. At the core of this ongoing transition we observe a continuous tension between elements of tradition and modernization. In order to better understand local food system functioning (main flows, and drivers), we focus on the socio-metabolic transition of diets, and explore opportunities of sustaining local food provision, in a modernizing setting.

## 9.2 Methods

### 9.2.1 Conceptual Framework

Rather than looking at dietary patterns in isolation, we affiliate with approaches that take a more holistic view of food systems (e.g. LaBianca et al. 1990) and attempt to incorporate our findings within a wider framework. In order to understand the socio-ecological system of Samothraki and assess the dynamics behind its self-reproduction, we use a model of interaction between cultural and natural spheres of causation (Fischer-Kowalski and Weisz 1999). We particularly focus on the central "hybrid" realm, the biophysical structures of society, where the two spheres interact. These are hybrid because they are governed both naturally and culturally. As long as the flows required for maintaining the stocks can be organized, the system can be reproduced, and the society sustained. When, on the contrary, critical stocks cannot be reproduced, the system risks collapsing (Petridis and Fischer-Kowalski 2016).

We use a comprehensive conceptual model of the socio-ecological system of Samothraki, adapted to the island's food system (Fig. 9.2), in order to qualitatively

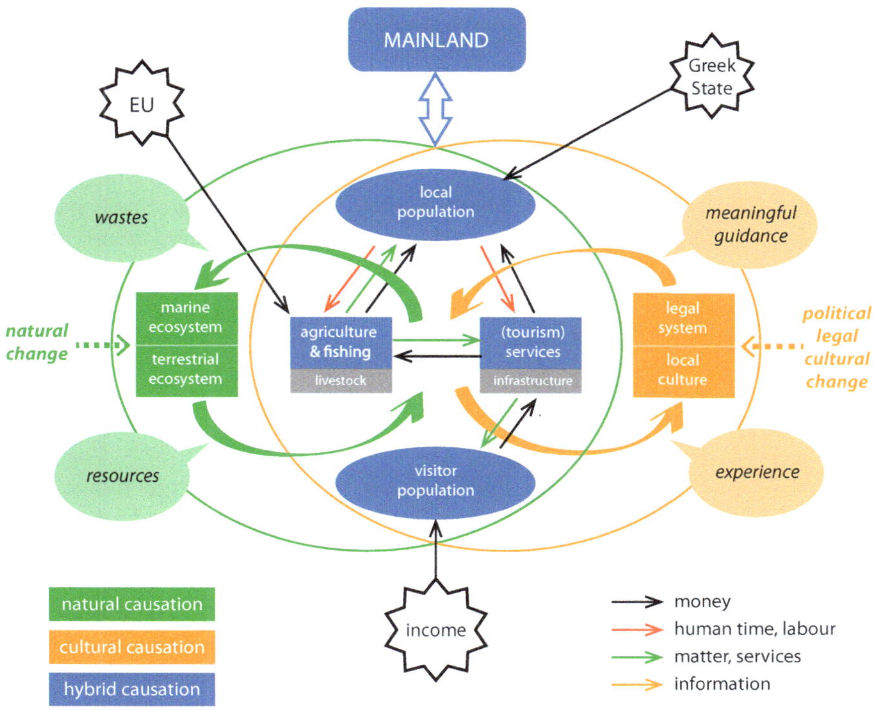


**Fig. 9.2** A conceptual model of the food system on Samothraki, described as a socio-ecological system *Source* adapted from Petridis and Fischer-Kowalski (2016)

identify the main flows associated with food, but also the natural and cultural drivers affecting food production and the consumption coevolving with them.

In the centre of the model, we find the core hybrid compartments of the socio-ecological system. These are the local population, the visitor population and the relevant economic sectors (agriculture and tourism), including the biophysical stocks they require and maintain. In addition, we can identify the major flows between the different compartments. The local population invests labour in agriculture, fishing, and tourism and receives agricultural produce and income in return. The visitors bring money from outside the system in exchange for food in restaurants and from shops. All sectors draw resources from the marine and/or terrestrial ecosystem—food grown on the island or caught in the seas surrounding it, and generate wastes in return as food waste. The behaviour of all actors is guided by the island's legal and cultural system, an example being the importance of fasting rules on people's eating habits, and this system in turn coevolves with it, by incorporating new experiences. For instance, novel imported food items may influence consumption preferences or perceptions about more traditional local products, especially among the young.

The system also strongly depends on the outside world. The connection to the mainland through the ferry is vital for the transportation of people, but also for food imports/exports. Natural changes influence marine and terrestrial ecosystems as well as most biophysical stocks: climatic conditions affect the island's agricultural production, while external drivers contribute to the depletion of fish stocks. Likewise, political, legal, cultural and economic changes affect the island's legal and cultural system: Notably, the Greek economic crisis might force people to cut back on food expenses and increase self-sufficiency. There are three regular major income flows originating from outside the system which are relevant for the food system: money visitors bring with them for purchasing local produce and spend in local gastronomy, European Union CAP subsidies for the agricultural sector, mainly livestock, and a (net) money flow from the Greek state to the local population in the form of payments such as pensions, a part of which is used to buy food.

### 9.2.2 Methodology Applied

The model assists us in describing the food system in socio-metabolic terms. First, we present some tentative results on food flows (left part of the model), utilising local statistics and complementing them with interviews conducted during fieldwork on several stays on the island (between 2014 and 2016). Then, in order to scrutinize tensions behind transition dynamics, we focus on the current food consumption patterns of the local population on Samothraki and complement this with exploring factors influencing food choices (right part of the model). For this purpose, we conducted an exploratory food consumption survey in May–June 2015 based on a food-frequency questionnaire adapted from Bountziouka et al. (2012) which included weighted quantities of food portions and food sources.

A paper-based questionnaire was used, along with a tablet showing food item pictures in different quantities to help respondents decide more correctly on food quantities consumed. The sample size consisted of 34 randomly selected individuals, aged between 13 and 81, with a balanced distribution across the different age groups, as well as education level: primary school (24%), secondary school (32%), and university degree (32%). Out of the total sample, 38% were female and 62% male—this slightly uneven ratio roughly corresponds to the sex distribution of the island (43.5% females, 56.5% males). In terms of occupational status, the greatest share of respondents was employed in gastronomy (38%), followed by agriculture and animal husbandry (17%) as well as public service (17%). In addition, 21% of the respondents were pensioners, 12% students and 9% unemployed.

The questionnaire assessed consumption of 39 typical food items (further grouped into 14 categories), which were elicited through exploratory interviews. In order to achieve higher accuracy, we focused on the summer season (May to October); the winter season was covered by a control question, as well as additional interviews, and requires further investigation. In addition, the questionnaire addressed factors influencing food choices. From the wide range of factors that influence food consumption patterns (Atkins and Bowler 2001), we focused on the role of health, price and localness (locally produced food). Interviews were also used to crosscheck categories, identify missing items and incorporate narratives about the past. Overall, results provided us with a satisfactory overall picture of the yearly average food consumption patterns and tendencies, sufficient enough to provide insights for our current research aims.

Finally, a methodological note: There is no universally accepted definition of local food (Martinez et al. 2010), and defining what is local is an ongoing debate (see: Born and Purcell 2006; Feagan 2007; Goodman et al. 2014; Hinrichs 2007). In part, it is a geographical concept related to the distance between food producers and consumers, but it can also be defined in terms of social and supply chain characteristics, such as production methods and provenance (Martinez et al. 2010). In traditional societies, these components were usually interconnected, but nowadays they are increasingly separated due to changes in food consumption and production processes (Wieser 2014). In our study, we define locally produced food as all food grown and processed within the island borders. This is relatively easy to identify, since there are clear physical borders, and currently a single access point. Consequently, we consider all raw and processed foods which enter the island via the ferry as imports, and all foods and food wastes which leave the island as exports.

## 9.3 Results

### *9.3.1 Notes on Food Metabolism*

A wide range of agricultural products grow on Samothraki all year around, but especially during the warmer months (from April to November), comprised of grains, pulses, olives, oil-bearing crops (sunflower, edible olives and olives for olive oil production) roots and tubers, vegetables, fruits and nuts (ELSTAT 2010). Moreover, there are currently about 50,000 goats and sheep, 9000 chickens and 1000 pigs on the island. The steep increase of ruminant populations has led to the need of switching to fodder production and consequently more feed imports. Currently, about a 1000 t/year of vetch and alfalfa are grown on the island for fodder use, while 70–90% of the feed for the goats and sheep on the island is imported (Fuchs 2014). Total local fish catch from small and large-scale fishermen is calculated at about 500 tons a year, of which 450 tons are exported (own data). The local agricultural production, including fish, but excluding livestock secondary products, accounts for about 3500 t/yr (own estimates).

Locally produced food includes goat and sheep meat, milk, yoghurt and cheese, honey, wine and distilled spirit (tsipouro). Traditional local food, which sometimes also involves imported ingredients (e.g. sugar), includes: traditional wheat pasta products (flomaria, trahanas), jams, and spoon sweet preserves. Food imports, including food for the supermarkets, drinks, meat for the butchers and flour for the bakeries, account for about 2100 t/yr. Food exports, including milk, yoghurt, cheese, honey, olives, wine, goats and sheep and fish account for an estimated 940 t/yr. Moreover, there is about 660 t/yr of food waste exported along with mixed waste (own estimates).

The current food supply infrastructure on the island is moderate and consists of four supermarkets, five mini markets, four bakeries, two butchers, two fishmongers, a women's agro tourism cooperative, and several taverns, cafés and bars (with more added during the tourist season). There is neither a regular farmers' market nor any form of community-supported agriculture (CSA). However, it is the widespread practice of many farmers, olive oil producers, wine makers and beekeepers all over the island to sell their agricultural products directly to consumers through informal networks or in rare cases export them in small amounts. In addition, sellers from the mainland also come irregularly to the island and offer their goods at the port.

### *9.3.2 Dietary Transition*

Our findings from the food frequency/food quantities questionnaire support the documented dietary transition. Whereas there is a relatively high adherence to the Mediterranean diet, there are also deviations in the form of higher consumption of white bread, sugar and coffee, as well as alcohol containing beverages, and to a lesser extent meat. The four products most *frequently* consumed ($\geq 2$ times daily) from May to October are: olive oil, white bread, coffee and sugar, albeit in small quantities. Figure 9.3 shows average *quantities* consumed for 14 food groups. We notice the consumption of large amounts of fruits and vegetables, as well as alcohol containing beverages. The main divergences in winter include the consumption of

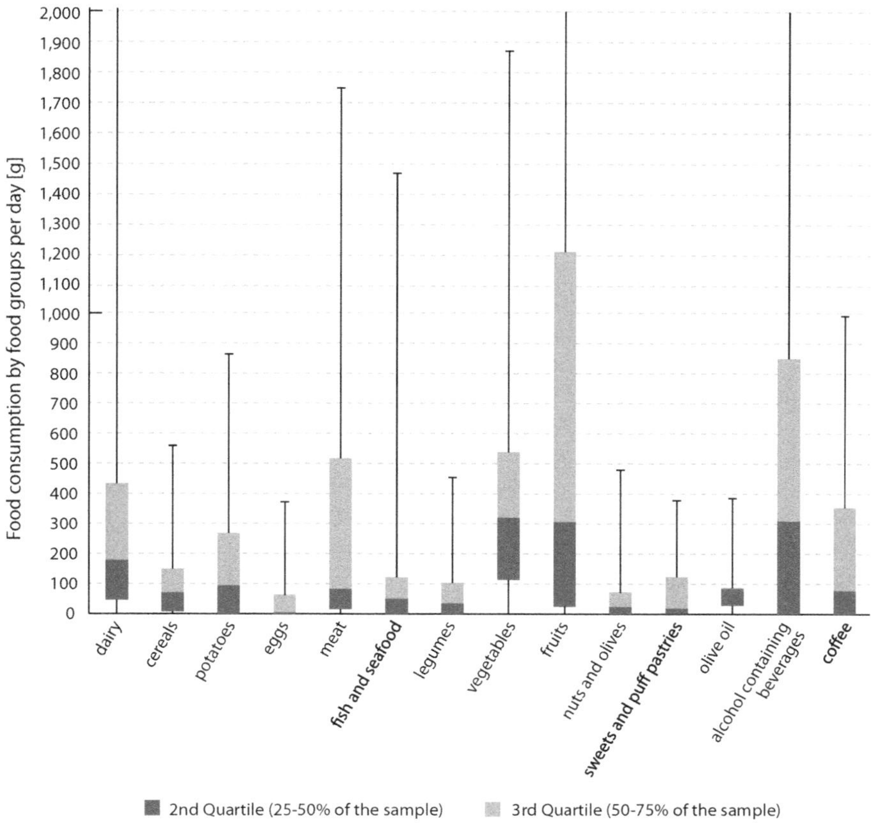


**Fig. 9.3** Median food consumption by food group per person per day (in g) for 14 food groups during May–October (N = 34)

more pulses, pasta, nuts, winter fruits and vegetables. Also, wild greens are collected mostly during winter, allegedly due to tasting better and having greater health benefits.

Comparing the data with the Greek average consumption values of 2004–2005 (Bountziouka et al. 2008) suggests that people on Samothraki appear to consume considerably more fruits and vegetables, slightly more meat, fish, legumes and olive oil, the same amount of potatoes, and less dairy products and cereals than the Greek average. Such comparison is perhaps indicative, but must be treated with care, taking into account the time difference, seasonality of our study and different methodological approaches.[2]

[2]Data on Greek average comes from household budget surveys (DAFNE V project).

### 9.3.3 Metabolic Transition

An analysis of the sources from which people obtain their food identified a wide range of options. People mostly obtain food from supermarkets, neighbours, friends or family, and grow/prepare food at home or eat out; to a lesser extent they get food directly from farmers, as well as in minimarkets, bakeries, butchers, fishmongers, beekeepers, the local dairy, and from sellers from the mainland; they also gather a small amount of food in their surroundings. In Fig. 9.4, we present the relative importance of different food sources and classify them as either local or imported from somewhere else.[3]

The results point to the ongoing metabolic transition. Overall, about half of all the food consumed in summer is imported. Supermarkets sell mostly imported foods, and only a small proportion of local produce, such as dairy products and some fruits and vegetables; the same applies to bakeries and butchers, apart from goat and sheep meat, which is local. As elderly respondents adhere significantly more to traditional food production and consumption patterns than young interviewees, this tendency may increase in the future. Still, if we read the results differently, half of the food consumed on the island still comes from local sources: there is an especially high share of food grown/gathered (15%), or obtained by family, friends and neighbours (22%). This covers all food groups, in particular olives/olive oil, fruit and vegetables, eggs, meat and dairy products. We will return to this important finding in the following sections.

### 9.3.4 Factors Influencing Food Choices

From the three main factors explored, *health* and *localness* were found to be of high significance in affecting food choices (Fig. 9.5): more than half of the respondents (especially the elderly and young parents) considered both factors as "very important" when choosing their food. Interestingly, most of the respondents linked the two concerns, stating they support local food production *because* it improves health, as well as creating more jobs and attracting more tourists. Respondents who consider health and localness as important also consumed significantly more local food, in most food categories, notably meat, dairy products, fruits and vegetables.

The picture was different regarding the question on *price*. Less than a quarter of the respondents noted price as 'very important' when choosing food, a finding that contradicts prevalent views on consumer behaviour (Atkins and Bowler 2001).

[3]There is a small amount of food, captured in the categories "homemade/own" and "neighbours, friends and family" that is imported, such as flour and sugar for traditional dishes, but the amount is negligible, as most food items in the questionnaire were single, raw food items. Conversely, a small amount of items sold at supermarkets come from local origin, but again this is considered negligible.

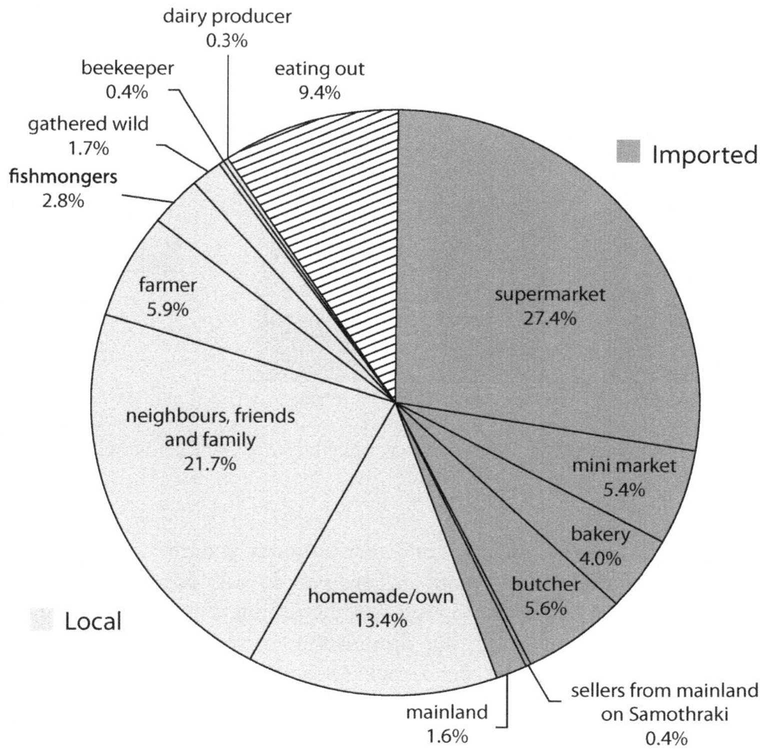


**Fig. 9.4** Food obtained in summer (May–October) (N = 34) according to 14 different sources, grouped according to their origin (local vs. imported). "Eating out" is treated as a split category, as cafés, bars, taverns and restaurants use a mix of imported and locally produced foods

Possible reasons for this may include the importance of informal food networks (which anyway provide a cheap option) as well as the prioritisation of food over other goods. Other factors mentioned during the interviews as affecting food choices were *quality*, *aroma* and *taste*. Again, interestingly, all those qualities were linked to locally produced food.

Looking at socio-economic factors, there appears to be a complementary relationship between young and old people, with the former obtaining more food from neighbours, friends and family, and eating out more, and the latter growing more food, but also buying more from supermarkets. This reflects a widespread tradition in Greece, where a part of the food prepared by the elderly goes to their childrens' households (Bountziouka et al. 2008). In addition, male respondents eat three times as much red meat (271 g/d) as females (90 g/d). Meat consumption remains below average for women and above average for men, regardless of whether health was perceived as of little or great importance.

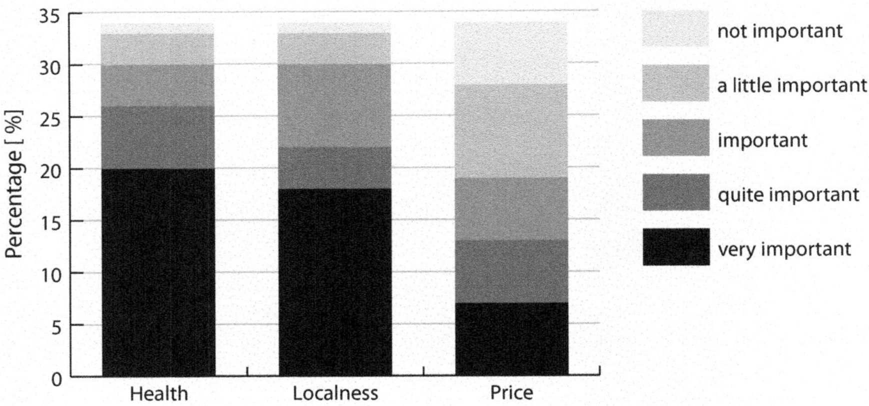


**Fig. 9.5** Relative importance of selected factors affecting food choices during summer (May–October) (N = 34)

Larger scale drivers influencing food consumption patterns include the legacy effects of labour workers in Germany, where people who have lived and worked abroad for several years continue to seek specific products consumed there, such as sausages. Still partially enduring, but diminishing in significance, are the dietary guidelines and fasting rules of the Greek Orthodox Church. According to the Church, it is advised to abstain from meat, olive oil and milk products on Wednesdays and Fridays throughout the year. Even though it is mostly elderly women who strictly adhere to this rule nowadays, it is nevertheless still quite common and customary that on those two days pulses and vegetable dishes are consumed. Moreover, many people do fast in the 48 days before Easter (abstinence from meat, fish and dairy products) and in the 40 days before Christmas (abstinence from meat and dairy products). It has been proposed that these rules may have indirectly promoted the use of wild plants and regulated exposure to certain foods throughout the year (Tourlouki et al. 2011).

The nursery school is a good example for pointing out some of the different tensions regarding food choices. There is at least a declared effort to offer a healthy, balanced diet and serve a wide variety of seasonal foods. Yet, almost all food the children consume at the nursery school is bought from supermarkets, and therefore is imported; it appears the effort to provide a healthy balanced diet is currently only possible this way. While there are increasingly westernized food consumption patterns observable, with numerous food items containing high amounts of sugar and/or fat served to the children, the dietary guidelines of the Church still exert some influence as there are two meat-free days per week, falling on Wednesdays and Fridays.

## 9.4 Discussion

Over the past decades, rural areas around the Mediterranean have been increasingly influenced by "supermarketization". This has led to a divergence from the traditional Mediterranean diet model (Simopoulos 2001; Trichopoulou et al. 2003), a fact that, apart from implications on health, has also economic, social and environmental repercussions (CIHEAM/FAO 2015). The renewed interest in the concept of sustainable diets in food studies is a welcome turn, as it proposes a more holistic view of diets than previous approaches based on health alone. We believe a socio-metabolic understanding of local food systems nicely complements and further adds to this emerging concept, by shedding light on additional aspects of sustainability. In our model we conceptualise the local food system in a holistic way, embedded in the socio-ecological system of the island, helping to understand the relation between material flows and their socio-economic drivers. This can be further elaborated by incorporating insights into the agroecological past of the island in order to guide future decisions.

The results of our study indicate that Samothraki finds itself in a dual, nutritional and metabolic transition. While we observe that the core of the traditional dietary pattern is still present, with a high consumption of fruits, vegetables and olive oil, our findings nevertheless indicate the gradual widespread diffusion of food items such as white bread and sugar (Alexandratos 2006; Scoullos and Malotidi 2007). In accordance with recent studies (Kontogianni et al. 2008; Papadaki et al. 2007; Tyrovolas et al. 2009; Yannakoulia et al. 2004), our findings show that the decline in adherence to the Mediterranean food culture is especially high among the young. Moreover, the local population has become increasingly dependent on food imports and thus vulnerable to market fluctuations. During summer, about 50% of all food consumed on the island is imported, a number likely to be higher during winter when agricultural production runs low. As elderly locals (51–81 years old) adhere significantly more to local food production and consumption patterns than young locals (13–30 years old), the share of imported food may likely increase further in the future. In order to evaluate this trend, we need a better understanding of seasonal population flows, as well as seasonal food availability. If there is not enough food in the winter, there may be a *need*, rather than a choice to import—of course this also depends on the subjective need for certain food items. Moreover, the availability of land and utilization of livestock demarcate the option space for increasing production or productivity. In addition, the ongoing shift away from agriculture to tourism means labour time on agriculture is reduced. This trend, unless reversed, will possibly further increase the need to import food.

### *9.4.1 Evidence of "Quiet Sustainability"*

However, our findings also show that certain traditional food production and consumption patterns have been preserved. Still, about 50% of food consumed, at least in summer, is obtained from local sources. One of the most interesting findings of our study (perhaps not surprising, but rarely documented), is the high share of food locally grown or wildly gathered (15%) and obtained through local informal networks of family, friends and neighbours (22%). These networks represent a widespread practice primarily motivated and utilized for reasons of better quality and taste, as well as solidarity and support for local products, and in some cases price, rather than explicit environmental reasons. This has, nevertheless, a huge impact on sustainability, and can be considered as an example of "quiet sustainability" (Smith and Jehlička 2013).

The concept of "quiet sustainability" has been used to describe a range of agricultural self-provisioning practices in Eastern Europe that:

> ...result in beneficial environmental or social outcomes, that do not relate directly or indirectly to market transactions, and that are not represented by the practitioners as relating directly to environmental or sustainability goals. Cultures of sharing, repairing, gifting and bartering characterise quiet sustainability. Everyday practices that have low environmental impacts, but that have not been pursued for that reason, are also features of the concept (Smith and Jehlička 2013, 155).

Similarly, widespread environmentally friendly food self-provisioning practices in Russia "...do not emerge from wider concerns about sustainability or contributing to an environmentally sound or localized agricultural system. They arise mostly from the *desire to grow healthy food* [...], the inability to buy expensive inputs, and self-interest in cultivating the small plots in a way that ensures longer-term fertility" (Visser et al. 2015, p. 7, emphasis added).

Our case fits the "quiet sustainability" discourse, as we are dealing with food self-provisioning practices that are environmentally more sustainable, involve smaller distances, use no or less synthetic fertilisers, as well as bringing community benefits (reciprocity, trust, sharing). The social dimension of quiet sustainability is crucial. The informal food networks on Samothraki can be viewed as examples of local collaboration in practice (as opposed to formal verbal agreements or trade) that achieve a better utilization of local produce, while reinforcing social cohesion, solidarity and a sense of community. Even though rarely studied as such, they in fact represent a practice widely performed in Greece, where it is typical that a significant part of agricultural and livestock produce serves to maintain a vast network of kin and client relations (Damianakos 1997). On Samothraki, these networks of food exchange are part of a complex nexus of reciprocal acts between individuals and families that go beyond food provisioning, into providing seasonal help in construction/agricultural activities, or offering financial or other services; receiving or buying food is often a way to pay back a previous favour (Kostakiotis 2015).

Moreover, this trend is associated with another characteristic of the Samothrakian community, that of pluriactivity. It is quite common for many local inhabitants to maintain a diversified household economy: most practice some degree of subsistence agriculture, own a piece of land and a few olive trees, grow their own vegetables, produce wine or own a few goats, while often seasonally utilising tourist opportunities. This pluriactivity is in fact the rule rather than the exception for most Greek farmers on marginal unproductive land (Damianakos 1997; Kasimis and Papadopoulos 2013). Family farming in Greece has been characterised as a survival/resistance strategy of flexibly using resources against a continuously changing setting (Kasimis and Papadopoulos 1997), as well as showing higher adaptive capacity to the recent crisis by displaying higher "rural resilience"(Kasimis and Papadopoulos 2013).

This subsistence/capitalist "hybrid" condition in a modernised and globalised world comes with its limitations: Currently on Samothraki most of the goat and sheep herders slaughter their animals on the farm and sell the meat directly to restaurants or households, or transport the animals to the mainland to achieve better selling prices. Only a small share of meat is currently processed in the island's newly renovated communal slaughterhouse (where supply chains are lacking). Those who do milk their animals sell their milk to the only local dairy on the south side of the island to produce yoghurt and cheese, which is then distributed locally, as well as exported to the mainland. For herders that reside further away, it is uneconomic to drive to the dairy every day, so they usually process smaller amounts of milk directly on their farm to produce cheese and sell it through informal networks.

## *9.4.2 Intervention Potential: Looking for Synergies*

In order to cope with a loss in income, and be less vulnerable to the ongoing economic stagnation, a common strategy is to maintain risk-averse behaviour and employ diverse agricultural and food self-provisioning practices. What is yet to be established in the local culture is that many of those practices are *also* more ecologically sustainable, and, if "rebranded" and made relevant and attractive for the young, also have the potential to bring employment opportunities to the island. Traditional knowledge on food production can be transformed from a "coping" strategy, or a return to the past, to a tool for promoting sustainable livelihoods within modern Greek society.

Somewhat contradicting similar studies (e.g. Bonaccio et al. 2014), our analysis has shown that factors of health and localness were found to be significantly more important in affecting food choices than prices, despite the acknowledged difficult economic situation. This is an interesting finding and points to the view that, despite the economic crisis, food in the Mediterranean culture seems to be one of the last refuges of quality for low-income people. Partly, this may be attributed to the fact that informal networks play an important role in food provisioning on Samothraki, and prices there are anyway low.

These findings highlight some promising ways to tackle the traditionalist vis-à-vis modernist tensions, and to evaluate the intervention potential of strategies to improve the sustainability of the food system on the island: The idea here is that existing sustainable parts of tradition (agricultural self-provisioning and informal networks of "quiet sustainability") can be reinforced by utilising arguments or *frames* that have a high resonance within the community, such as health concerns or pride in food quality. In other words, there is the potential for promising novel arguments supporting local food production and consumption, beyond established but threatened beliefs based on religion. Our results identified a strong preference for local and healthy food; a shift towards providing more local and healthier food, for local people as well as tourists, could help strengthen the (currently weak) synergy between the two main economic sectors of the island, agriculture and tourism, by promoting a more efficient utilization of local produce while providing economic support for local farmers. Moreover, instead of looking at the adherence to the Mediterranean diet and culture as backwardness, or a response to falling incomes, a narrative that offers a fresh look on localisation could make it more attractive for the young and shift the current trend of importing processed goods to the branding, marketing and exporting of quality local products. In this vein, it is worthwhile exploring the possibilities of utilising traditional knowledge, which is already occurring, for example in the offer of cooking classes and attempts to re-establish some of the old grain mills.

Our research points to several ways in which interventions can take place. People consume many fruits and vegetables, but a large proportion of those are imported. At the same time, most people value health and explicitly support local food because it is "more healthy", while there is increased interest in organic food, especially from young parents, but also from the elderly. Moreover, there is also the widespread belief that locally produced food could create more jobs on the island, as well as attracting more tourists. All this calls for bringing local producers and consumers closer together, and is an opportunity to upscale high quality organic production (branding for organic products, such as cheese, meat and olive oil) by new agricultural cooperatives, to improve value chains of local agricultural products (e.g. packaging and exporting of goat meat), and to develop the agro-tourism sector (Petridis 2012; Petridis et al. 2013). Evidence from other rural areas in Greece shows that producing "quality" products is the most significant factor motivating farmers to switch to organic agriculture (Koutsoukos and Iakovidou 2013). The pending inclusion of Samothraki in the World Network of Biosphere Reserves (Fischer-Kowalski et al. 2011; Petridis 2016) provides an excellent framework to further support this transition, and increase its visibility.

An additional research direction warranting further investigation is the possible association between dietary patterns and agricultural practices, in particular, whether there are signs of a coevolution between the modernisation of agriculture and westernisation of diets. It has been argued that the abandonment of widespread practices like terracing, that supported agricultural production on mountainous islands, can affect diets, in the sense that a decline in local cultivation may intensify incorporation to the market, where new food products challenge "traditional" ones (e.g. Tourlouki et al. 2011). If such an assumption holds true, then could this association be intentionally reversed? Can the "gradual re-evaluation of older farming systems" (Kizos 2008), be used synergistically with an increased dietary demand for local organic produce? Could it be that such an association is easier to conceptualise and achieve on Samothraki, where, like in other "marginal" rural areas in Greece, transition to modernisation only appeared later, and is slower and still incomplete? Evidence from dietary patterns are "promising"; however, there needs to be a more thorough evaluation of agricultural practices and local traditional knowledge, and if/how this can be linked to changing diets. Socio-metabolically speaking, a key parameter that needs to be taken into account is the role of energy input. Lack of fossil fuels in the past forced people to make a virtue out of necessity. Could recognition of the sustainability of local traditional practices guide people to making a virtue out of choice?

Samothraki, being simultaneously insular and mountainous, exemplifies Horden and Purcell's (2000) concept of the "marginal". Throughout Mediterranean history, this "omnipresence of the marginal has enforced diversity, flexibility and opportunism in managing the environment, re-examining the relationships between extraction of food from the uncultivated environment and the formal procedures of agriculture *sensu stricto*" (ibid. p. 180). In this line, it has been argued that an opening to the markets was one of the many response strategies to uncertainty and risk. "From the beginning of the historical record onward, we hear of the association of some of the islands with specialized production of certain high-value commodities destined for the wide market. [why?] The answer can only be connectivity" (ibid. p. 225). Seen from this perspective, a wise use of markets (or the creation of new ones) that link to the dietary demands and preferences of locals and visitors, could be a re-visiting of a centuries-old strategy of flexibility and resilience. At the same time, it is also a vivid reminder of the importance of connectivity with higher scales: The possibility of maintaining a sustainable local food system on Samothraki, centred on high quality organic products, may only be possible if there is a larger system providing tourists and monetary surplus that can buy those products!

### 9.4.3 A Latent Transformation

The transformative potential of alternative food networks has been vividly discussed, more so in the context of crisis (Benessaiah 2015; Calvário and Kallis 2017; Rakopoulos 2014). But can we look at the local food system on Samothraki as a "real utopia" (Cucco and Fonte 2015)? We argue that, so far, an explicit emphasis on emancipatory food politics on Samothraki is not in sight, though the potential may be there. Nevertheless, we believe that, in a way, certain current practices do implicitly challenge the overall food system. Considered as practices of "quiet sustainability", informal food networks "…allow parallel and overlapping narratives about families, networks, competencies and relations with nature. They are not a replacement or an alternative to the market economy of food, or a response to its environmental or social failings, but rather a vivid demonstration that that is only part of life" (Smith and Jehlička 2013, 155). In particular, due to the emphasis on solidarity and the social dimension of sustainability, we can perhaps also describe them as examples of "quiet food sovereignty", i.e. food sovereignty without a movement (Visser et al. 2015).

> Quiet food sovereignty "…does not challenge the overall food system directly through its produce, claims, or ideas, but focuses on individual economic benefits and ecological production for personal health, as well as a culturally appropriate form of sociality, generated by the exchange of self-produced food. Furthermore, a rights discourse—which is so central in food sovereignty—while not absent, is rather implicit. It is grounded in the longstanding tradition of self-provisioning, and taken for granted" (Visser et al. 2015, 13).

While dominant food policies globally are aiming at achieving more sustainable production and consumption, here we present a case of already existing practices (e.g. growing food, barter exchange) that, while they are both "sustainable" and "sovereign", are mainly performed for reasons such as coping with insecurity, rather than environmental sustainability or explicit resistance. This, again, resonates the case of smallholders in Russia that "…see cultivating their own food as an important right. However, this right is implicit. It is rarely expressed, as it is a longstanding tradition, and a substantial degree of control over their own production is seen as the natural order of things. Why would smallholders state the obvious?" (Visser et al. 2015, 10).

Our story does not fit the narrative of a purposive transition, as current patterns of food provisioning, although "sustainable", mainly refer to practices already sustained for decades (Smith et al. 2015). Nevertheless, the deadlocks presented by the ongoing economic stagnation, together with a sustained and informed socio-ecological research presence on the island (Petridis et al. 2016), may provide an opportunity to speed up social dynamics and help identify institutional reforms in support of local practices (e.g. the setting up of social cooperatives or seed banks), that would implicitly prefigure emancipatory possibilities (Kloppenburg 2014; Petridis et al. 2015). This may present farmers on Samothraki with the chance of "leapfrogging" to producing (and consuming) high quality organic products, and making a decent living out of it, without experiencing a further westernisation of dietary patterns. They may do so by taking advantage of the 'food sovereignty countermovement' (McMichael 2014), of the emerging 'third food regime' (Burch and Lawrence 2009), manifested in the desire for quality, 'healthy' foods. Possible indirect gains for Samothraki could be the enlargement of regional markets for organic products and perhaps even a reverse mobility of young farmers to the island.

To sum up, in order to fully develop the intervention potential towards more sustainable local food systems on Samothraki, a socio-metabolic reading of food provision and dietary patterns allows us to identify (and utilize) local opportunities, and promote a more efficient utilization of local produce. After outlining existing tensions, we propose ways to use the traditionalist vis-à-vis modernizing tendencies and to support "social innovations" that are built upon sustainable parts of existing traditions. In particular, this would mean utilizing existing frames around health, quality and taste in order to "rebrand" traditional food patterns, prevent a loss of traditional knowledge on food production by making it more attractive, and explore marketing strategies that would generate employment opportunities while reinforcing already existing environmentally sustainable solutions.

## References

Alexandratos, N. (2006). The Mediterranean diet in a world context. *Public Health Nutrition, 9* (1A), 111–117.

Aounallah-Skhiri, H., Traissac, P., El Ati, J., Eymard-Duvernay, S., Landais, E., Achour, N., et al. (2011). Nutrition transition among adolescents of a South-Mediterranean Country: Dietary patterns, association with socio-economic factors, overweight and blood pressure. A cross-sectional study in Tunisia. *Nutritional Journal, 10/38,* 1–17.

Atkins, P., & Bowler, I. (2001). *Food in society. Economy, culture, geography*. London: Hodder Education.

Bach-Faig, A., Berry, E. M., Lairon, D., Reguant, J., Trichopoulou, A., Dernini, S., et al. (2011). Mediterranean diet pyramid today. Science and cultural updates. *Public Health Nutrition, 14* (12A), 2274–2284.

Baroni, L., Cenci, L., Tettamanti, M., & Berati, M. (2007). Evaluating the environmental impact of various dietary patterns combined with different food production systems. *European Journal of Clinical Nutrition, 61*(2), 279–286.

Benessaiah, K. (2015). Transforming our food system: Let's enable the back-to-land movement in Greece. *Newsletter für Engagement und Partizipation in Europa*, 05/2015. Bundesnetzwerk Bürgerschaftliches Engagement (BBE).

Bonaccio, M., Di Castelnuovo, A., Bonanni, A., Costanzo, S., De Lucia, F., Persichillo, M., et al. (2014). Decline of the mediterranean diet at a time of economic crisis. Results from the moli-sani study. *Nutrition, Metabolism and Cardiovascular Diseases, 24*(8), 853–860.

Born, B., & Purcell, M. (2006). Avoiding the local trap. Scale and food systems in planning research. *Journal of Planning Education and Research, 26,* 195–207.

Bountziouka, V., Bathrellou, E., Giotopoulou, A., Katsagoni, C., Bonou, M., Vallianou, N., et al. (2012). Development, repeatability and validity regarding energy and macronutrient intake of a semi-quantitative food frequency questionnaire: methodological considerations. *Nutrition, Metabolism & Cardiovascular Diseases, 22*(8), 659–667.

Bountziouka, V., Tsiotas, K., Economou, H., Naska, A., & Trichopoulou, A. (2008). Trends in food availability in GREECE—The DAFNE V Project. Department of Hygiene and Epidemiology, Medical School, University of Athens. http://ec.europa.eu/health/ph_projects/2003/action1/docs/greece_en.pdf.

Burch, D., & Lawrence, G. (2009). Towards a third food regime: Behind the transformation. *Agriculture and Human Values, 26*(4), 267–279.

Burlingame, B., & Dernini, S. (2011). Sustainable diets: the Mediterranean Diet as an example. *Public Health Nutrition, 14*(12A), 7–2285.

Calvário, R., & Kallis, G. (2017). Alternative food economies and transformative politics in times of crisis: Insights from the Basque Country and Greece. *Antipode, 49*(3), 597–616.

CIHEAM. (2012). *Mediterra 2012. The Mediterranean diet for sustainable regional development.* International Centre for Advanced Mediterranean Agronomic Studies (CIHEAM). Paris: Presses de Sciences Po.

CIHEAM/FAO. (2015). *Mediterranean food consumption patterns: Diet, environment, society, economy and health.* A White Paper Priority 5 of Feeding Knowledge Programme, Expo Milan 2015. CIHEAM-IAMB, Bari/FAO, Rome.

Cucco, I., & Fonte, M. (2015). Local food and civic food networks as a real utopias project. *SOCIO. HU, 2015*(3), 22–36.

da Silva, R., Bach-Faig, A., Quintana, B. R., Buckland, G., de Almeida, M. D. V., & Serra-Majem, L. (2009). World variation of adherence to the Mediterranean diet, in 1961–1965 and 2000–2003. *Public Health Nutrition, 12*(9A), 1676–1684.

Damianakos, S. (1997). The ongoing quest for a model of Greek agriculture. *Sociologia Ruralis, 37*(2), 190–208.

De Marco, A., Velardi, M., Camporeale, C., Screpanti, A., & Vitale, M. (2014). The adherence of the diet to Mediterranean principle and its impacts on human and environmental health. *International Journal of Environmental Protection and Policy, 2*(2), 64–75.

Dernini, S., & Berry, E. M. (2015). Mediterranean diet: From a healthy diet to a sustainable dietary pattern. *Frontiers in Nutrition, 2,* 15.

Dernini, S., Meybeck, A., Burlingame, B., Gitz, G., Lacirignola, C., Debs, P., et al. (2013). Developing a methodological approach for assessing the sustainability of diets: The Mediterranean diet as a case study. *New Medit, 12*(3), 28–36.

Duchin, F. (2005). Sustainable consumption of food: A framework for analyzing scenarios about changes in diets. *Journal of Industrial Ecology, 9*(1–2), 99–114.

EC. (2015). The greek food basket—Athens. European Commission. http://ec.europa.eu/social/main.jsp?catId=738&langId=en&pubId=7836&type=2&furtherPub.

ELSTAT. (2010). Agricultural production on Samothraki. http://www.statistics.gr/en/home/. Accessed February 17, 2016.

ELSTAT. (2012). 2011 population-housing census. Interactive Map. http://www.statistics.gr/en/interactive-map. Accessed February 12, 2016.

EUROSTAT. (2013). Analysis of EU-27 household final consumption expenditure—Baltic countries and Greece still suffering most from the economic and financial crisis. http://ec.europa.eu/eurostat/documents/3433488/5585636/KS-SF-13-002-EN.PDF/a4a1ed61-bac7-4361-a3f0-4252140e1751?version=1.0.

Evros S. A. (2004). *Study on the grazing capacity of Samothraki Grazelands*. Greece: Dimosineteristiki Evros. (in Greek).

FAO/Bioversity. (2010). Report of the international symposium on biodiversity and sustainable diets. Rome (available at www.fao.org/ag/humannutrition/28506-0efe4aed57af34e2dbb8dc578d465df8b.pdf).

FAO/Bioversity. (2012). Sustainable diets and biodiversity. Directions and solutions for policy, research and action. Rome (available at www.fao.org/docrep/016/i3004e/i3004e00.htm).

Feagan, R. (2007). The place of food: Mapping out the 'local' in local food systems. *Progress in Human Geography, 31*(1), 23–42.

Fischer-Kowalski, M., & Weisz, H. (1999). Society as hybrid between material and symbolic realms. *Advances in Human Ecology, 8,* 215–251.

Fischer-Kowalski, M., Xenidis, L., Singh, S. J., & Pallua, I. (2011). Transforming the Greek island of Samothraki into a UNESCO biosphere reserve. An experience in transdisciplinarity. *GAIA, 20,* 181–190.

Fraňková, E., & Johanisová, N. (2012). Economic localisation revisited. *Environmental Policy and Governance, 22*(5), 307–321.

Fuchs, N. A. (2014). *Effekte der EU-Agrarsubventionen auf das Extensive Weidehaltungssystem der Griechischen Insel Samothraki. Sozialökologische Fallstudie im Hinblick auf Umweltrelevante Veränderungen (Master Thesis).* Alpen-Adria University Klagenfurt.

Gomiero, T., Pimentel, D., & Paoletti, M. G. (2011). Is there a need for a more sustainable agriculture? *Critical Review in Plant Science, 30,* 6–23.

Goodman, D., Dupuis, M. E., & Goodman, M. K. (2014). *Alternative food networks. Knowledge, practice, and politics*. London, New York: Routledge.

Gussow, J. D. (1995). Mediterranean diets: Are they environmentally responsible? *The American Journal of Clinical Nutrition, 61*(suppl), 1383S–1389S.

Gussow, J. D., & Clancy, K. L. (1986). Dietary guidelines for sustainability. *Journal of Nutrition Education, 18*(1), 1–5.

Hinrichs, C. (2007). Introduction. Practice and place in remaking the food system. In C. Hinrichs & T. A. Lyson (Eds.), *Remaking the North American food system. Strategies for sustainability*. Lincoln, London: University of Nebraska Press.

Horden, P., & Purcell, N. (2000). *The Corrupting Sea: A study of Mediterranean History*. Wiley-Blackwell.

Kasimis, C., & Papadopoulos, A. G. (1997). Family farming and capitalist development in Greek agriculture: A critical review of the literature. *Sociologia Ruralis, 37*(2), 209–227.

Kasimis, C., & Papadopoulos, A. G. (2013). Rural transformations and family farming in contemporary Greece. In D. Ortiz-Miranda, A. Moragues-Faus & E. A. Alegre (Eds.), *Agriculture in Mediterranean Europe: Between old and new paradigms. Research in rural sociology and development* (Vol. 19, pp. 263–294). Emerald Group Publishing.

KEPKA. (2006). Greek consumers' dietary habits, pan-hellenic research for the diet. KEPKA (Greek Consumers' Protection Center) (in Greek).

Keys, A. (1995). Mediterranean diet and public health: Personal reflections. *The American Journal of Clinical Nutrition, 61*(6), 1321S–1323S.

Keys, A., Menotti, A., Aravanis, C., Blackburn, H., Djordevič, B. S., Buzina, R., et al. (1984). The seven countries study: 2289 deaths in 15 years. *Preventive Medicine, 13*(2), 141–154.

Kizos, T. (2008). Rural environmental management in Greece as a cultural frontier between the "occident" and the "orient". *Arbor, 729,* 127–142.

Kloppenburg, J. (2014). Re-purposing the master's tools: The open source seed initiative and the struggle for seed sovereignty. *Journal of Peasant Studies, 41*(6), 1225–1246.

Kontogianni, M. D., Vidra, N., Farmaki, A. E., Koinaki, S., Belogianni, K., Sofrona, S., et al. (2008). Adherence rates to the Mediterranean diet are low in a representative sample of Greek children and adolescents. *The Journal of Nutrition, 138*(10), 1951–1956.

Kostakiotis, G. (2015). *Politics of care: Challenges of aging on a small island (PhD Thesis).* University of the Aegean (In Greek).

Koutsoukos, M., & Iakovidou, O. (2013). Factors motivating farmers to adopt different agrifood systems: A case study of two rural communities in Greece. *Rural Society, 23*(1), 32–45.

LaBianca, O. S., Hubbard, L. E., & Running, L. G. (1990). *Sedentarization and Nomadization: Food System Cycles at Hesban and Vicinity in Transjordan*. Institute of Archaeology, Andrews University Press.

León-Muñoz, L. M., Guallar-Castillón, P., Graciani, A., López-García, E., Mesas, A. E., Aguilera, M. T., et al. (2012). Adherence to the Mediterranean diet pattern has declined in Spanish adults. *The Journal of nutrition, 142*(10), 1843–1850.

Martinez, S., Hand, M. S., Da Pra, M., Pollack, S., Ralston, K., Smith, T., et al. (2010). *Local food systems: Concepts, impacts, and issues*. ERR 97, U.S. Department of Agriculture, Economic Research Service.

McMichael, P. (2014). Historicizing food sovereignty. *Journal of Peasant Studies, 41*(6), 933–957.

Moisides, A. (1986). *Rural society in modern Greece, 1950–1980*. Athens: Mediterranean Studies Foundation. (In Greek).

Nestle, M. (1995). Mediterranean diets: Historical and research overview. *American Journal of Clinical Nutrition, 61*(suppl.), 1313S–1320S.

Papadaki, A., Hondros, G., Scott, J. A., & Kapsokefalou, M. (2007). Eating habits of university students living at, or away from home in Greece. *Appetite, 49*(1), 169–176.

Petridis, P. (2012). Perceptions, attitudes and involvement of local residents in the establishment of a Samothraki biosphere reserve, Greece. *Eco.mont—Journal on Protected Mountain Areas Research, 4*(1), 59–63.

Petridis, P. (2016). Establishing a biosphere reserve on the island of Samothraki, Greece: A transdisciplinary journey. *Sustainable Mediterranean, 72,* 39–41.

Petridis, P., & Fischer-Kowalski, M. (2016). Island sustainability: The case of Samothraki. In H. Haberl, M. Fischer-Kowalski, F. Krausmann, & V. Winiwarter (Eds.), *Social ecology: Society-nature relations across time and space* (pp. 543–557). Cham: Springer International Publishing.

Petridis, P., Fischer-Kowalski, M., Singh, S. J., & Noll, D. (2017). The role of science in sustainability transitions: Citizen science, transformative research, and experiences from Samothraki Island, Greece. *Island Studies Journal, 12*(1), 115–134.

Petridis, P., Hickisch, R., Klimek, M., Fischer, R., Fuchs, N., Kostakiotis, G., et al. (2013). *Exploring local opportunities and barriers for a sustainability transition on a Greek Island.* Social Ecology Working Paper. Nr. 142, Vienna.

Petridis, P., Muraca, B., & Kallis, G. (2015). Degrowth: Between a scientific concept and a slogan for a social movement. In J. Martinez-Alier & R. Muradian (Eds.), *Handbook of ecological economics* (pp. 176–200). Edward Elgar.

Popkin, B. M. (2004). The nutrition transition: An overview of world patterns of change. *Nutrition Reviews, 62*(Suppl 2), S140–S143.

Rakopoulos, T. (2014). The crisis seen from below, within, and against: From solidarity economy to food distribution cooperatives in Greece. *Dialectical Anthropology, 38*(2), 189–207.

Rosenbloom, J. I., Nitzan-Kaluski, D., & Berry, E. M. (2008). A global nutrition index. *Food and Nutrition Bulletin, 29,* 266–277.

Scoullos, M., & Malotidi, V. (2007). *Mediterranean food: Historical, environmental, health & cultural dimensions. Educational material.* Athens: MIO-ECSDE.

Simopoulos, A. P. (2001). The Mediterranean diets: What is so special about the diet of Greece? The scientific evidence. *The Journal of Nutrition, 131*(11), 3065S–3073S.

Smith, J., & Jehlička, P. (2013). Quiet sustainability: Fertile lessons from Europe's productive gardeners. *Journal of Rural Studies, 32,* 148–157.

Smith, J., Kostelecký, T., & Jehlička, P. (2015). Quietly does it: Questioning assumptions about class, sustainability and consumption. *Geoforum, 67,* 223–232.

Tello, E., & González de Molina, M., Chapter 2 in this volume. Methodological challenges and general criteria for assessing and designing local sustainable agri-food systems: A socio-ecological approach at landscape level. In E. Fraňková, W. Haas & S. J. Singh (Eds.), *Socio-metabolic perspectives on sustainability of local food systems.* Springer.

Tourlouki, E., Matalas, A. L., & Panagiotakos, D. (2011). Cultural, social, and environmental influences on surviving dietary patterns of the past: A case study from the Northern Villages of Karpathos. *Nature and Culture, 6*(3), 244–262.

Trichopoulou, A., Costacou, T., Bamia, C., & Trichopoulos, D. (2003). Adherence to a Mediterranean diet and survival in a Greek population. *New England Journal of Medicine, 348* (26), 2599–2608.

Tsilimigkas, G., & Kizos, T. (2014). Space, pressures and the management of the Greek landscape. *Geografiska Annaler: Series B, Human Geography, 96*(2), 159–175.

Tyrovolas, S., Polychronopoulos, E., Bountziouka, V., Zeimbekis, A., Tsiligiani, I., Papoutsou, S., et al. (2009). Level of adherence to the Mediterranean Diet among elderly individuals living in Mediterranean islands: Nutritional report from the MEDIS study. *Ecology of Food and Nutrition, 48*(1), 76–87.

UNESCO. (2013). Mediterranean diet. Representative list of the intangible cultural heritage of humanity. Paris. http://www.unesco.org/culture/ich/en/RL/mediterranean-diet-00884.

Vareiro, D., Bach-Faig, A., Quintana, B. R., Bertomeu, I., Buckland, G., de Almeida, M. D. V., et al. (2009). Availability of Mediterranean and Non-mediterranean foods during the last four decades: Comparison of several geographical areas. *Public Health Nutrition, 12*(9A), 1667–1675.

Varela-Moreiras, G., Avila, J. M., Cuadrado, C., Del Pozo, S., Ruiz, E., & Moreiras, O. (2010). Evaluation of food consumption and dietary patterns in spain by the food consumption survey: Updated information. *European Journal of Clinical Nutrition, 64,* S37–S43.

Visser, O., Mamonova, N., Spoor, M., & Nikulin, A. (2015). 'Quiet food sovereignty' as food sovereignty without a movement? Insights from post-socialist Russia. *Globalizations, 12*(4), 513–528.

Wahlqvist, M. L., Kouris-Blazos, A., & Wattanapenpaiboon, T. (1999). The significance of eating patterns: An elderly Greek case study. *Appetite, 32,* 23–32.

Wieser, H. (2014). *Zooming in and zooming out on the practice of sustainable food shopping: Evidence from Austria, Hungary and The Netherlands (Master Thesis)*. WU Vienna University of Economics and Business.

Yannakoulia, M., Karayiannis, D., Terzidou, M., Kokkevi, A., & Sidossis, L. S. (2004). Nutrition-related habits of Greek adolescents. *European Journal of Clinical Nutrition, 58*(4), 580–586.

## Author Biographies

**Panos Petridis** has an educational background in biology, and since 2010 has been a researcher at the Institute of Social Ecology of the Alpen-Adria University in Vienna. He works within the broad family of "sustainability sciences", focusing on society-nature interactions. For several years he has been involved in the management of a transdisciplinary research project on the Greek island of Samothraki, a UNESCO Biosphere Reserve candidate, exploring future transition pathways in more sustainable directions. Lessons learnt from this socio-ecological research formed part of his recently completed doctoral study on the role of science in sustainability transitions. He has also published on the notion of "Degrowth", a collective endeavour at redefining prosperity, outside the economic imperative frame.

**Julia Huber** holds a Bachelor's degree in International Development from the University of Vienna and a Master's degree in Social Ecology from the Alpen Adria University. Her research interests and passions include sustainable food systems, local food initiatives, the Mediterranean diet as well as alternative food networks. In her master's thesis she analysed the food consumption patterns of the local population on the Greek island of Samothraki from a social-ecological perspective. Currently, she works for an NGO called respACT—Austrian business council for sustainable development supporting Austrian companies in their sustainability efforts.

# Chapter 10
# Local, Mixed and Global Organic Tomato Supply Chains: Some Lessons Learned from a Real-World Case Study

**Gonzalo Gamboa, Sara Mingorria, Marina Di Masso and Mario Giampietro**

**Abstract** This chapter presents the evaluation of three organic tomato supply chains in Catalonia (Spain): the Local and Global supply chains, and a Mixed supply chain. The evaluation is based on a set of multidimensional indicators derived from different narratives: Commodity, Environmental and Livelihoods. In so doing, we identified the main challenges when implementing the supply chain approach, and propose some recommendations for its application. The supply chain approach proves to be very useful in identifying the critical points at which to intervene and/or improve the performance of food supply chains, but presents some disadvantages when it is used to compare the sustainability of food supply chains. Finally, we show that organic production is not enough to achieve a sustainable food provision system, and the context and the food supply chain in which organic food is commercialised also determine the sustainability of organic food production.

---

G. Gamboa (✉)
Departament d'Economia i d'Història Economica, Facultat d'Economia i Empresa, Universitat Autònoma de Barcelona, Office B3-112, Building B, Campus UAB, 08193 Bellaterra, Spain
e-mail: gonzalo@riseup.net

G. Gamboa · S. Mingorria · M. Giampietro
Institut de Ciència i Tecnologia Ambientals (ICTA), Universitat Autònoma de Barcelona, Campus UAB, 08193 Bellaterra, Spain
e-mail: sara.mingorria@gmail.com

M. Giampietro
e-mail: Mario.Giampietro@uab.cat

M. Di Masso
Internet Interdisciplinary Institute (IN3), Universitat Oberta de Catalunya (UOC), Av. Carl Friedrich Gauss, 5, 08060 Castelldefels, Spain
e-mail: madimasso@gmail.com

M. Giampietro
Institució Catalana de Recerca i Estudis Avançats (ICREA), Passeig Lluís Companys, 23, 08010 Barcelona, Spain



E. Fraňková et al. (eds.), *Socio-metabolic Perspectives on the Sustainability of Local Food Systems*, Human-Environment Interactions 7,
https://doi.org/10.1007/978-3-319-69236-4_10

**Keywords** Food supply chains · Multi-scale integrated assessment · Narratives · Biophysical and Socio-economic indicators

## 10.1 Introduction

Recent decades have witnessed an increasing interest in the development and study of local food supply chains (LFSC) or networks. According to Marsden et al. (2000), these food supply chains can be *face-to-face*, *spatially proximate* or *spatially extended*, and these refer respectively to schemes in which products are sold directly by the producer to the consumer, products are sold in local outlets, and products are sold outside the local area (including information on the production process and origins), respectively. The critical characteristic of these food supply chains is that the "product reaches the consumer embedded with information…, which enables the consumer to confidently make connections and associations with the place/space of production, and, *potentially, the values of the people involved and the production methods employed*" (Marsden et al. 2000, p. 425. Emphasis in the original). The diversity identified by these authors reflects the wide range of food supply chains that can be grouped under the same umbrella term.[1]

According to Forssell and Lankoski (2015), the characteristics used to describe LFSCs in the literature can be classified as *background*, *core* and *outcomes*. Background characteristics refer to the unconventional values and aims of the actors participating in the network (e.g. commitment to sustainability and the non-industrial logic operating in LFSC). Core characteristics are those that refer to the quality requirements of products (e.g. organic, traditional, healthy) and production processes (e.g. territorially embedded). In this sense, the value of food production goes beyond commodity value to include the cultural and territorial characteristics of the product (Sonnino and Marsden 2006). Core characteristics also refer to the reduced distance between producers and consumers (in terms of physical distance, reduced intermediaries and shared information) and the new forms of market governance that would redistribute power and share economic risk along the chain. Outcome characteristics are those characteristics that refer to changes in the nature of relationships between the actors along the chain. As networks are based on trust and social embeddedness, and information is shared among actors, relationships within the network are supposed to be strengthened (Hinrichs 2000; Marsden et al. 2000; Morgan and Murdoch 2000; Renting et al. 2003; Ilbery and Maye 2005).

[1]These food supply chains are referred to as local or short food supply chains, alternative food networks or localised food systems. We use the term "Local Food Supply Chain" to refer to them.

Forssell and Lankoski (2015) analyse the potential contribution of LFSCs to sustainable food provision. According to them, there is a general perception that local food supply networks are more sustainable than their conventional counterparts. The same authors conclude that there is little evidence based on empirical work that explores and compares the sustainability of different food supply chains. The comparison of different food supply chains usually focuses on specific stages of the FSC (Matopoulus et al. 2015), such as transport (e.g. Jones 2002), purchasing (e.g. Sanye et al. 2012) or from post-harvest to retail (e.g. Blanke and Burdivck 2005). However, sustainability in food provision must be scrutinised in relation to the whole food chain (Cobb et al. 1999), including up-stream inputs of agricultural production, production practices and distribution-related steps of food supply (e.g. transport, commercialisation) (Ilbery and Maye 2005).

On this basis, the objective of this chapter is to propose and apply a framework with which to evaluate and compare food supply chains. The evaluation and comparison of food supply chains is undertaken using a set of quantitative indicators, which enables the analyst to look for strengths and weaknesses in the different organic tomato supply chains and to search for sustainable paths in food production and consumption.

As suggested by Tello and González de Molina (Chap. 2 in this volume), we must explore the capacity of self-reliant organic farming to provide a healthy and sufficient diet for the global population. Accordingly, this chapter evaluates and analyses three organic tomato supply chains operating in Spain. We use the term "organic" to refer to production systems that rely on ecological processes, biodiversity, cycles adapted to local conditions, and ecosystem management rather than systems which rely on the use of synthetic agricultural input, veterinary drugs, genetically modified seeds and breeds, preservatives, additives or irradiation. Organic agriculture is part of an extensive supply chain, which includes food processing, distribution and retailing. It pursues consumer confidence, environmental protection, food quality and animal welfare.[2]

The supply chains have been classified as Local, Mixed and Global, according to four criteria: geographical distance between producer and consumer, territory and identity, knowledge and technology, and governance of the supply chain. Farms operating in each supply chain differ from one another in terms of the technology used for production, which allows us to explore the degree of self-reliance of different types of organic production and their effects on sustainability. Also, we analyse how different configurations of the supply chains (i.e. physical and relational distance between producers and consumers) affect their sustainability.

Finally, the main problems found when evaluating and comparing food supply chains under the proposed methodology are presented, and then some lessons which have been learned from this exercise are briefly discussed.

[2]Based on the definitions presented in http://www.ifoam.bio/en/organic-landmarks/definition-organic-agriculture, and http://www.fao.org/organicag/oa-faq/oa-faq1/en/.

## 10.2 Methods

As mentioned in the introductory section, this chapter presents an analytical framework, with which we evaluate and compare food supply chains. The evaluation is based on a case study of three organic tomato supply chains, whose performance is evaluated in economic and biophysical terms—that is, using quantitative indicators.

The following sections present the case study: the organic tomato supply chain in Catalonia, Spain. We then discuss the following: the criteria used to differentiate between Local, Mixed and Global supply chains; the methodological framework used to define narratives, attributes and indicators, including the Multi-Scale Integrated Assessment of Societal Metabolism (MuSIASEM) approach (Giampietro 2003; Giampietro et al. 2009); and finally, the data collection methods.

### *10.2.1 Case Study*

This chapter draws on research undertaken in the context of the GLAMUR project. The GLAMUR project (*Global and Local food chain Assessment: a MUltidimensional performance-based approach*)[3] aimed to advance scientific knowledge about the impact of food chains in order to increase food chain sustainability through public policies and private strategies. In order to advance our knowledge, several case studies were carried out to evaluate, compare and analyse different food supply chains from a multidimensional perspective: i.e. using attributes and indicators from economic, social, environmental, ethical and health dimensions.

As part of the GLAMUR project, several case studies were carried out across Europe. The case studies covered five types of food supply chains (wine, cereals, fruits and vegetables, cheese and dairy products, and pork). The supply chains of each selected product were analysed in two different countries. Fourteen partners participated in the project, and the team from the Universitat Autònoma de Barcelona (UAB) carried out the case study of organic tomato supply chains presented in this chapter.

We selected the organic tomato supply chain because of the growing importance of this sector in Spain and Catalonia. The tomato is one of the most important vegetables produced in Spain. Over the course of the last five years, Spain has produced approximately 13% of the total fresh vegetables produced in Europe (EU28) and about 17–20% of total tomatoes produced.[4] According to the FAO food balance sheets, as of 2011, approximately 90% of the fresh vegetables consumed in Spain were domestically-produced (12.6 thousand tons), and the rest were

[3]For more information, visit http://glamur.eu.

[4]Data obtained from FAOSTAT web site.

imported. Tomatoes account for about 30% of the total Spanish domestic production of vegetables; about half of the domestic tomatoes produced were exported and the rest consumed internally. Thus, the importance of the vegetable and tomato sector in both the Spanish and European agricultural sectors is evident.

Spain is the European country with the largest surface dedicated to organic production (about 17%). Regarding the production of organic fresh vegetables, Spain is located in fourth position and approximately 10% of the total European organic crop area is located there. Within Spain, organic tomato production represents about one-fifth of organic fresh vegetable production (MAGRAMA 2015). In Catalonia, organic tomato production acquires even greater importance when considering the increasing participation of organic products in the Catalan market. For instance, the land cultivated with organic vegetables increased by 30% from 2000 to 2013 in Catalonia, and currently covers approximately 500 ha (GENCAT 2014). Also, according to the Centre d'Estudis d'Opinió (CEO 2015) about one-third of Catalan households buy some organic products monthly, and fresh food (i.e. fruits and vegetables) the most frequently consumed organic food. On the other hand, unlike the global character of the Catalan agricultural and agro-industrial sectors, organic production in Catalonia is local in terms of scope (Badal et al. 2011); about 45% of organic production is commercialised within Catalonia, about one-third is sold to the rest of the Spanish state, and about one-quarter of the organic production is exported (CCPAE 2015).

### *10.2.2 Differentiating Between Local, Mixed and Global Organic Tomato Supply Chains*

The GLAMUR project has brought together experts and academics in the field of food systems and food provision from around Europe. Based on their knowledge and experience of food systems and food supply chains, the following four criteria for classifying food supply chains as local or global have been defined:

1. The geographical distance between producers and consumers
2. The type of governance and organisation of the supply chain
3. The kind of resources, knowledge and technologies used
4. The role of the territory, and modes of production used, in shaping the identity of the product.

First, the geographical distance relates the number of kilometres food has to travel from production to consumption. Second, the type of governance and organisation of the supply chain relates to the degree of control exerted by local and global factors over the supply chain, expressed, for example, through the number of intermediaries between producers and consumers.

Third, the types of resources, knowledge and technologies used are linked to the presence of traditional and local know-how within the food supply chain, since standardisation processes may lead local chains to incorporate globalised logic and dynamics (Wiskerke 2003).

Fourth, the role of the territory and modes of production in shaping product identity relate to the links between production systems and cultural aspects of the territory, as expressed e.g. through protected geographical indications.

Table 10.1 illustrates the organic tomato supply chains analysed in this study, which are differentiated according to the four criteria defined by the GLAMUR project. The selection of the organic tomato supply chains was based on purposive sampling, i.e. a non-probabilistic sampling of individuals with some characteristics relevant to the research questions. This research involved producing a non-statistically representative sample. In contrast, the selection of supply chains was based on extreme or deviant sampling—a sample that included the purest or clearest-cut instances of the studied phenomenon (see Given 2008). In this chapter, we have chosen a local organic tomato supply chain, based on direct sales (e.g. vegetable basket scheme). In global terms, we have chosen a supply chain that includes production in greenhouses and two intermediaries, and this is a supply chain in which producers and consumers are geographically distant. Moreover, we have included an intermediate organic tomato supply chain, which is situated in between the local and global chains.

**Table 10.1** Criteria used to differentiate between local, mixed and global organic tomato supply chains

| Criteria | Local | Mixed | Global |
|---|---|---|---|
| Geographical distance between producer and consumer | Grown and sold in Catalonia | Grown in Catalonia and sold both in Catalonia and elsewhere | Grown in Almeria and sold both in Catalonia and elsewhere |
| Governance and organisation of the supply chain | Direct sales schemes;<br>Producer is a member of an agro/ecological farmers' network | More than 2 intermediaries (wholesale market and retailers);<br>Producer is a partner of a wholesale cooperative | More than 2 intermediaries (wholesale market and retailers);<br>Producer is a collaborator in the wholesale cooperative |
| Resource, knowledge and technologies used | Organic production outdoors;<br>Local natural resources used (e.g. water, manure) | Organic production outdoors;<br>Local natural resources used (e.g. water, manure) and organic certified fertilisers and pesticides | Organic production in greenhouses;<br>Fertigation systems and organic-certified pesticides |
| Territorial aspects shaping the identity of the product | Local and traditional varieties | Local and hybrid varieties | Hybrid varieties |

### *10.2.3 Defining Attributes and Indicators to Evaluate Food Supply Chains*

As mentioned in Gamboa et al. (2016), the pre-analytical adoption of different narratives on the food supply chains leads to non-equivalent assessments of the performance of food networks. That is, for instance, the endorsement of an economic narrative leads to a formal representation of FSCs based on productivity, prices, and supply and demand; the endorsement of an environmental narrative leads to a formal representation of FSCs based on the use of energy and resources, and the generation of waste and contamination. In this way, different narratives use different attributes to describe food supply chains. In order to measure and monitor the state of the system, attributes are mapped into formal categories, or indicators. Indicators are thus a means of representing an attribute of the system (see Table 10.2).

The definition of performance indicators has been guided by the MuSIASEM approach (Giampietro et al. 2009), which is an operationalisation of the fund-flow model developed by Georgescu-Roegen (1971). According to Georgescu-Roegen (1971), any metabolic system can be represented by using *fund* and *flow* categories. In terms of the time scale of the representation, *fund* categories transform *inflows* into *outflows*, and *flows* are either consumed or generated in order to reproduce the *funds*. Therefore, *fund* categories remain "the same" throughout the duration of the representation (e.g. capital, people, Ricardian land). *Flow* categories refer to elements that appear and/or disappear at some point during the representation (e.g. added value, water, energy, matter). What we call production is in reality a transformation process of resources into useful products and waste products: a transformation of some materials into others (the flow elements) by some agents (the fund elements).

Figure 10.1 presents the fund-flow model, as applied to a farm with three crops. In the analysis of the farm as a whole, the biophysical and economic inflows are transformed into outflows of products, waste and added value. This transformation is undertaken using the funds total human activity and total land consumed (measured in hours of human activity and hectares of land).

**Table 10.2** Definitions of narratives, attributes and indicators

| | |
|---|---|
| Narrative | Stories identifying the relations of causality used to structure the perception of the observed system and to define what is relevant among its infinitely rich dynamics (Magrini 1995; Allen and Giampietro 2006). A story-teller is an actor who chooses a given narrative to organise the information about the perceptions of external events in order to support action |
| Attribute | An observable quality used to characterise a system, which allows us to describe and evaluate its behaviour and characteristics |
| Indicator | An operational representation of attributes, defined in terms of a specific measurement or observation procedure (Gallopín 1997). The value (i.e. the state of the variable) provides information on the condition and/or trend of an attribute (or attributes) of the system considered |

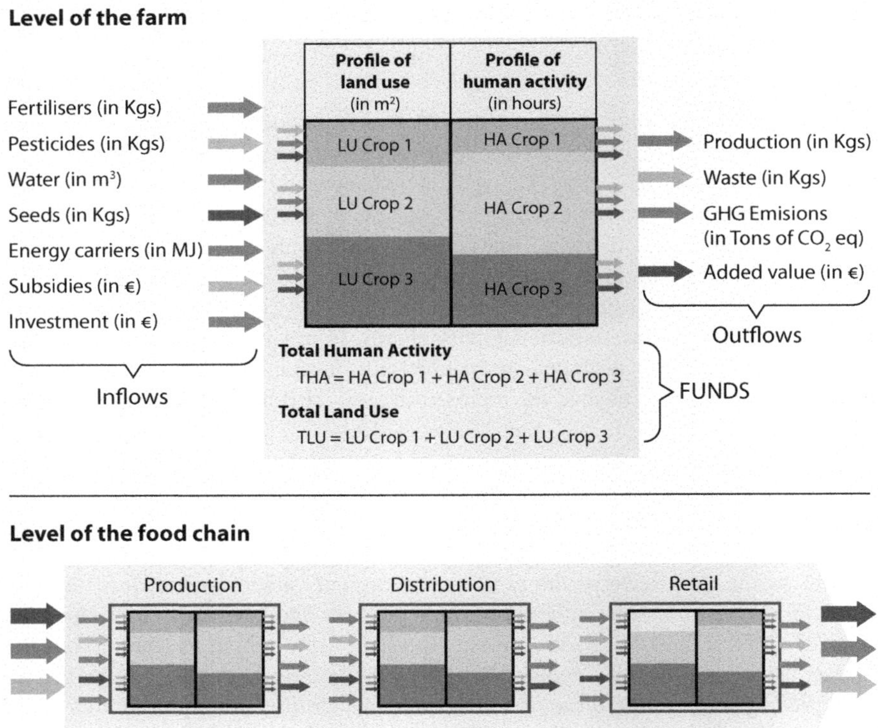


**Fig. 10.1** Representation of the fund-flow model applied to a farm and a food supply chain *Source* Own elaboration

By combining fund and flow categories in flow/fund ratios, one can describe *how the system does what it does*. That is, the flow/fund ratios represent the speed and intensity of the system's metabolic processes, i.e. the speed at which flows are consumed or produced per unit of fund category. In other words, the intensity of the use of fund categories to use/produce flows (e.g. flow of production per $m^2$ per year calculated by Production/Total Land Use and measured in $kg/m^2$ per year, or the flow of added value created per hour of human activity, calculated by Added Value/Total Human Activity and measured in €/h).

This evaluation can be performed at lower or higher levels (e.g. at the level of different crops or at the level of the food supply chain, respectively). At a lower level, biophysical and economic flows must be disaggregated by crop, as well as the *funds* of total human activity and total land use. The calculation of flow/fund ratios considers the economic and biophysical flows involved in the production of a defined crop, and the land and human activity allocated to cultivate that crop. At the level of the food supply chain, one can include the rest of the steps in the chain (e.g. distribution, retail), as shown in the lower part of Fig. 10.1. In this case, the evaluation of the performance of the food supply chains must consider the aggregation of biophysical and economic flows, as well as the human activity and land use in all the steps of the chain.

Farms and food supply chains are very different in terms of size and activities (i.e. requirements of land and human activity, volume of production, use of resources, generation of waste). Therefore, indicators must allow for the comparison of different farms. The intensive indicators mentioned above refer to one unit of two productive factors: human time and land. It is possible to compare the productivity of both farms in terms of kilograms per unit of land and/or per unit of human time allocated to produce tomatoes (measured in kg/m$^2$ and kg/h respectively). Also, it is possible to compare the energy consumption per unit of land surface or human time, or the net income per unit of land or per unit of familiar work.

## *10.2.4 Selection of Performance Indicators*

As mentioned in Sect. 10.2.3, the performance of the organic tomato supply chains is assessed using a set of economic and biophysical indicators that represent different non-equivalent narratives. The detailed process of identifying and defining narratives, attributes and indicators is presented in Gamboa et al. (2016). For the purposes of this chapter we have selected three narratives in order to assess the organic tomato supply chains: the commodity, livelihoods, and an environmental narrative. The story-tellers of the *commodity* narrative view food as a commodity, and the food supply chains are mainly described and evaluated in economic terms. The story-tellers of the *livelihood* narrative have a complex view of food, which is related to identity and culture, rural development issues and working conditions; the story-tellers of the *environmental* narrative prioritise the ecological aspects related to food supply chains, focusing on climate change impact, animal welfare and ecosystems; more information on the identification and definition of narratives can be found in Gamboa et al. (2016).

Table 10.3 presents the narratives, attributes and indicators used to evaluate the performance of the farms and the organic tomato supply chains.

## *10.2.5 Data Collection*

### 10.2.5.1 Primary Data Collection

Primary data collection encompasses two main methods to record time and resource allocation: (i) activity log and (ii) in-depth interviews.

**Activity Log**. When analysing the time allocated to different agricultural tasks, the low number of crops cultivated in small and diversified farms makes it very

**Table 10.3** Narratives, attributes and indicators used for the evaluation and comparison of organic tomato supply chains

| Narrative | Attribute | Scale | Indicator | |
|---|---|---|---|---|
| Commodity | Added value | Chain | Economic labour productivity | Amount of added value generated per hour of human activity<br>Added value is calculated by adding the salaries, the net income of the farmer/wholesaler/retailer and the taxes. The added value of the chain is calculated by aggregating the added value of the chain stages<br>The results are divided by the human activity allocated to each stage of the chain |
| | Productivity | Farm | Land productivity | Total tomato production per unit of land use allocated to produce tomatoes |
| | Job creation | Chain | Number of jobs | Number of jobs per unit of land use is calculated by dividing the total number of hours required to produce, distribute and retail tomatoes. One job equals one person working 172 h/month over 11 months. This value is divided by the surface of land use allocated to tomato production, distribution and retail |
| Livelihoods | Distribution of income | Chain | Distribution of income | Share of added value created along the chain retained by the producer<br>Added value is calculated by adding the salaries, the net income of the farmer/wholesaler/retailer and the taxes. The added value of the chain is calculated by aggregating the added value of the chain stages |
| | Farmers' income | Farm | Net income | Net income of farmers, per unit of family labour<br>Calculated by subtracting the costs of production and the |

(continued)

**Table 10.3** (continued)

| Narrative | Attribute | Scale | Indicator | |
|---|---|---|---|---|
| | | | | taxes to the gross income. The result is divided by the total number of family hours used for production |
| | Job creation | Farm | Number of rural jobs | Number of jobs per unit of land use ($m^2$)<br>It is calculated by dividing the total number of hours required to produce tomatoes by the surface of land use allocated to tomato production. One job equals one person working 172 h/month over 11 months |
| Environmental | Energy consumption | Farm | Consumption of energy carriers | Direct consumption of electricity, per unit of land use allocated to tomatoes |
| | $CO_2$ emissions | Farm | Direct $CO_2$ emissions | Tonnes of $CO_2$ equivalent directly emitted in the production stage of the chain, per unit of land use allocated to tomato production<br>It is calculated by multiplying the total burned fossil fuels by the corresponding emission factor, and dividing by the total land use allocated to produce tomatoes |
| | Biodiversity | Farm | Number of crops | Number of crops on the farm. It is a proxy indicator of agro-biodiversity, which refers to the variety and variability of animals, plants and microorganisms that are important to food and agriculture. It includes practices undertaken to manage water and land uses, as well as local knowledge and culture that shape agro-biodiversity (FAO 1999) |

difficult to estimate the time allocated to different products. This is not a case of semi-diversified farms with few crops, where the time allocated to different tasks and crops is quite easy to estimate. In the case of monoculture farms, human time is allocated to only one crop. Therefore, the activity log has been applied to one small farm involved in the Local supply chain. An activity log is a written record of how, in this case, farmers spend their production time.

Data collection took place from April to November 2013. Farmers were asked to record all activities related to tomato production, including time and resources (e.g. land, machinery, equipment) allocated to the activity. For instance, when applying agricultural treatments to the plants, farmers had to record the time spent in preparing and applying treatments, the quantities of the chemical compounds and their cost.

The information collected by means of the activity log encompasses time allocated to production and commercialisation tasks, the number and the costs of resources (e.g. agricultural inputs, gas oil, electricity) used in production and commercialisation, as well as the use of machinery and equipment, and distance travelled for commercialisation.

After collecting the information, certain production parameters were calculated (e.g. land productivity, direct and indirect costs) and compared with figures in the literature in order to check the reliability of the gathered data.

**In-Depth Interviews** Interviews were carried out with two farmers (Local and Mixed), two wholesalers and one retailer involved in the different tomato supply chains. A common objective of the interviews was description of the operation of the different stages in which the actors participate. The main aspects considered in the interview were: the description of productive, distribution and retail activities (depending on the actor interviewed), and the description of operational aspects of the enterprise, as well as the advantages and obstacles of the supply chain in which they participate.

Interviews allowed us to obtain additional data on salaries, purchase and sale prices, and the quantity of product produced/distributed/sold. Also, wholesalers and retailers were asked about labour costs and labour requirements, the costs of equipment, machinery, transport and rent of the premises, consumption of energy carriers (i.e. electricity, natural gas and diesel), total turnover and turnover per product, the total weight of handled products and by-products, the share of products' sales to different customers (e.g. retail shops, school canteens, small distributors) and the origin of products.

In the case of the medium-size farmer involved in the Mixed supply chain, data correspond to 2014. The information provided by the organic wholesaler corresponds to the period between July 2013 and June 2014.

#### 10.2.5.2 Secondary Data Sources

Secondary data sources have been used to both check the quality of data obtained by primary data collection techniques and to complement that information. The works of Moleres (2009), Capgemini Consulting (2009), Guzmán et al. (2008), Montero et al. (2011), Sanyé et al. (2012) and Sanyé-Mengual et al. (2013) were used to check the quality of information obtained in primary data collection techniques. Data for the farms involved in the global organic tomato supply chain were obtained from Moleres (2009).[5] Economic information was updated according to the consumer price index. Sanyé et al. (2012) and Capgemini Consulting (2009) were also used to provide complementary data for the distribution stage.

## 10.3 Results

The results presented here include a description of the organic tomato supply chains, a description of the economic and biophysical attributes as well as the indicators, and a comparison of three organic tomato supply chains in relation to three non-equivalent narratives. In this way we can identify the expected trade-offs of integrated assessment and associate them with the legitimate-but-contrasting views on food supply chains.

### *10.3.1 The Organic Tomato Supply Chains*

In the following sections, the three organic tomato supply chains are described. More specific information regarding the costs and revenues, as well as the energy consumption of the supply chains, are presented in the Appendix.

#### 10.3.1.1 Local Organic Tomato Supply Chain

The local organic tomato supply chain focuses on direct sales schemes (i.e. a vegetable basket scheme), in which the farmer sells tomatoes directly to consumers. In this case, we have considered a small farm of less than half a hectare, of which about 15% is allocated to tomato production. There are four workers (including the farmer) on the farm, which is rented and mostly irrigated with dripping systems. The farmer produces seasonal vegetables using organic principles.[6]

The farmer produces as many as 20 different products during the year. Most of the seeds are reproduced on the farm or obtained from an organic seed bank. Production tasks are performed mostly manually or using small machinery. Manure is used as fertiliser, which is obtained locally from organic livestock.

After production and harvesting, tomatoes are processed on-farm. Processing consists mainly of cleaning the product and preparing individual boxes and/or bulk

[5]Human activity was calculated using the labour costs reported in Moleres (2009). A workload of 45 h/week and a net salary of €1000/month were considered.

[6]The farmers are members of a network of agroecological farmers. The network has developed a Participatory Warranty System (SPG) by means of a participatory process with producers and consumers. The SPG involves more than just production issues, and is guided by the aim of developing social transformation projects, promoting fair relations and small farmers' autonomy.

products. This task is performed one day per week, in the morning just after harvesting and before distribution takes place. Products are then transported by one or two workers to the distribution points, which include consumption cooperatives, organic grocery stores and school canteens.

#### 10.3.1.2 Mixed Organic Tomato Supply Chain

This tomato supply chain considers a medium-size Catalan farmer, producing tomatoes on a semi-diversified farm (i.e. between 4 and 5 crops on a family farm). The farm size is 1.8 ha, of which about 15% is occupied by tomatoes. Other products occupy approximately 60% of the land used, and green fertilisers are used on approximately 25% of land. Production is carried out outdoors according to organic principles, and seeds are mostly commercial hybrids. Some paid labour is used during the summer. After harvesting, processing takes place on-farm and consists of the cleaning and packaging of the tomatoes.

The tomatoes are then sold to the main organic wholesaler of Catalonia, who distributes tomatoes to retailers, small wholesalers, farmers, consumption cooperatives, restaurants and school canteens. The organic wholesaler is a cooperative, of which the farmer is a partner. Members of the wholesale cooperative plan their production jointly and supply about 20% of the commercialised products. Approximately 10% of the products are supplied by close collaborators, with whom the cooperative plans the supply of products at the beginning of the year. The remaining 70% of products are supplied by non-members.

The wholesale cooperative then sells the tomatoes (among other fruits and vegetables) to organic shops specialising in fruits and vegetables, and these shops are considered the final stage of this chain.

#### 10.3.1.3 Global Organic Tomato Supply Chain

In this supply chain, tomatoes are produced in greenhouses in Southern Spain (Almeria). Greenhouses are 3 ha on average and are equipped with ventilation systems. Cultivation is carried out following organic principles, and the automated dripping system delivers nutrients to the plants. Inputs are purchased in the market and seeds are mostly commercial hybrids. In most cases, processing takes place on-farm and consists of cleaning and packaging the tomatoes.

For the evaluation of this chain, only a small fraction of the production is considered as being sold to the organic wholesaler in Barcelona (i.e. about 41 Tonnes). The rest of the production is sold through different distribution and commercialisation channels. The effect of this issue will be noted when performing the evaluation of the global supply chain.

Tomatoes are then transported about 800 km from the production site to the main organic wholesaler in Barcelona. As already mentioned, about 60% of the tomatoes sold by what is considered an organic wholesaler come from the province of Almeria (Andalusia). Then, the wholesaler supplies tomatoes (and other vegetables and processed products) to local supermarkets, retailers and local distributors. The wholesale cooperative then sells tomatoes (among other fruits and vegetables) to organic shops specialising in fruits and vegetables, which are considered the final stage of this chain.

#### 10.3.1.4 The Organic Wholesale Cooperative

Eleven people work in the organic wholesaler.[7] The 2000 $m^2$ premises is located in Mercabarna, the main wholesale market in Catalonia. The turnover of the cooperative reaches about €5 million. Tomatoes represent 2.8% in terms of value with respect to total sales, and 2.5% in terms of volume. About 60% of the tomatoes are supplied by Almeria and the remaining 40% are supplied by Catalonia. Therefore, tomatoes from the Mixed as well as from the Global supply chains pass through the organic wholesaler.

About 70% of tomatoes handled by the organic wholesaler are sold to organic shops; about 13% are sold to small wholesalers; 13% are sold to other farmers; and the rest of the tomatoes are sold to consumer cooperatives, restaurants and school canteens. According to these figures, we have considered the commercialisation in organic shops the last step in both the Mixed and Global chains.

#### 10.3.1.5 The Organic Retail Shop

As a final stage of the mixed and global chains we have considered sales in small grocery stores that specialise in fresh organic products. According to CEO (2015), about half of organic buyers purchase food in shops specialising in organic products.

For this study, we have collected data from a shop specialising in organic fresh fruits and vegetables; a small portion of its products are packed (preserves, juices and dried pulses). The 80 $m^2$ store is located in Barcelona. The turnover is about €90 thousand a year. More than 80% of the fruit and vegetables they sell come from the main organic wholesaler in Barcelona. The shop is managed by two owners who work 6 h a day, 6 days a week.

### *10.3.2 Performance of the Organic Tomato Supply Chains*

As mentioned in Sects. 10.2.3 and 10.2.4, the comparison of organic tomato supply chains was carried out according to different narratives: the commodity, the

[7]A workload of 40 h/week and 48 weeks/year is assumed.

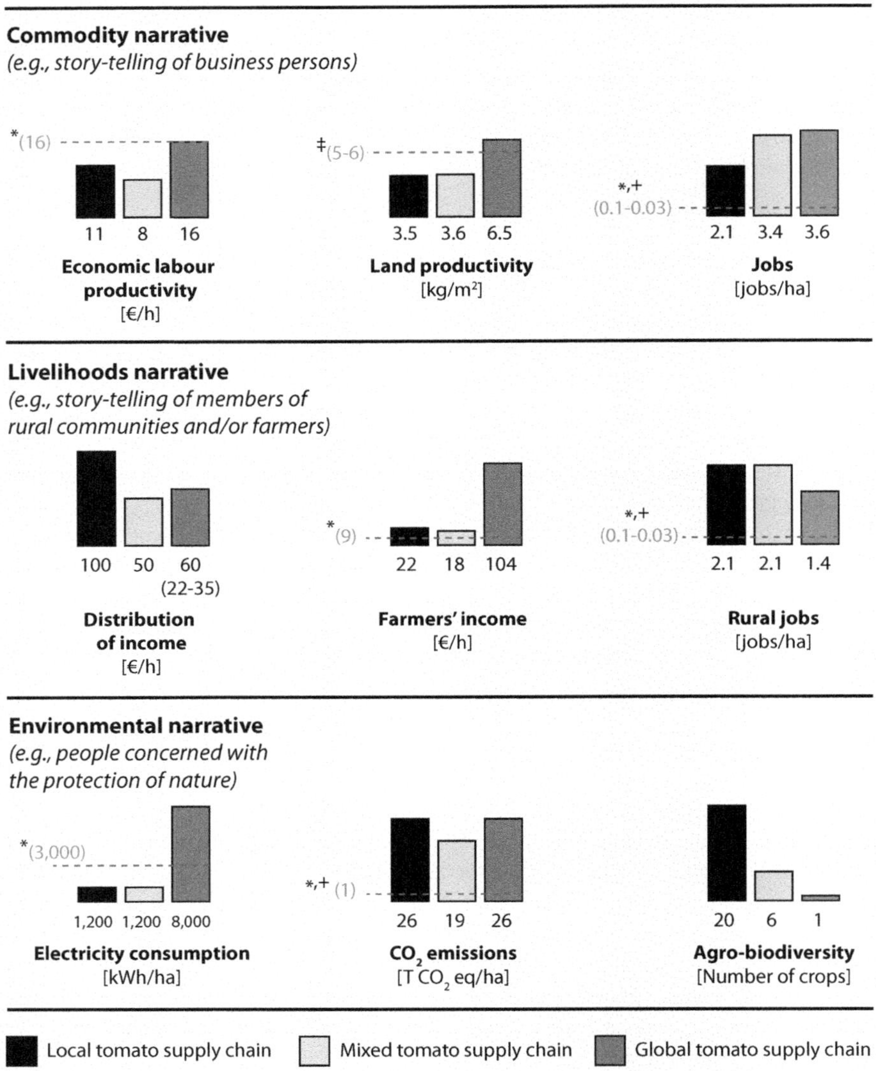


**Fig. 10.2** Characterisation of the performance of the organic tomato supply chain, reflecting the existence of different story-tellers (i.e. narratives). *Source* Own elaboration, adapted from Gamboa et al. (2016)

livelihoods and the environmental narratives. Figure 10.2 presents the performance of the organic tomato supply chains according to indicators derived from these narratives.

The values in brackets refer to benchmarks based on (*) the average of the Spanish agricultural sector; (‡) average of the Spanish organic tomato sector (+) the average of the Spanish agricultural sector without considering emissions and jobs in distribution and retail activities. To calculate the indicator "number of jobs", the total number of hours required in each chain is calculated by assuming that one worker works 172 h/month, over 11 months.

According to the indicators selected in the *Commodity story-telling*, the global supply chain creates greater added value per hour of paid work allocated to producing, distributing and retailing tomatoes. The farms producing tomatoes for the global chain also present higher productivity, which is due to cultivation in controlled environments in greenhouses and the fertigation system.[8] In fact, the global farm is the only one that has higher productivity than the Spanish average for organic tomato production, which is about 5–6 kg/m$^2$ (in the last few years; MAGRAMA 2012, 2013, 2014, 2015). These larger productivities are at the expense of the higher consumption of energy carriers (see below).

The global and mixed supply chains create a higher number of jobs per m$^2$ of land use. This is basically due to the contribution of activities performed in wholesaling and retailing (where land use is negligible compared to production).

Story-tellers of the *Livelihoods* narrative usually argue that local food supply chains favour producers in terms of distribution of the added value created along the supply chains. This is in line with the results of the evaluation: the producer involved in the Local supply chain retains a higher share of the added value created along the chain due to the absence of intermediaries. Farmers participating in Mixed and Global supply chains also retain a large share of the added value created along the chain. It should be noted that the farmer of the Global chain has to sell their production via other commercialisation channels, which usually include more intermediaries, and thus the farmer retains a much lower share of added value (number in brackets in Fig. 10.2).

When it comes to comparing the net income per hour of familial labour allocated to producing and commercialising tomatoes, the Global supply chain performs much better than the other two supply chains. In the three supply chains, farmers obtain a higher level of income per hour of familial labour than the average farmer operating in the Spanish agricultural sector. It should be noted that when considering the whole production of tomatoes of the farmers involved in the Global organic tomato supply chain, of which an important part is sold at half of the price considered in this study, the farmer's income is reduced to one-third, and thus is quite close to the income of farmers operating in the local and mixed chain (number in brackets in Fig. 10.2). According to Renting et al. (2003), the elimination of intermediaries would allow farmers to retain a higher share of the value created along the supply chain, which would ameliorate the reduction of farmers' income due to price squeeze. However, the time and resources needed to commercialise in SFSC may decrease the profitability of farmers (Jarosz 2008; Ilbery and Maye 2005), and this should be linked with the results obtained here: e.g. farmers

[8]Fertigation refers to the injection of soil amendments, fertilizers and other water-soluble products into the irrigation system.

involved in the Local supply chain also carry out production tasks linked with other crops and the commercialisation tasks of the whole production.

When comparing the jobs created in rural areas, the Local and Mixed supply chains perform better than the Global supply chain. This is in line with the argument that LFSC helps maintain the population in rural areas and helps people deal with ageing issues (De Roest and Menghi 2000; Knickel and Renting 2000; European Parliament 2016). In the case of the Local supply chain, production and distribution is performed by the same people who are living and working in rural areas. In contrast, the Mixed and Global chains create jobs in rural areas only in the production stage of the chain, and the farm operating in the Mixed supply chain is more labour intensive than the farm operating in the Global supply chain. The differences in labour requirements at the farm level stem from the different stages of the production cycle. Local and Mixed farms require more labour to complete tasks related to production (e.g. preparing the land, seeding and/or weeding) than the Global farm. In contrast, and due to higher productivities, the Global farm requires more labour than Local and Mixed farms during the period of harvesting. In the case of the Local and Mixed supply chains, labour is required to produce other seasonal vegetables, which increases the amount of labour generated by these farms.

Finally, the indicators in the *environmental* story-telling, concerned with the preservation of ecosystems and the integrity of natural processes, support the Local and Mixed tomato supply chains. Farms working in the Global chain consume greater levels of electricity, in order to control the temperature inside greenhouses, which increases productivity and net income at the expense of higher biophysical costs. In fact, electricity consumption more than doubles the average electricity consumption per hour of human activity in the Spanish agricultural sector. It is worth noting that the farm operating in the Mixed supply chain has similar electricity consumption to the farm operating in the Local supply chain, which is basically used to power the irrigation system. However, the Mixed supply chain increases its electricity consumption in the wholesale and retail stages of the chain.

The Local supply chain has higher $CO_2$ emissions per unit of land use. However, this evaluation considers only direct $CO_2$ emissions, i.e. emissions generated by the combustion of fossil fuels. This higher level of $CO_2$ emissions is due to the use of small trucks, which are not fully loaded (used in the distribution of the vegetable baskets in the city of Barcelona). These results indicate how performance could be improved in the distribution system of local food supply chains: coordination between producers and the implementation of storage and distribution centres across the city are but two actions that would make distribution tasks in direct selling schemes easier. In the case of the farm involved in the Global supply chain, $CO_2$ emissions are produced indirectly: that is, emissions are generated in the energy sector when electricity is generated.

Agro-biodiversity is much higher on the Local farm, which is based on a diversified system of production. Here, farmers cultivate as many as 20 different crops during the year to supply consumption groups, school canteens or the local market with their own products. This is an expression of the different nature of the tomato supply chains analysed. The local chain aims to produce and supply fresh seasonal vegetables, forming a model that is based on relations of trust between producers and consumers. The global chain, on the other hand, aims to supply tomatoes all year round, using a system which is detached from natural cycles. In the case of the farmers participating in the Mixed supply chain, agro-biodiversity is medium: they produce between 5 and 6 different crops. This is a result of belonging to a cooperative whose members have reached an agreement to reduce the number of crops (in order to simplify the work on the farm), but have also agreed to continue producing a wide range of products between all members of the cooperative. Also, farmers are not dependent on one product, and thus they avoid economic shocks due to crop failures.

## 10.4 Discussion

As mentioned in the introduction, there is little evidence based on empirical work that evaluates and compares the performance of different food supply chains under the lens of sustainability (Forssell and Lankoski 2015). Most of the studies focus on specific steps in the food supply chains for their comparison (see Matopoulus et al. 2015). Thus, the main aim of this paper has been to propose and implement a framework with which to evaluate and compare food supply chains.

One of the first difficulties when analysing food supply chains is the fact that most producers, wholesalers and retailers handle a diverse range of products. Hence, it is very difficult to study and evaluate the performance of the food supply chain of one single product.

Primary data collection was possible only in the case of the Local and Mixed supply chains. In the Local case, the use of time and resources specific to tomato production were collected by means of the activity log. In the case of the farm participating in the Mixed supply chain, time and resources allocated to the production of tomatoes was obtained via in-depth interviews and the estimations were provided by the farmer. In such cases, the use of new technologies can make data collection easier. For instance, the use of a time-tracker application on a smartphone may allow the farmer to fill in the activity log on-site and collect the information on time spent on, for example, tasks and crops respectively. This was not the case for the farmer operating in the Global tomato supply chain, since all activities are geared towards the production of tomatoes.

The overall use of resources and activities performed by the wholesaler and retailer was obtained from an in-depth interview. The economic and biophysical resources used by these actors were then allocated according to the weight of the different products. That is, for instance, the time (and resources) required to handle 1 kg of tomatoes is equal to the time (and resources) required to handle 1 kg of another product. These assumptions may entail biases in the evaluation of the performance of the food supply chains. For instance, fresh products are stored in a refrigerated chamber. There, the requirement for electricity to cool depends on the quantity of products, their volume and weight, the length of time they spend being stored, as well as the initial temperature of the product, among other factors. This situation should also be noted in relation to the human time and effort needed to handle products. Here, weight also plays an important role in the allocation of human labour (as has been taken into account in this study), but volume also plays an important role in determining labour requirements: e.g. the time needed to handle boxes of lettuces and tomatoes is not necessarily proportional to the weight of the boxes. In this sense, smartphones and time-tracker applications can facilitate the implementation of activity logs for different products.

Another difficulty with which we were faced when analysing food supply chains is that some actors participate in more than one distribution channel. In order to deal with this issue, we have presumed that the total number of tomatoes produced by the farmer consists of those sold directly to consumers. In the Mixed supply chain, all tomatoes produced by the farmer are sold to the wholesaler. Therefore, no assumptions were needed. However, in the case of the farms participating in the Global chain, 15% of total production is sold to the wholesaler considered in this study, and the rest of the production is sold to other intermediaries, at lower prices (€0.8/kg this is the actual value) rather than to the organic wholesaler cooperative located in Barcelona (€1.6/kg). In this case, certain indicators (e.g. farmers' income) are presented using two values: one that takes into consideration only a fraction of the total product sold through the organic wholesale cooperative, and a second value (in brackets in Fig. 10.2) that indicates the performance of the farm after tomato production as a whole has been taken into consideration. For the purpose of this evaluation, it has been hypothesised that the wholesaler sells all these tomatoes to organic shops, which are responsible for buying about 70% of the tomatoes sold by the wholesaler.

Therefore, on the basis of this study, the supply chain approach is useful in evaluating the performance of a food supply chain and looking for aspects to improve, rather than comparing different food supply chains. As mentioned above, the assumptions required for the calculation of the indicators may affect the results of the evaluation and tip the scale towards either side. This is the case for small and diversified horticultural farms as well as a wholesaler, as the adequate allocation of resources and costs across products is very difficult.

It is interesting to recall that the farm operating in the Local supply chain is part of the Catalan Network of Agro-ecological Farmers, and these farmers mostly produce on small diversified farms. In these cases, one of the main strategies of farmers is to offer a diversified portfolio of products. This means that some crops are produced because of their attractiveness to the consumer rather than for their profitability—something which is compensated for by including in the vegetable basket products which are more profitable but less attractive to the consumer. Therefore, tomatoes produced and commercialised through the Local supply chain acquire a different meaning: they are part of a diversified vegetable basket and the flagship of the summer. In this context, the comparison of different food supply chains must be based on the comparison of vegetable shopping baskets, rather than on individual products. Sanye et al. (2012) is an example of this sort of evaluation, but it lacks the evaluation of stages of the chain prior to purchasing.

As mentioned in the Sect. 10.2, the evaluation of the Local and Mixed organic tomato supply chains are based on specific farms that operate in each organic tomato supply chain, and thus is not meant to be representative. However, information is very useful for stakeholders involved in the supply chains, in order that they can identify critical points to intervene and/or to improve the different phases and stages of the studied supply chains. In order to facilitate the use of information by stakeholders, it is of fundamental importance that biophysical indicators are accompanied by qualitative information. This information not only complements quantitative figures, but also gives meaning to the numbers. For instance, when evaluating either the number of "jobs" or "rural jobs", further considerations related to the quality of jobs must be taken into account (e.g. salaries, workload, working conditions, vacations, social security). Moreover, when evaluating the distribution of added value along the chain, there are additional factors (other than economic) that influence the results, such as the distribution of power and the role in defining prices along the chain. For instance, in the Global supply chain, market dynamics play a key role in defining prices. Over the last few years, tomatoes from Morocco and the Netherlands have increased competition because of their low prices, which in turn has forced Spanish producers to sell at even lower prices. In the case of the Local supply chains considered in this case study, price is a matter of agreement between producers and consumers, and market prices are considered to be reference values which are used to define prices. In the case of the Mixed supply chain, prices are defined according to the market, but the lack of intermediaries gives farmers the ability to obtain relatively good prices compared to the average prices paid by intermediaries. The producers are partners of the wholesale cooperative, and production is planned between the partners of the cooperative in order to ensure that products will be sold through the wholesale cooperative.

The results relating to electricity consumption question the self-reliance of organic production under greenhouses, which is highly dependent on the external input of electricity. Moreover, it also highlights the fact that the organic production is not always self-reliant and its sustainability is also related to the supply chain via which products are commercialised. In fact, the post-harvest stages of the supply chain increase electricity consumption by about 15 times in the Mixed chain and by more than 4 times in the Global supply chain, whereas the electricity consumption of the Local supply chain is almost equal to that of the farm operating in that supply chain. In terms of $CO_2$ emissions, the results obtained in this present study reinforce the idea that inefficiencies in transport may counteract the potential benefits of a short distance between production and consumption Coley et al. (2009). Other factors can also counteract the potential benefits of producing locally, such as the requirement to store food in a cold environment or the requirement that a specific food be produced in greenhouses (Gomiero et al. 2011). These results indicate that the context heavily influences the energy and emission performance of the food supply chains.

## 10.5 Conclusions

This chapter has presented the evaluation of three organic tomato supply chains under different narratives. Story-tellers of the Commodity narrative support the global organic tomato supply chain since it generates higher added value per hour of human activity allocated to produce, distribute and retail organic tomatoes, creates more jobs along the chain, and provides higher levels of productivity of tomatoes per hectare. The story-tellers of the Livelihood narrative consider the local and mixed chains to be more sustainable since the farmer obtains a higher share of the value generated along the chain, and these supply chains help to maintain the rural population by providing more rural jobs. Income for farmers is important and must be improved if people are to stay in, and be attracted to, rural areas. People of the Environmentalist narrative favour local and mixed supply chains. The lower consumption of electricity and higher agro-biodiversity support this choice. The $CO_2$ emissions from fossil fuel consumption are similar for all supply chains, but provide some indications of where to improve the operation of the Local supply chain. Information is very useful for stakeholders involved in the evaluation, as it can be used either to identify critical points at which to intervene and/or to improve the different phases of the chain.

Based on the knowledge acquired through this study, we suggest that the comparison of local and global organic tomato supply chains is somewhat flawed because chains are not interchangeable. These chains are of different natures and have different purposes. Seasonal production is a differentiating element: local and mixed supply chains produce tomatoes on a seasonal basis and the global chain supplies tomatoes mainly when the open-air production season in Spain has ended. If we accept that tomatoes can be consumed all year round, we would be looking at complementary organic tomato supply chains rather than competing ones.

This chapter has also demonstrated the usefulness of the food supply chain approach, as it can be used to identify the critical points at which to intervene and/or improve the performance of food supply chains. Also, we have presented the main difficulties inherent in implementing this analytical framework and have proposed some strategies to deal with them. The most salient obstacles in implementing the analytical framework are the allocation of human activity, that of farmers, wholesalers and retailers, and resources to produce and handle different crops. Also, owing to the differences in the purpose and strategies of the farmers involved in the different food supply chains, the evaluation and comparison of food supply chains must be based on the evaluation of a purchasing basket, rather than on a single product.

Finally, we have demonstrated that organic production is not enough to achieve a sustainable food provision system. Self-reliance is questioned when production is carried out under greenhouses in order to provide food out of season. Moreover, we have noted that the context and the food supply chain in which organic food is commercialised also determine the sustainability of organic food production.

## Appendix

See Tables 10.4, 10.5 and 10.6.

**Table 10.4** Characteristics of the Local organic tomato supply chain

| | Units | Production | Distribution | Chain |
|---|---|---|---|---|
| Total land use | $m^2$ | 3000 | 80 | 3080 |
| Land use for tomatoes | $m^2$ | 400 | 3 | 403 |
| Human activity | Hours | 126 | 35 | 161 |
| Production/purchases | kg | 1385 | 1316 | 1385 |
| Sale price | € | | 2.5 | 2.5 |
| Sales | kg | 1316 | 1316 | 1316 |
| *Costs* | | | | |
| Agricultural inputs/bought tomatoes | € | 499 | 50 | 549 |
| Cost machinery | € | 39 | 2 | 41 |
| Labour costs | € | 672 | 153 | 825 |
| Transport | € | 0 | 62 | 62 |
| Direct costs | € | 1210 | 267 | 1477 |
| Rent | € | 34 | 10 | 44 |
| Other indirect costs | € | 6 | | 6 |
| Indirect costs | € | 40 | 10 | 50 |
| Total costs | € | 1250 | 277 | 1527 |
| *Family labour* | | | | |
| Family labour | € | 142 | 78 | 220 |
| Family labour | Hours | 63 | 17 | 80 |
| *Energy consumption* | | | | |
| Diesel consumption | Litres | 10 | 30 | 40 |
| Energy consumption—diesel | kWh | 103 | | 103 |
| Energy consumption—diesel transport | kWh | | 295 | 295 |
| Electricity consumption | kWh | 47.8 | 0.3 | 48 |

**Table 10.5** Characteristics of the mixed organic tomato supply chain

| | Units | Farm | Distribution | Retail | Chain |
|---|---|---|---|---|---|
| Total land use | $m^2$ | 18,000 | 2000 | | 20,000 |
| Land use—tomatoes | $m^2$ | 2600 | 6.6 | 33 | 2639 |
| Human activity | Hours | 1044 | 69 | 591 | 1704 |
| Production/purchases | kg | 9300 | 8835 | 8197 | 9300 |
| Sale price | € | 1.59 | 2.2 | 3.5 | 3.5 |
| Purchasing price | €/kg | 1.59 | 1.59 | 2.20 | |
| Sales | kg | 8835 | 8197 | 7958 | 7958 |
| *Costs* | | | | | |
| Agricultural inputs/Bought tomatoes | € | 1274 | 14,048 | 17,512 | 32,834 |
| Cost machinery | € | 230 | | | 230 |
| Labour costs | € | 4658 | 1683 | 5031 | 11,372 |
| Transport | € | 643 | 403 | 597 | 1643 |
| Direct costs | € | 6805 | 16,134 | 23,140 | 46,079 |
| Rent | € | 266 | 307 | 1671 | 2244 |
| Other indirect costs | € | 471 | 478 | 723 | 1672 |
| Amortisation | € | 735 | | 1033 | 1769 |
| Indirect costs | € | 1472 | 785 | 3428 | 5685 |
| Total costs | € | 8277 | 16,919 | 26,567 | 51,763 |
| *Family labour* | | | | | |
| Cost family labour | € | 2177 | | | 2177 |
| Family labour | Hours | 324 | | | 324 |
| *Energy consumption* | | | | | |
| Diesel consumption | Lts | 70 | 118 | 3 | 191 |
| Energy consumption—diesel | kWh | 707 | | | 707 |
| Energy consumption—diesel transport | kWh | 0 | 1193 | | 1193 |
| Electricity consumption | kWh | 311 | 654 | 3619 | 4584 |

**Table 10.6** Characteristics of the global organic tomato supply chain

| | Units | Farm | Distribution | Retail | Chain |
|---|---|---|---|---|---|
| Total land use | $m^2$ | 6710 | 2000 | | 8710 |
| Land use—tomatoes | $m^2$ | 6710 | 29 | 145 | 6884 |
| Human activity | Hours | 1825 | 306 | 2616 | 4747 |
| Production/purchases | kg | 43,616 | 41,147 | 36,265 | 43,616 |
| Sale price | € | 1.59 | 2.2 | 3.5 | 3.5 |
| Purchasing price | €/kg | | 1.59 | 2.20 | |
| Sales | kg | 41,147 | 38,267 | 35,209 | 35,209 |
| *Costs* | | | | | |
| Agricultural inputs/bought tomatoes | € | 8298 | 65,424 | 77,479 | 151,201 |
| Cost machinery | € | 680 | | | 680 |
| Labour costs | € | 9130 | 7838 | 22,259 | 39,228 |
| Transport | € | 1679 | 1878 | 2641 | 6197 |
| Direct costs | € | 19,787 | 75,140 | 102,379 | 197,306 |
| Rent | € | 854 | 1431 | 7392 | 9677 |
| Other indirect costs | € | 408 | 2224 | 3201 | 5833 |
| Amortisation | € | 5564 | | 4573 | 10,137 |
| Indirect costs | € | 6,826 | 3655 | 15,166 | 25,646 |
| Total costs | € | 26,613 | 78,794 | 117,545 | 222,952 |
| *Family labour* | | | | | |
| Cost family labour | € | 2344 | | | 2344 |
| Family labour | Hours | 373 | | | |
| *Energy consumption* | | | | | |
| Diesel consumption | Lts | 36 | 551 | 12 | 599 |
| Energy consumption—diesel | kWh | 361 | | | 361 |
| Energy consumption—diesel transport | kWh | 889 | 5554 | | 6443 |
| Electricity consumption | kWh | 5368 | 3047 | 16,011 | 24,426 |

# References


Allen, T. F. H., & Giampietro, M. (2006). Narratives and transdisciplines for a post-industrial world. *Systems Research and Behavioral Science, 23,* 1–21.

Badal, M., Binimelis, R., Gamboa, G., Heras, M., Tendero, G. (2011). Arran de terra, indicadors participatius de Sobirania Alimentària a Catalunya. Resum de l'informe. Associació Entrepobles & Institut d'Economia Ecològica i Ecologia Política.

Blanke, M., & Burdick, B. (2005). Food (miles) for thought—Energy balance for locally-grown versus imported apple fruit. *Environmental Science and Pollution Research, 12*(3), 125–127.

Capgemini Consulting. (2009). Estudio de la cadena de valor y formación de precios del tomate. Observatorio de precios de los alimentos, Ministerio de Medio Ambiente y Medio Rural y Marino.

Centre d'Estudis d'Opinió (CEO). (2015). *Baròmetre 2015 de Percepció i Consum d'Aliments Ecològics. Generalitat de Catalunya*. Available at: http://pae.gencat.cat/web/.content/al_alimentacio/al01_pae/05_publicacions_material_referencia/arxius/15_Barometre.pdf. Accessed on July 11, 2016.

Cobb, D., Dolman, P., & O'Riordan, T. (1999). Interpretations of sustainable agriculture in the UK. *Progress in Human Geography, 23*(2), 209–235.

Coley, D., Howard, M., & Winter, M. (2009). Local food, food miles and carbon emissions: A comparison of farm shop and mass distribution approaches. *Food Policy, 34,* 150–155.

Consell Català de Producció Agraria Ecològica (CCPAE). (2015). *Recull d'estadístiques del sector ecològic a Catalunya 2000–2015*. Available at: http://www.ccpae.org/docs/estadistiques/2015/00_2015_ccpae_recull-estadistiques.pdf. Accessed on January 15, 2016.

De Roest, K., & Menghi, A. (2000). Reconsidering "Traditional" food: The case of Parmigiano Reggiano Cheese. *Sociologia Ruralis, 40,* 439–451.

European Parliament. (2016). *On how the CAP can improve job creation in rural areas.* (2015/2226(INI)). Committee on Agriculture and Rural Development.

FAO. (1999). *Agricultural biodiversity, multifunctional character of agriculture and land conference, background paper 1*. Maastricht, Netherlands, September 1999.

Forssell, S., & Lankoski, L. (2015). The sustainability promise of alternative food networks: An examination through "alternative" characteristics. *Agriculture and Human Values, 32,* 63–75.

Gallopín, G. (1997). Indicators and their use: Information for decision-making. In B. Moldan & S. Billharz (Eds.), *Sustainability indicators*. Report of the project on Indicators of Sustainable Development. Scope 58.

Gamboa, G., Kovacic, Z., Di Masso, M., Mingorría, S., Gomiero, T., Rivera-Ferré, M., et al. (2016). The complexity of food systems: Defining relevant attributes and indicators for the evaluation of food supply chains in Spain. *Sustainability, 8,* 515.

Generalitat de Catalunya (GENCAT). (2014). *Superfícies i produccions dels conreus agrícoles any 2014*. Available at: http://agricultura.gencat.cat/web/.content/de_departament/de02_estadistiques_observatoris/02_estructura_i_produccio/02_estadistiques_agricoles/01_llencols_definitius/fitxers_estatics/produccions_catalunya/Llencols_Catalunya_2014.pdf. Accessed on July 11, 2016.

Georgescu-Roegen, N. (1971). *The entropy law and the economic process*. Cambridge, MA, USA: Harvard University Press.

Giampietro, M. (2003). *Multi-scale integrated analysis of agroecosystems*. Boca Raton, FL, USA: CRC Press.

Giampietro, M., Mayumi, K., & Ramos-Martin, J. (2009). Multi-scale integrated analysis of societal and ecosystem metabolism (MuSIASEM): Theoretical concepts and basic rationale. *Energy, 34,* 313–322.

Given, L. M. (Ed.). (2008). *The SAGE Encyclopedia of qualitative research methods*. US: SAGE Publications Inc.

Gomiero, T., Pimentel, D., & Paoletti, M. G. (2011). Environmental impact of different agricultural management practices: Conventional vs organic agriculture. *Critical Reviews in Plant Sciences, 30*(1), 95–124.

Guzmán, G., García, A., Alonso, A., & Perea, J. (2008). *Producción Ecológica: Influencia en el Desarrollo Rural*. Ministerio de Medio Ambiente y Medio Rural y Marino Secretaría General Técnica. Centro de Publicaciones.

Hinrichs, C. C. (2000). Embeddedness and local food systems: Notes on two types of direct agricultural market. *Journal of Rural Studies, 16,* 295–303.

Ilbery, B., & Maye, D. (2005). Food supply chains and sustainability: Evidence from specialist food producers in the Scottish/English borders. *Land Use Policy, 22,* 331–344.

Jarosz, L. (2008). The city in the country: Growing alternative food networks in Metropolitan areas. *Journal of Rural Studies, 24*(3), 231–244.

Jones, A. (2002). An environmental assessment of food supply chains: A case study on dessert apples. *Environmental Management, 30*(4), 560–576.

Knickel, K., & Renting, H. (2000). Methodological and conceptual issues in the study of multifunctionality and rural development. *Sociologia Ruralis, 40,* 512–528. doi:10.1111/1467-9523.00164.

Magrini, T. (1995). Ballad and gender: reconsidering narrative singing in Northern Italy. *Ethnomusicology Online*. Available online: http://www.umbc.edu/eol/magrini/magrini.html. Accessed on January 15, 2016.

Marsden, T., Banks, J., & Bristow, G. (2000). Food supply chain approaches: Exploring their role in rural development. *Sociologia Ruralis, 40*(4).

Matopoulos, A., Barros, A. C., & van der Vorst, J. G. A. J. (2015). Resource-efficient supply chains: A research framework, literature review and research agenda. *Supply Chain Management: An International Journal, 20*(2), pp. 218–236. doi:http://dx.doi.org/10.1108/SCM-03-2014-0090.

Ministerio de Agricultura, Alimentación y Medio Ambiente (MAGRAMA). (2012). *Agricultura Ecológica*. Estadísticas 2011.

Ministerio de Agricultura, Alimentación y Medio Ambiente (MAGRAMA). (2013). *Agricultura Ecológica*. Estadísticas 2012.

Ministerio de Agricultura, Alimentación y Medio Ambiente (MAGRAMA). (2014). *Agricultura Ecológica*. Estadísticas 2013.

Ministerio de Agricultura, Alimentación y Medio Ambiente (MAGRAMA). (2015). *Agricultura Ecológica*. Estadísticas 2014.

Moleres, J. (2009). *Estudio y análisis de costes de producción de explotaciones de tomate ecológico bajo invernaderos en Almería*. Final degree project, Technical School of Agricultural Engineers, Public University of Navarra.

Montero, J., Antón, M., Torrellas, M., Ruijs, M., & Vermeulen, P. (2011). *Environmental and economic profile of present Greenhouse production systems in Europe*. Deliverable 5, EUPHOROS project.

Morgan, K., & Murdoch, J. (2000). Organic vs. conventional agriculture: Knowledge, power and innovation in the food chain. *Geoforum, 3*(2), 159–173.

Renting, H., Marsden, T. K., & Banks, J. (2003). Understanding alternative food networks: Exploring the role of short food supply chains in rural development. *Environment and Planning A, 35,* 393–411.

Sanyé, E., Oliver-Solà, J., Gasol, C. M., Farreny, R., Rieradevall, J., & Gabarrell, X. (2012). Life cycle assessment of energy flow and packaging use in food purchasing. *Journal of Cleaner Production, 25,* 51–59.

Sanyé-Mengual, E., Cerón-Palma, I., Oliver-Solà, J., Montero, J. I., & Rieradevall, J. (2013). Environmental analysis of the logistics of agricultural products from roof top greenhouses in Mediterranean urban areas. *Journal of Science of Food and Agriculture, 93,* 100–109.

Sonnino, R., & Marsden, T. (2006). Beyond the divide: Rethinking relations between alternative and conventional food networks in Europe. *Journal of Economic Geography, 6,* 181–199.

Wiskerke, J. S. C. (2003). On promising niches and constraining socio-technical regimes: The case of Dutch wheat and bread. *Environment and Planning, 35*(3), 429–448.

## Author Biographies

**Gonzalo Gamboa** is a visiting professor at the Department of Economy and Economic History and associate researcher at the Institute of Environmental Science and Technology (ICTA), at the Autonomous University of Barcelona (UAB). Drawing on the multi-scale integrated assessment of societal metabolism approach, his research focuses on the study of the metabolic patterns of agri-food systems, from production to consumption. He also collaborates with the association Arran de Terra (At ground level) to promote agroecology and food sovereignty, by focusing on food systems and local development.

**Sara Mingorría** is a postdoctoral researcher at the Institute of Environmental Science and Technology at the Autonomous University of Barcelona (ICTA-UAB). Her Ph.D. was focused on oil palm and sugarcane conflicts in Guatemala. Currently, she is working in the EnvJustice project (envjustice.org) mapping and analysing environmental conflicts around the world (ejatlas.org). Among her many interests, she is involved in understanding the role of Traditional Ecological Knowledge (TEK) in resilient systems and the functioning of global and local food systems. At the same time, she is reporting social and environmental injustices through documentaries and Human Rights' evaluations. She is also co-founder of the educational and agroecological cooperatives: Mardetierras and La Noguera Medinaceli in Spain.

**Marina Di Masso** is an environmental sociologist, and postdoctoral researcher at the Sustainable Communities, Social Innovation and Territory research group (SoPCI) of the Universitat de Vic-Universitat Central de Catalunya (UVic-UCC), Spain. Her major research interests are transformative processes of social change, with a focus on the agri-food system. Her research focuses particularly on critical perspectives of the global agri-food system, alternative food networks, and food sovereignty as a transformative paradigm. Recent research and personal interests include wider alternative socio-economic practices, with a focus on feminist economics and social and solidarity-based economy. She is a member of the Agroecology and Alternative Food Systems Chair of the UVic-UCC.

**Mario Giampietro** is ICREA Research Professor at the Institute of Environmental Science and Technology (ICTA) at the Autonomous University of Barcelona (UAB), Spain. He works on integrated assessment of sustainability issues using new concepts developed in complex systems theory. He has developed an innovative scientific approach—Multi-Scale Integrated Analysis of Societal and Ecosystem Metabolism (MuSIASEM)—that integrates biophysical and socio-economic variables across multiple scales, thus establishing a link between the metabolism of socio-economic systems and the potential constraints of the embedding natural environment. His recent research focuses on the nexus between land use, food, energy and water in relation to sustainable development goals.

# Chapter 11
# Connecting Local Food and Organic Waste Management Systems: Closing Nutrient Loops in the City of Madrid

**Marian Simon-Rojo and Barbora Duží**

> *If it were practicable to collect, with the least loss, all the solid and fluid excrements of the inhabitants of the town, and return to each farmer the portion arising from produce originally supplied by him to the town, the productiveness of the land might be maintained almost unimpaired for ages to come, and the existing store of mineral elements in every fertile field would be amply sufficient for the wants of increasing populations.*
> (Von Liebeg, founder of artificial fertilisers, 1863).

**Abstract** Cities stand out as the main destination of processed and consumed resources coming from all over the world. The urban metabolism approach warns us against the prevailing linear processes that move from production to consumption generating huge amounts of waste, which are not reintegrated back into the system. A transition towards a circular urban metabolism is a fundamental issue. In the case of the food system, the goal of returning organic matter and nutrients to the soil and closing nutrient cycles poses an important societal challenge. Considering that our food system is globalised, means with which to find feasible ways of closing nutrient cycles at the local level are not evident. This chapter explores the potential of addressing simultaneously the issues of food production and organic waste and their re-connection, so that the transition to a more re-localised urban food system is complemented by a revisited model of local organic waste management. We present an empirical case study of the city of Madrid (Spain), an experience that reintegrates organic waste into regional Alternative Food Networks. It was initiated as a bottom

M. Simon-Rojo (✉)
Surcos Urbanos and Technical University Madrid, Madrid, Spain
e-mail: m.simon@upm.es

B. Duží
Department of Environmental Geography, Institute of Geonics of the Czech Academy of Sciences, Brno, Czech Republic
e-mail: arobrab@centrum.cz

 
E. Fraňková et al. (eds.), *Socio-metabolic Perspectives on the Sustainability of Local Food Systems*, Human-Environment Interactions 7,
https://doi.org/10.1007/978-3-319-69236-4_11

up approach by the civic platform Madrid Agroecologico, which demanded new public policies and the definition of a sustainable urban food strategy. Local farmers became responsible for composting organic waste from selected schools, residential areas and municipal markets within a pilot project called Madrid Agrocomposta, financed by the local municipality. The project was instrumental in raising awareness of the issue concerning waste and its potential re-use as compost to amend soils. We also explore the potential and the implications for public policies to accommodate a fundamental shift towards recycling organic solid waste into compost for urban agriculture or green areas. The analysis of the material and physical factors from the perspective of metabolism flows is applied to identifying ways to close the nutrient cycles by community composting inside the city or in peri-urban farms and distributed small scale waste processing plants. They would result in the integration of different waste management systems and introducing new players into a sector dominated by large companies, willing to retain a tight control of the urban waste management business.

**Keywords** Urban metabolism · Organic waste · Closing nutrient cycles · Urban waste management · Local food farm · Madrid · Community compost

## 11.1 Introduction: Moving from Linear to Circular Food Systems Through Waste Composting for Peri-Urban Agriculture

Cities stand out as the main destination of resources coming from all over the world and are therefore key areas in addressing the urgent need to reduce the inflows of matter and energy and the outflows of wastes and emissions, that have increased over time (Barles 2007; Clift et al. 2015). To that end, it is worth adopting a system-wide socio-metabolic approach that relates the movement, circulation and dislocation of material resources in the city to the social, political and economic circumstances that influence these flows (Tornaghi et al. 2015; Kennedy et al. 2010). The main challenge is to overcome the prevailing linear model of production, consumption and final disposal of materials which are not reintegrated into the system. To reduce the overuse of resources and the unsustainable rates of generation of waste, it is essential to move to a circular approach, and redesign the system of production and consumption as biological models with closed loops and zero waste. This approach seems to be more coherent with a planet of finite resources in which there is no place for dead ends of worthless garbage (Hottle et al. 2015; Hawken et al. 1999).

In the case of the food system, finding feasible ways of returning organic matter to the soil and closing the nutrient loop is crucial. This chapter explores the potential of re-connecting the issues of food and waste. We suggest that the transition to a more re-localised urban food system should be performed in parallel with a revisited model of local waste management that returns organic matter and nutrients to the soil, where food will once again be cultivated.

Urban agriculture and the overall food system are at a turning point. Local food initiatives have started to be reflected in the European agenda, as Friends of the Earth, Agricultural and Rural Convention (ARC2020) and others had been demanding for years. Arguments in favour of the relocalisation of food cover a wide range of issues, like improving energy efficiency, reducing food miles and emissions as well as dependency on fossil fuels, strengthening local communities, their culture and identity and reconnecting the urban with the rural (Renting et al. 2003; Holloway et al. 2007). We argue that local food should not be tackled only as a matter of shorter supply chains, energy and distance, but that the whole process of food production should be brought into the wider picture, a process that starts with the soil where the seeds will be sown. As explained below, no sustainable local food system will be achieved if soil amendment and closing loops are not considered together. Thus, the main idea of this chapter is to connect local food systems with local composting to return nutrients to peri-urban land.

The chapter describes an empirical case study of the city of Madrid (Spain) and its hinterland, in order to understand the challenges faced when trying to introduce new organic waste management models. We focus on a project for recovering urban organic waste for composting and use in peri-urban farms. It was a project initiated from bottom up, that gained support from the local municipality. Techniques for composting organic urban waste are well-developed, while mechanisms to facilitate their implementation and ensure good source separation with the engagement of the population, are lagging behind. Pilot projects are a recurrent approach to identifying key cultural, social, governance and economic factors that are determinant for the success of introducing changes in a system that affects the whole population. How can we move from pilot projects to consistent public policies? What are the possibilities of scaling it up and of combining different waste management systems in one city? Given the short duration of the project and its limited geographical scope, rather than providing generalisations we offer early suggestions and reflections.

The chapter is divided into five sections. After this introduction, the Soil Sect. (11.2) reviews its importance for agricultural production and describes the main threats posed by unsustainable agricultural practices that lead to loss of organic matter and key nutrients. The next Sect. (11.3) explains the most relevant food localisation concepts and examines the potential of connecting local food system and organic waste management. A methodology Sect. (11.4) then describes the case study of project Madrid Agrocomposta in Madrid and provides initial analysis. It is followed by the Results Sect. (11.5) that highlights preliminary results given the short duration of the project. Finally, the Discussion part (11.6) touches on questions dealing with the potential of actively engaging the population in a different waste management system, making explicit the idea of closing loops in proximity, and nurturing their re-connection with local farmers.

## 11.2 Soil: A Key Factor to Re-Localising the Food System

### *11.2.1 Healthy Soil—Essential for Food Production*

Soil is the basis for food production,[1] as it is the foundation for vegetation used for feed, fibre, fuel and medicinal products. Besides, it hosts biodiversity, helps to mitigate and adapt to climate change by playing a key role in the carbon cycle, and improves our resilience to floods and droughts by storing and filtering water (FAO 2015a, b).

Soil organic matter is considered the most significant single indicator of soil quality and agricultural productivity (Šarapatka et al. 2011). From an agroecological perspective, a holistic approach is needed to deal with food incorporating recycling of biomass, optimising nutrient availability and securing favourable soil conditions (Altieri 2002). Cycles of nitrogen, phosphorus and potassium are particularly relevant in achieving fertile soils to grow healthy crops. This represents a global challenge endangering food security in the future, as agriculture has become increasingly dependent on artificial fertilisers as a substitute for proper soil management (Duncan 1990). Environmental scientists predict that conventional phosphorus supply derived from phosphate rock will achieve its peak around 2030 and will then begin to decline as a result of increasing demand for this non-renewable natural resource (Cordell et al. 2009).

Despite its relevance, the current quality of soils is far from being optimal. At global scale, at least 33% of land is moderately to highly degraded due to the erosion, salinisation, compaction, acidification, and chemical pollution of soils (FAO 2015a, b), along with the application of inadequate farming techniques (Šarapatka et al. 2011). In many areas of Europe, long-term intensive cultivation of soils has resulted in a decline in its organic matter content that is equivalent to a 'mining' operation (Rusco et al. 2001). The authors show the results of the European Soil Database, in which organic matter in soil is estimated on the basis of values of organic carbon. The current decline in organic matter contents should be a serious source of concern, especially in Southern Europe, where "74% of the land has a surface soil horizon (0–30 cm) that contains less than 2% organic carbon, i.e. 3.4% organic matter" (Rusco et al. 2001, p. 6). Moreover, in some parts of the Mediterranean region, erosion has already reached the stage of irreversibility (Grimm et al. 2002).

Adopting sustainable farming practices would increase organic matter contents in soils, which is essential in order to maintain water and nutrient function in the soil, as well as to enhance soil biodiversity (European Commission 2006).This is critical, given that eroded soils can be responsible for 15–30% lower crop yields as

[1]With exceptions like aquaponics, which in global terms have a relatively minor role in food provision, that do not require soil to grow plants.

compared to uneroded soils (Schertz et al. 1989; Langdale et al. 1992), and low levels of soil organic matter jeopardise the good functioning of agricultural systems (Mäder et al. 2002).

### *11.2.2 Compost, or How to Feed the Soil After Having Fed the City*

As Marmo et al. (2004) point out, a large part of the potential pool of organic matter for soils does not leave the farm. This pool of organic matter is made up of roots, crop residues, agricultural wastes (manure spread to land) and the existing soil organic matter (SOM) itself. Management of this direct pool is the easiest and the cheapest way to implement strategies to maintain or improve SOM and should be given priority.

Von Liebig's quotation at the beginning of the chapter explains how this direct pool of organic matter benefits from its combination with organic urban waste, returning to the peri-urban farmers the food scraps from what they have previously supplied to the town. The use of compost is a beneficial farming practice to recover depleted soil organic matter, and to reduce the dependency from external inputs with high environmental impact that entails substantial costs to farmers. Several studies have analysed how organic waste could contribute to soil regeneration and improvement of its biochemical parameters. For example García et al. (1994) studied the effects of adding the municipal organic fraction of solid waste to a degraded soil in Spain. After forty months, they observed increases in organic matter, available mineral substances and other compounds (including phosphorus) and enzymatic activity in all the treated plots. Similar beneficial effects were observed by Plošek et al. (2013) who examined the optimal usage level of compost in soil reclamation substrates for degraded soils in the Czech Republic. Dulac (2001) summarised the main benefits of using compost, based on experiences from the Urban Waste Expertise Programme (1995–2001) held in Brazil and Argentina. While its nutrient levels are low in comparison to chemical fertiliser, compost is a long term source of valuable mineral and organic materials, including slow-release nitrogen, so that it serves as kind of 'soil bank'. Moreover composting contributes to improvement of water balance and resilience to drought (Dulac 2001).

Besides these environmental benefits, composting organic urban waste to feed the soil contributes to fulfilling at least in part the European Union waste management strategy, aiming at reducing the volume of waste that ends up in a landfill by making use of organic waste (Dumitrescu et al. 2014, p. 229).

## 11.3 Bringing Local Food Plans Down to Earth

In 2016, 129 cities signed the Milan Urban Food Policy Pact (MUFPP) to develop sustainable food systems that are inclusive, resilient, safe and diverse. Besides that, a steadily increasing number of cities have been developing sustainable urban food plans or strategies and setting up local food councils (Moragues et al. 2013). At the same time, social movements linked to food sovereignty understand the re-localisation of food as a way to counter the globalised system and the inequalities and exploitation on which it is based.

### *11.3.1 The Right Scale for Closing the Food Loop*

Local food entering the political agenda offers an opportunity to coordinate strategies from waste and food plans, that traditionally belong to different domains. If the goal of closing the cycle of nutrients is to be achieved, it is not enough to work on composting food waste to feed peri-urban soils. The origin of that food, the matter and nutrients, is of relevance, otherwise we cannot say that nutrients and organic matter returns to the soil, as it will be amended with soil coming from somewhere else which will not be able to recover what was deployed in the cultivation process. It is not possible to return organic matter to exactly the land it came from, so at what scale should this connection of re-localizing food and closing cycles be considered?

Food and waste management systems extend far beyond the official domains of a city, raising a critical question: at what scale does it make sense to work on closing the nutrients loop? The answer varies with each context, although we should bear in mind that the goal of closing the loop will be only partially achieved, given the impossibility of large cities becoming self-dependent in terms of food provision.

Whereas local is usually explained as the opposite of global, there is no generally accepted definition of local food, neither in legal, nor in geographical terms (Kneafsey et al. 2013). Many studies aim at defining what 'local food' means and covers, and subsequently examine, measure and evaluate them through various qualitative and quantitative methods (Marsden et al. 2000; Hiroki et al. 2016; Schmitt et al. 2016; Mundler and Laughrea 2016). Such analysis is beyond the scope of this chapter. For the purpose of this chapter, we consider the relevant work of Hiroki et al. (2016) who have summarised the main methodological concepts and suggest three main approaches to define local food. The first approach focuses on geographical proximity, which is framed mainly by administrative boundary or metric distance; the second focuses on geographical origin that stresses the source of food, while the last concentrates on distribution methods and consumer-producer relationships and supply chains. Along with these, several attributes can be associated with local food, especially benefits to society through environmental sustainability, local development or food security. We assume that geographical proximity and benefits to society are key factors to address the recovery of organic waste so as to integrate it into the local food circuit.

The European Commission (2013a, b) relates "local farming" to a relatively small geographical area, "but acknowledges that there is no uniform definition of the term local area" and that "it is essentially the consumer who decides whether a product comes from a local area or not" (Johnson 2016). Therefore, at the European level, local food is not associated with any official administrative boundary. On the contrary, waste management services have to be designed for specific areas, with clear administrative boundaries and a hierarchy from local to regional governments with different responsibilities and competencies. Therefore, connecting local food and local waste implies that technicians and public officers, with different planning and management cultures, should align their goals and work together on common measures. Other relevant stakeholders are engaged in the issue, contributing to shape a complex context. Connections among top-down and bottom-up approaches, local food networks and local food programmes have been developed in many countries (Kirwan et al. 2013). When we look closer at selected programmes trying to combine food and waste, an interesting example comes from Rotterdam, the first Dutch city with a Food Policy Council. It faces the challenge of improving nutrient and waste water management and providing resources for community-based and entrepreneurial economic activities. The city plan deals with the composting of householdorganic wasteat the community level as well as at the regional level with peri-urban farms (de Baat et al. 2014).

### 11.3.2 Feeding the Soil and Eating Pollution?

A frequent objection regarding the amendment of soil with compost obtained from urban agriculture's organic waste is connected with its potential hazard due to pollution of soils. Warnings in this regard were first associated with specific areas where soil may be contaminated to some extent, especially brownfields, vacant soil or places located close to industrial areas or near roadways loaded with heavy traffic (Nehls et al. 2015; Säumel et al. 2012; Schwarz et al. 2016). Organic waste might then contain some traces of health and environmentally risky substances. The complex relationships between the use of contaminated municipal compost and the risk for the environment and health have been the subject of many studies. Some anticipate accumulation of pollutants after several years of disposal, which might lead to future hazards (Déportes et al. 1995). As for chemical contamination, mainly heavy metals or products of organic chemistry (for example long term prevailing polychlorinated biphenyls) have been detected, as well as biological contamination in the form of pathogenic microorganisms (Déportes et al. 2015). The main danger consists of human exposure to these contaminations through the setup of urban circular food chains. On the other hand, some scientists found that in cases of soil contamination the addition of organic matter dilutes the concentration of the contaminant, particularly in the case of lead, in soil (Schwarz et al. 2016).

To sum up this discussion, we should take into account these potential risks and not underestimate some environmental and health circumstances connected with a specific urban environment. Nevertheless, we should also be aware that these concerns do not apply to all urban agriculture, a concept which covers agricultural activity which goes far beyond brownfields or informal gardens in no-man's land at roadsides. It should also be noted that local food does not equal urban agriculture developed within the city or in its fringe. On the contrary, the bioregional and the city-region approaches are gaining strength (Dubbeling et al. 2015; Moragues et al. 2013). If these points are clear, many of the concerns related to the levels of contamination of urban soil are only of relevance for relatively small areas, which play a minor role in food provision. This is at least the case in countries such as Spain that can rely on growing local food in open agrarian areas, not too far from urban agglomerations, which would indeed benefit from re-connecting to the city by means of food.

### *11.3.3 Competing Interests and the Right to Waste*

So far, urban organic waste is no longer seen as something to get rid of, but as a valuable resource, among other goals, to close nutrients and head toward sustainable local food systems, and is also seen as a part of the strategy of how to mitigate climate change.

At the same time, in economic terms, composting organic wastecould result in household savings in taxes for waste disposal, as well as in energy and operating savings for the city council. Stimulating a circular economy related to waste management would boost local jobs (European Commission 2015). If local farmers are integrated in the process with compost facilities in their farms, it would improve their economies, through extra incomes as waste managers.

Despite all the potential benefits, the path is full of pitfalls. The fraction of urban solid waste that goes for composting is likely to increase in the coming years, because the Waste Framework Directive[2] sets the target of recycling 50% of household and similar wastes by 2020. Waste treatment is highly regulated and local authorities have the responsibility to take decisions on operators and treatment options.

---

[2]The Waste Framework Directive 2008/98/EC sets the basic concepts and definitions related to waste management, such as definitions of waste, recycling, recovery. It explains when waste ceases to be waste and becomes a secondary raw material (so called end-of-waste criteria), and how to distinguish between waste and by-products. The Directive lays down some basic waste management principles: it requires that waste be managed without endangering human health and harming the environment, and in particular without risk to water, air, soil, plants or animals, without causing a nuisance through noise or odours, and without adversely affecting the countryside or places of special interest. Waste legislation and policy of the EU Member States shall apply as a priority order the following waste management hierarchy: prevention, preparation for re-use, recycling, recovery and finally, disposal.

Composting has to compete with incineration for energy production, since the EU still considers it as part of its strategy to achieve the goal of 20% renewable energy by 2020. It collides with the Waste Framework Directive, which establishes a hierarchy, giving priority to composting over burning for energy recovery. In different official documents it is easy to find arguments pro-incineration because, according to the same hierarchy, energy recovery from waste is a preferred option to landfill/disposal, though not "at the expense of waste reduction or recycling" (DEFRA 2013).

In practice, many city councils are bounded by contract with the incineration company to deliver a minimum quantity of waste each week, so waste cannot be diverted to other management systems, even by the individual decisions of each citizen (Sage et al. 2015). Ultimately, it is essential to bear in mind that waste management involves large amounts of money through public contracts, and therefore it is at the core of potential conflict of interests. The evolution of urban waste management systems under the paradigm of the economies of scale has led to major contracts and large companies have taken precedence. Nonetheless, new actors are questioning this model and trying to gain a foothold in the waste sector. Some of them come from the Food Sovereignty Movement, which also contests the globalisation of an industrial and neoliberal model of agriculture (Wittman et al. 2010) by claiming their right to waste. They call for the abolition of legal impediments that hinder the possibilities of developing alternative channels for waste recovery through compost systems connected to local/regional farms. That is the case of Madrid (Spain), that will be analyzed in the next section, exploring the possibilities of coexistence of different management systems operating in the same city, one of them aligned with a strategy to support local farmers engaged in alternative food networks.

## 11.4 Methodology of Analysis

This chapter explores a pilot project in Madrid, promoted by a civic initiative and aimed at competing with the hegemonic urban waste management system by integrating local farmers as new participants with a socio-metabolic perspective. This experience is analyzed to address two basic research questions: What are the main contributions, obstacles and outputs of the project dealing with the integration of local farmers into the urban waste management system? On what conditions is it feasible to scale up this small-scale project?

In order to make the concept of local food operational and to enable a quantitative analysis of available public data on food, waste and agriculture, we adopt the municipality and the region as basic geographic units of analysis. Disaggregated data are readily available when working with official administrative boundaries, therefore the case study focuses on the municipality of Madrid, and the region in which it is embedded (Comunidad de Madrid, regional government which corresponds to NUTS II).

The case study unfolds on two levels, first a pilot project is analyzed and then a basic model is applied to the city and the region, to explore the possibilities of introducing different waste management systems. We start by analyzing Madrid Agrocomposta, an innovative composting initiative implemented at a verysmall scale. The factors considered in understanding this initiative as regards the connection between organic waste and food production for the recovery of nutrients and the amendment of soil have been: stakeholders, management and organisation system, technology and resources involved, and outcomes. The analysis of these factors is based on open dialogues with key players and on the processing of technical reports.

The analysis at a regional scale relates to indicators about food consumption, generation of organic waste and brown material from parks, the availability of agricultural land and needs of compost. In this case we use data from official statistics (see Table 11.1) as well as accepted values of reference: we consider an average of 15 kg pruning in 1 $m^2$ of green area (ENT 2006); the dose of compost varies largely, from 2–3 ton/ha per year for rainfed crops to 20–30 ton/ha for horticultural crops and 4 ton/ha for rainfed woody crops and 15 ton/ha for irrigated woody crops (FAO; Peña-Turruella et al. 2002).

## 11.5 Feasibility—How to Make a New Model Operational. The Case Study of Madrid

Here we present some recent experiences of community composting and agro-composting in Madrid. These experiences were intended to explore new ways of dealing with urban organic waste, connecting it to the food process. They may illustrate alternatives in governance and waste management that contribute to achieve more resilient urban regions through the re-localisation of the food system and returning nutrients to the soil, reducing the dependency upon chemical fertilisers and reconnecting consumers and farmers.

### *11.5.1 Agriculture and Food Dependence on the Global Market: Geographical Context*

Some basic data illustrate the dimensions of the food security and waste management challenges that Madrid is facing. The capital city of Spain has 3.1 million inhabitants[3] and the functional urban area 6.5 million inhabitants. The region of Madrid aspires to consolidate as a large service hub, and farming has become irrelevant in terms of its contribution to the GDP (0.10%) and to the workforce

[3]Source of data are those listed in Table 11.1.

**Table 11.1** Needs and flows in the urban metabolism of the agri-food system at the regional level, indicators, and sources of data

| Factor | Description | Unit | Data availability | Source of data |
|---|---|---|---|---|
| *Production* | | | | |
| Land uses | Existing area for food production according to the agrarian census: Agricultural land and pasture land | Land per cap. Ha/ inhab | At regional (C. M) and municipal level | Instituto de Estadistica Comunidad de Madrid |
| Land requirements for food production | Surface of land required to meet food consumption needs, according to population diet and yield of land | Land required per capita $m^2$/ inhab. year | Based on average values at national level | Technical report Carpintero (2006) |
| Flows. Use of fertilisers | Consumption of fertilisers in the region | K, P, N per Ton/ha of agricultural land | Based on absolute values at regional level | National Association of Fertilisers |
| Organic food production | Existing agrarian area under organic food production, registered in official statistics | ha of agricultural land | Data available at regional (C.M) and Data available at municipal level | Instituto de Estadistica Comunidad de Madrid |
| *Food consumption* | | | | |
| Population (R, M) | Basic data to assess demand of food and generation of waste | Population n. inhab. | At regional (C. M) and municipal level | Demographic Census-padrón Instituto Nacional Estadistica |
| Needs. Food consumption | Average amount of food according to surveys on household's budget | Food intake per inhab Kg/inhab. year | Based on household average statistics at national | MAGRAMA |
| Flows. Origin of food | Percentage of food consumed in the region (C.M) which is imported from other Spanish regions or from other countries | % food imported (thousands of euros) | Data available at regional level | Inter-regional Input Output Tables. Institute Statistics Comunidad Madrid |
| *Organic waste generation and management* | | | | |
| Needs. Urban organic waste | The organic fraction of urban waste generated annually, on average | Organic waste per capita Kg/ inhab. year | Data available at municipal level for the city of Madrid | Area of statistic information. Municipality of Madrid |
| Flows. Management of organic urban waste | The amount of organic solid waste sent to the landfill and the amount composted annually | Tons per year | Municipal level | Eurostat |

(0.75%). The figures on the origin of the food entering the region are eloquent: by 2003 food imports accounted for 2330.60 Mill €, by 2010 imports accounted for 98% of the total,[4] a proportion that gives an idea of the regional dependency of the food system, both on external supply areas and on global chains.

Looking at the "waste side" of the equation, we find that in 2014 the city of Madrid produced 1.2 million tons of total urban waste.[5] This implies 385 kg waste per capita which is under the average 475 kg/cap in the E U 28.[6] Almost half of them correspond to organic waste (0.5 millions tons). Madrid's solid urban waste mostly ended up in the main dump of the region and only 14,462 tons were composted, that means only 3% of the total, far short of the 42% of solid municipal waste being composted in Italy (CIC 2014). In short, we are far from closing the nutrient cycles of organic matter, a goal that goes beyond returning organic waste from food produced and consumed within the Madrid region, but should indicate the re-localisation of a higher rate of food.

Simultaneously addressing the issues of food and waste is not a new goal. Two of the main inputs for making fertilisers—potash and phosphates—are extracted from rock beds through mining. In the 80s, local experts explained that alternatives to these mined products, which were becoming scarce, were to be found inside the urban systems: "the landfills at the metropolitan functional area of Madrid are the biggest source of fertilisers in Spain" (Ballesteros et al. 1983).

Nevertheless, the warning was not heeded and, 30 years later, the region is still stuck with an industrialised agriculture which depends heavily upon mined products. In the agricultural season 2014–2015, the consumption of fertilisers in the Madrid region was not negligible: 4270 Tons N (there is a strong upward trend with 6510 Tons by 2015–16), 2314 Tons $P_2O_5$ and 1916 Tons $K_2O$. These figures imply that, on average, the productive land in the region received 30, 16 and 13 kilogrammes per hectare respectively, an amount which is largely below the average use of fertilisers in Spain: 72, 29 and 23 kg/ha.

The presence of organic farming is increasing in the Madrid region, although it still represents a small part of the total. The surface of 5000 ha registered as organic farming in the region by 2005, has almost doubled in less than 10 years: 9400 ha by 2013, which nonetheless, is barely 4% of the total agrarian land in the region. The lower levels of fertilisers are not attributable to a higher presence of organic agriculture, as this is not mainstream in the region, but can indicate that the intensity of agrarian exploitation is relatively low. The way in which this situation could be a factor that reduces the obstacles for a transition into agroecological practices is an open and key question that challenges public policy approaches. Some clues can be found in the recent evolution of new farmers: officially there are 355 organic farmers and processors, joined in the last years by a myriad of small agroecological

---

[4]Imports refer to goods obtained from other Spanish regions as well as from outside the country.
[5]Source of data: Area of statistic information. Municipality of Madrid. http://www.madrid.es.
[6]Source of data: EUROSTAT http://ec.europa.eu/eurostat/statistics-explained/index.php/Municipal_waste_statistics (retrieved 10. 07. 2016).

projects that have emerged around Madrid and which do not follow the institutionalised process of being certified or registered (del Valle 2013). This emergent group of small agroecological farmers was the one integrated in the pilot project of agro-composting that will be analyzed.

### 11.5.2 Socio-Political Context

In some areas of the Madrid region, nowadays the number of new organic farmers is almost as large as those registered officially by 2009 at the agrarian census, in a process that has been propelled by the severe financial crisis. They usually start operating in an informal way, many of them driven by the goal of creating alternatives to the dominant economic system (Simon-Rojo et al. 2017).

For a long time, alternative farmers and committed consumers engaged in Community Supported Agriculture and Alternative Food Networks (AFN), had very low expectations of having an impact on public policies. A new political landscape is now on its way: political candidatures emanated from grass-root social movements, gained considerable power in 2015's municipal elections, and are now governing in Madrid and other municipalities. These new local governments are—in principle—more favourable to integrating demands from social movements (Simon-Rojo et al. 2015).

Different activists engaged in food sovereignty movements assumed then that time had come to take a qualitative step forward and to generate a political impact. In January 2015, a significant number of farmers, consumers, cooperatives, researchers and environmentalists launched a civic platform called Madrid-Agroecologico.[7] They launched a participatory process to define a collaborative strategy to put the ideas of food sovereignty and agroecology into the political agenda. One of the specific demands that was formulated called for the introduction of a system of agro-composting, based on a new paradigm of waste management connected to local farmers, which requires firm support from local governments.

In October 2015, the city of Madrid signed the Milan Urban Food Policy Pact[8] and a monitoring committee was set up in which the Madrid Agroecologico civic platform participates. The proposal of waste management was shaped as a pilot project that, along with other plans about farmer's market or public procurement, is the groundwork of a future plan for a sustainable urban food strategy.

---

[7]Madrid Agroecologico: http://madridagroecologico.org/.

[8]The Pact is an international protocol, engaging the cities for the development of food systems, based on the principles of sustainability and social justice, which has been ratified by 130 cities from all over the world ( http://www.milanurbanfoodpolicypact.org/).

### *11.5.3 The Pilot Project. Madrid Agrocomposta*

According to previously established research questions, here we provide an analysis of the whole civic initiative process and explain its main contributions, obstacles and outputs.

In May 2015, members of Madrid Agroecologico decided to advance the cause of closing the nutrient cycles and addressing the issue of unsustainable organic waste in Madrid. A dialogue with the city council was opened, based on a proposal for improving the efficiency of the recovery of urban organic matter to be composted in peri-urban farms, i.e. connecting the recovery of food waste to local food production.

**Scale and scope**. This dialogue resulted in a pilot project, called Madrid Agrocomposta which started in March 2016. Implementing a distributed organic waste management system in a sector dominated by a few huge corporations is not conflict free. The annual municipal budget of 170.5 million euros goes to large companies from the building sector.[9] Therefore, it was conceived as a very small-scale initiative to test the operability of the proposal, the citizens' engagement and the outcomes. It was launched as a pilot project operating between March and June, in the framework of the Department of Environmental Education, whose support was fundamental. A second pilot project was implemented during November and December 2016. And the third stage, on a larger scale, is expected to start in 2017.

**Participants**. A steering committee was established within the civic platform, which addressed the demand to the Municipal Department of Environment. The municipality financed an initial four-month pilot project with 16,000 €. Because of legal and technical obstacles, the project did not fall under the department of "Waste Management", but under the "Environmental Education" one, and the awareness campaign did play a key role in the project. A wide range of stakeholders became involved including:

- EconomiasBioRegionales (EBR), a social economy organisation commissioned by the city council to coordinate the project and to develop a dissemination plan.
- Four peri-urban farms located within 30 km around the city of Madrid (see Fig. 11.1), all of them met the criterion of being very small agricultural holdings with less than 2 ha and a business model based on direct sales (mainly through vegetable box-schemes, to which consumers subscribe in order to receive a regular supply of local vegetables) or short supply chains.
- Eight nodes for collecting waste, including three schools canteens, two municipal markets and three community gardens. Others nodes joined the project afterwards.

[9]FCC, Ferrovial, Urbaser, OHL, Sacyr, Ascan and Acciona.

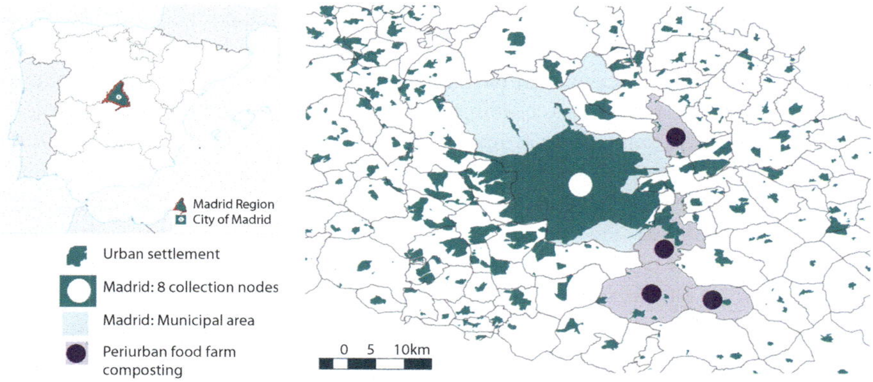


**Fig. 11.1** Map of the Madrid region and the areas involved in the Madrid Agrocomposta project. *Source* Own elaboration, based on fieldwork data collection

- The transport of food was arranged with the small "El Olivar" company, which is devoted to social inclusion.
- A scientific committee monitored the project through monthly meetings.

**Technology and Resources**. The project adopted low-cost solutions. The bins for collecting organic waste had to be easy to move and adequate to avoid unpleasant odours and nuisances. EBR chose a cheap and standard plastic model available at the market. The farmers built the composters in situ re-using common materials in the farm.

The amount of waste of each delivery was weighed and scrutinised manually on the farm, to control and discard pieces of plastics and other foreign materials. Periodically farmers checked the temperature and turned the compost pile over to promote aeration. Once the compost was mature, it was used to amend the soil.

**Outcomes**. During the first month, only one ton of organic waste was gathered in the collection nodes. Households' participation was weaker than expected, the awareness campaign was reinforced and in the following months the amount increased to 3 tons/month, meeting the target set for the project. The project has elicited public sympathy and attention from the media and from the general public. The quality of the compost was exceptional, the amount of foreign materials to be discarded was lower than 0.1%.

The project worked only with organic farmers, therefore there has been no substitution of synthetic inputs. The impact such a system could have in terms of economic savings for a conventional farmer is not negligible: according to RECAN statistics, in the horticultural sector fertilisers account for 8% of total costs, 12% in the case of herbaceous crops.

The monetary compensation assigned to the small agricultural holdings participating in the project was a valuable supplement for their often limited incomes. At this stage, they received 250 €/month that, besides the waste management activities, covered the self-construction of the compost facilities, training (with external experts and internal sessions of mutual learning and exchange of knowledge) and organising visits to each farm for members of the general public included in the awareness campaign. According to the steering committee of the project, when the system is stable, it can operate at the relatively low cost of 50 €/ton, slightly over the upper bound of operating costs estimated in the Waste Management Strategy of Madrid (25–45 €/ton). The advantage of composting with farmers, compared to the standard centralised compost system, is that their investment costs are almost negligible compared to the 150 €/ton in the latter, as well as the wider social benefits derived from their educational activities, anchoring the importance of food cycles into the collective imagination of those urban dwellers engaged in the project and/or visiting the farm.

To sum up, even at this very early stage of the project, we found several benefits linked to agro-composting in peri-urban farms, that contribute to the move towards a more sustainable waste management system in Madrid.

Although the amount of waste diverted from landfill has been necessarily marginal, given the micro scale of the project, the high quality of the compost achieved with an insignificant amount of foreign material (less than 0.1%) that is discarded manually by the farmers before adding the waste to the compost piles, shows the potential of this type of initiative.

In social terms, it is a successful bottom-up practice with high social value that encourages local development. It has proven to be a space for social innovation, with the implication of civic movements and a wide range of participants, including people at risk of social exclusion. It confirms the interest of supporting small and micro local food farmers who should play a key role in creating close ties between consumers and producers (Llobera Serra and Simon-Rojo 2014), in doing so becoming a symbolic relevant player in the re-localisation of the food system.

## 11.6 Discussion: Recovering the Logic of Interweaving Food and Waste

The Madrid Agrocomposta pilot project, small as it was, points to possible ways of dealing with waste and food. Lessons derived from the pilot project, together with a basic overview of general conditions in Madrid, offer a global perspective to understanding the potential and the implications for public policies to accommodate a fundamental shift towards recycling organic solid waste into compost for urban agriculture or green areas. To this end, we first analyze the material and physical factors from the perspective of metabolism flows, with the aim of identifying ways to close the nutrient cycles by composting in situ, be it inside the city or in peri-urban farms.

### 11.6.1 Flows, Needs and Resources

The analysis of the potential flow of materials in closed cycles is twofold, as it includes a rough quantitative characterisation of needs and of resources available and also a basic spatial assessment of the different waste treatment processes that could shape a management plan. The analysis of resources considers urban organic waste on the one hand and local farmers on the other. Problems with data availability hinder a detailed and precise assessment, but they provide a basic overview to help making informed choices about which ways to explore when planning future urban waste management systems.

The city of Madrid generates 500,000 tons of organic waste annually, with another 500,000 tons produced at regional scale. Kitchen organic waste is mainly green, which means nitrogen-rich materials, in contrast to brown carbon-rich material. In order to be transformed into good quality compost, the proportion of brown and green materials should be such as to ensure a 30:1 ratio of Carbon-Nitrogen. A relevant source for brown material is to be found in forest areas—whose surface accounts to almost 200,000 has in the Madrid region—and in the farms. After all, "management of this direct pool is the easiest and the cheapest way to implement strategies to maintain or improve soil organic matter [with adequate agricultural and forestry practices]" (Marmo et al. 2004).

Local food produced in the region is not enough to meet the food needs of the population. Even if all the agricultural land production was aimed at the local market, it would satisfy only 30% of food needs (Moran 2015). In terms of agriculture and soil needs, there are 230,000 ha of agricultural land in the region, most of them in a poor condition, with less than 2% of organic matter in topsoils. Given a standard concentration of 2–3% of nitrogen (*N*), phosphorus (*P*) and potassium (*K*) and, according to practice on different crops explained in the Methodology section, we calculated the amount of compost required[10] as being a total of 700,000 tons per year, for which almost 350,000 tons of organic waste would be needed annually. That sum is less than the total amount of organic waste coming from the city, or in other words, the compost needed to mend soils for all the surface of agrarian land in the region can be obtained by reintegrating the organic urban waste into the loop. The rest of the cities, towns and villages in the region generate another 500,000 tons of organic waste, in locations which are closer to the agricultural land. To reduce energy consumption and emissions associated with transport, priority should be given to recovering waste for composting in farms as close as possible.

In parallel, the amount of organic waste is to be reduced if the goal of moving into more sustainable food systems is assumed, as it implies dealing with the problem of waste at the origin and renders it imperative to reduce the generation of food waste. It should be noted that on average 30% of the food produced ends up as

[10]Calculations: 3 tons/ha for 69,000 ha rainfed crops, 20 tons/ha for 19,000 ha of irrigated crops, 4 tons/ha for 37,000 ha of rainfed olive groves and vineyards and 15 tons/ha for 1.200 ha of irrigated olive groves and vineyards.

waste (MAGRAMA 2013). The volume of food loses generated daily in Mercamadrid—the main food logistics platform of the city—would be enough to feed a city of 200,000 inhabitants.[11]

### *11.6.2 Re-Organisation of Flows to Introduce Diversity in Waste Management*

Calculations applying general average data about consumption and waste generation to specific spatial data provide a starting point to assess the potential of re-connecting food waste and agriculture by means of compost and soil regeneration. Nonetheless, defining the process and management system or, in other words, who will play a role and how, is also critical. The implications of different possibilities are briefly summarised: domestic and community composting, distributed and centralised compost plants and agro-composting in situ. Sludge is not included in the range of possibilities, as it is not considered compost in Directive 2008/98/EC of the European Parliament, but rather biosolids or stabilised organic residues.[12] Figure 11.2 depicts the arrangement of collection and treatment of waste. On the left (1) we can see current centralised management system, and on the right, two scenarios of decentralised systems that integrate community composting in green areas (2.1) and peri-urban farms (2.2) are displayed.

In the transition to organic waste management systems intertwined with local food policies, we analyze different compost systems which can be operated in a de-centralised framework. Domestic and community composting are included in the second model and both, along with composting in farms, are considered in the third model (Fig. 11.2).

**Domestic Composting**. By domestic composting we refer to small-scale processes that take place at the household level. It is mostly associated with detached or semi-detached houses with gardens, where yard waste is integrated in the process. Domestic composting is the most decentralised system that deals with organic waste as close as possible to the places where it is generated: the household level. In Madrid, the relevance of domestic composting is low, as the fraction of this type of houses is small. It can also be argued that, insofar as they are not food producers (or only anecdotally), this could not be considered part of a proposal to intertwine

[11]The volume of food distributed through Mercamadrid could feed roughly 7 million inhabitants, therefore the food discarded would be less than the 3% of the total, which is a very conservative estimate.

[12]It has to be noted that in Spain, 80% of anaerobically digested sewage sludge is used as an agricultural fertiliser (Source: Registro Nacional de Lodos. Ministerio de Agricultura, Alimentación y Medio Ambiente), which poses a problem because of the presence of heavy metals, hormones, pharmaceutical residues and other pathogens. A stricter regulation of chemicals and heavy metals in the environment should be a priority.

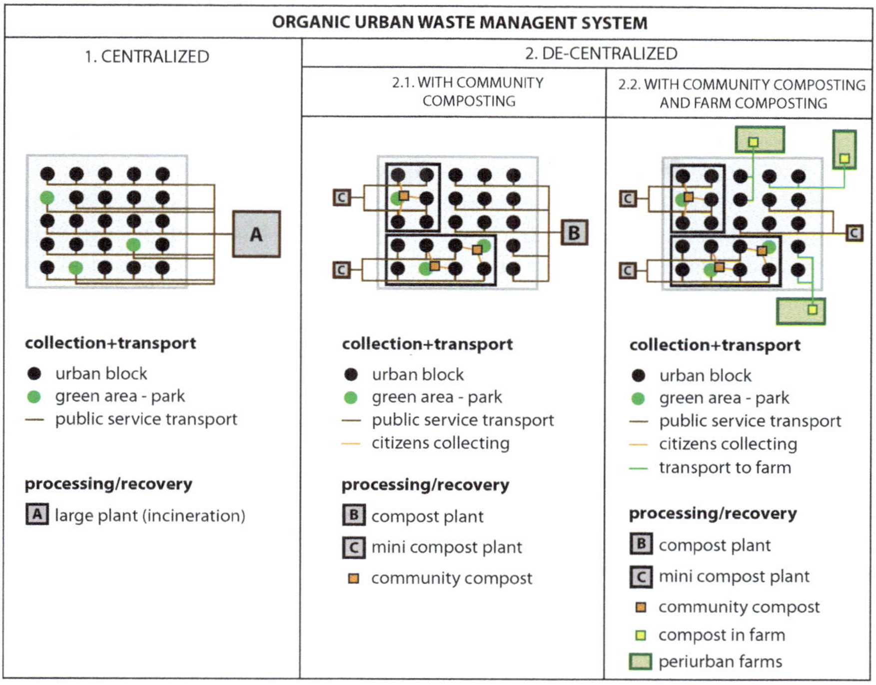


**Fig. 11.2** Different models for urban organic waste management. *Source* own elaboration

local food and waste management systems. Some cities, like Victoria (Canada),[13] have developed a public service that periodically takes the organic and yard waste from the household container to a community composter and, with the same frequency, returns a bag of soil compost.

**Community Composting in Parks**. There are better prospects for other forms of community composting, a term that encompass a diverse range of groups and activities related to local processing of organic material, in which the composting operation is open to a community, not done individually. Ideally, the compost obtained is used in the same area and/or by the same community (Slater et al. 2010). So far, there are only a few examples in Madrid, linked to community gardens and emphasising the educational goals of working together as growers and composters on sustainable waste management and local food systems.

There are 4200 ha of public green areas and urban parks, 61% of them in the periphery of the city. Considering a standard rate of 15 kg of pruning material per $m^2$ of green areas (ENT 2006), 245,000 tons of brown material would be available in green areas dispersed throughout the city. This brown material could be

[13]http://www.communitycomposting.ca/.

combined with 80,000 tons of kitchen waste from immediate residential and tertiary buildings and transformed in-situ into compost. As a result, all the pruning from public green areas would be diverted from incineration. The amount of organic waste integrated in this process of compost varies considerably between districts within the city (Fig. 11.3): in the central ones it would absorb barely 1–2% of food waste, but in others this percentage rises to 29 and 30%.

In theory, brown materials from public parks and green areas could be combined with almost 10% of the organic fraction of municipal solid waste for compost in a decentralised network of composting facilities distributed close to parks in each neighbourhood. At first sight this compost is not really returned into the local food circuit, but the situation may change by developing in parallel edible gardens in public parks or public open spaces. Another possibility is to deliver the compost, once it is mature, to peri-urban fields.

**Agro-Composting in Peri-Urban Farms**. Moving one step further, we find the concept developed in Madrid Agrocomposta of connecting community collection of waste to a decentralised network of local food organic farms that become waste managers, composting it. Within a regional perspective, most of these farms should have closer areas of organic waste supply (see previous section), but if we consider other issues beyond transport efficiency and logistics, this type of waste management draws on fostering different relationships towards food and farmers and has the potential to induce behavioural changes in consumers which is a critical aspect in moving to sustainable food systems.

Their quantitative impact in absolute terms would be more relevant in those districts without green areas for community compost (Fig. 11.3). In all cases they can be the flagship of an agroecological transition in Madrid, given their condition of being part of alternative food networks and committed to food sovereignty. Payments for acquiring the compost service would help these projects to consolidate. At the same time, as they develop educational activities to raise awareness about food and waste, they may act as a reference for the re-connection of consumers with producers.

**Distributed Plants for Composting**. Institutional inertia tends to favour continuity in management, therefore future plans are likely to follow a scheme based on a centralised system in which only major companies would be eligible to provide such a public service. Expectations of achieving compost of good quality are low with this system that hinders affective ties, the engagement of citizens, and the possibilities of feedback to redirect the situation if the level of discards is high.

A more nuanced approach should be adopted, instead of assuming that centralisation, specialisation and large scale are more efficient. It is worth opening a discussion about a distributed model that associates different districts[14] in the city to different composting plants and they, in turn, to different agricultural areas, complemented with strategies to promote food production for local markets.

[14]Population in Madrid's districts range from 45,000 to 240,000 inhabitants, on average a district has 150,000 inhabitants.

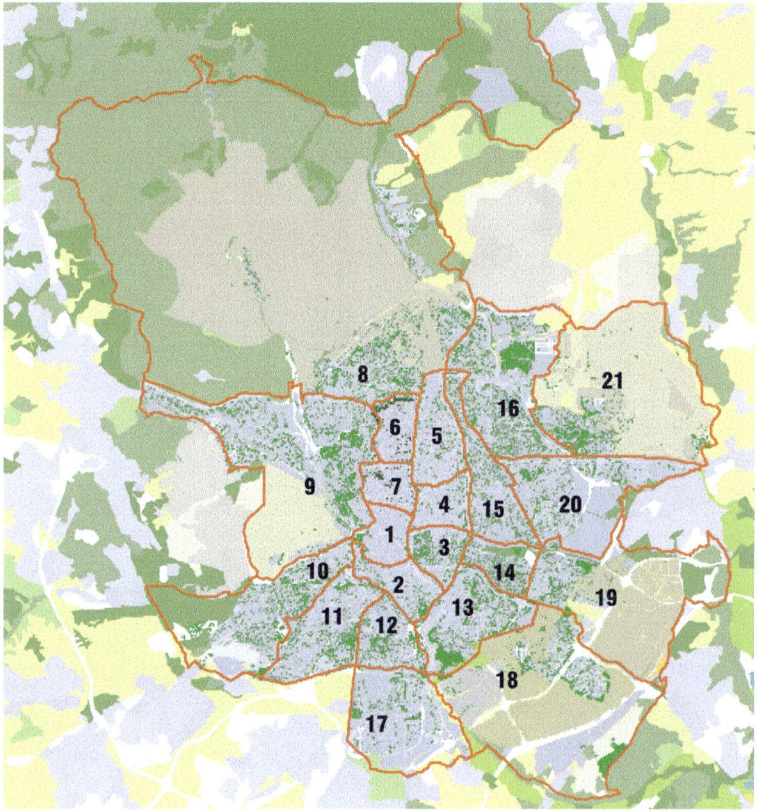


| District | Population | Green Area (ha) | Food waste composted (%) |
|---|---|---|---|
| 01. Centro | 132,644 | 28.49 | 6% |
| 02. Arganzuela | 151,520 | 16.48 | 3% |
| 03. Retiro | 118,559 | 10.72 | 2% |
| 04. Salamanca | 143,244 | 8.41 | 2% |
| 05. Chamartín | 142,610 | 30.87 | 6% |
| 06. Tetuán | 152,545 | 26.94 | 5% |
| 07. Chamberí | 137,532 | 6.71 | 1% |
| 08. Fuencarral-El Pardo | 235,482 | 140.17 | 16% |
| 09. Moncloa-Aravaca | 116,689 | 99.93 | 23% |
| 10. Latina | 234,015 | 146.39 | 16% |
| 11. Carabanchel | 242,000 | 148.78 | 16% |
| 12. Usera | 134,015 | 152.02 | 30% |
| 13. Puente de Vallecas | 227,195 | 236.08 | 27% |
| 14. Moratalaz | 94,607 | 71.73 | 20% |
| 15. Ciudad Lineal | 212,431 | 66.52 | 8% |
| 16. Hortaleza | 177,738 | 121.45 | 18% |
| 17. Villaverde | 141,442 | 113.87 | 21% |
| 18. Villa de Vallecas | 102,140 | 51.15 | 13% |
| 19. Vicálvaro | 69,800 | 76.45 | 29% |
| 20. San Blas-Canillejas | 153,411 | 70.5 | 12% |
| 21. Barajas | 462,64 | 12.42 | 7% |

**Fig. 11.3** Potential for community compost based in public green areas. *Source* Own elaboration

### 11.6.3 Final Reflections

Some final remarks could orientate future discussion. We propose that the transition to a more re-localised urban food system should be complemented by a revisited model of local organic waste management so that they can reinforce each other. They are both essential parts of a strategy of an agroecological transition into a more sustainable urban food system. If the amendment of degraded soils and the closing of the nutrient cycles are done using scraps from food that come from far away, we have to ask ourselves what happens to the soil where those vegetables were grown and how do they recover the missing nutrients? An ethical issue comes to the forefront. In the forthcoming years, the management of humanure should also be brought into the picture.

Simultaneously fostering local food production and implementing a new model of waste management should take into account new actors, and public policies are decisive in supporting local and agroecological food to strengthen the position of an emergent sector of agroecological family farms. To do so, it will be necessary to work with decision-makers and adapt the regulations, but also to persuade technicians to waive their objections and make it possible that diverse systems of bio-waste recovery operate in parallel, performing different functions in connection with the re-localisation of urban food systems.

**Acknowledgements** We would like to thank Franco Llobera, from Madrid Agroecologico, who inspired the project Madrid Agrocomposta and directed our attention to more sustainable urban waste management systems; and Pedro Almoguera, master of bio-intensive agriculture who invites us to introduce a social justice perspective (whose soil are we eating?) when thinking on closing nutrient and material loops. The author from the Institute of Geonics, Academy of Sciences of the Czech Republic, would like to thank for support of the presented chapter by the long-term conceptual development of research organisation, RVO: 68145535.

## References

Altieri, M. A. (2002). Agroecological principles for sustainable agriculture. In N. Uphoff (Ed.), *Agroecological Innovations: Increasing Food Production with Participatory Development*, (pp. 40–46). Sterling, Va.: Earthscan Publications.

Ballesteros, G., Gaviria, M., Baigorri, A., & Domingo, E. (1983). *Agricultura periurbana (periurban agriculture)*. Madrid: Technical Report.

Barles, S. (2007). Feeding the city: Food consumption and flow of nitrogen, Paris, 1801–1914. *Science of the Total Environment, 375,* 48–58.

Carpintero, Ó. (2006). La huella ecológica de la agricultura y la alimentación en España, 1955–2000. *Áreas. Revista Internacional de Ciencias Sociales, 25,* 31–45.

CIC. (2014). *Country report on biowaste collection and recycling in Italy*. ECN Country Report on Italy: Italian Composting and Biogas Association.

Clift, R., Druckman, A., Christie, I., Kennedy, C., & Keirstead, J. (2015). *Urban metabolism: A review in the UK context. Future of cities: Working paper* (p. 82). London, Government Office for Science.

Cordell, D., Drangert, J.-O., & White, S. (2009). The story of phosphorus. Global food security and food for thought. *Global Environmental Change, 19*, 292–305.

DEFRA. (2013). *Incineration of Municipal Solid Waste*. Waste Management Technology Brief.

de Baat, P., Valstar, A., & Renting, H. (2014). Harvesting nutrients in the cities of Rotterdam and Tamale. *Urban Agriculture Magazine, 28*, 24–25.

del Valle, J. (2013). Dime quien eres y te dire como vendes. *Canales de comercializacion del sector hortofrutícola de la Comunidad de Madrid.*

Déportes, I., Benoit-Guyod, J. L., & Zmirou, D. (1995). Hazard to man and the environment posed by the use of urban waste compost: a review. *Science of the Total Environment, 172*(2), 197–222.

Dubbeling, M., Renting, H., Hoekstra, F., Wiskerke, J. S. C., & Carey, J. (2015). City Region Food Systems. *Urban Agriculture Magazine, 29*.

Dulac, N. (2001). *The organic waste flow in integrated sustainable waste management tools for decision-makers* (p. 49). Experiences from the Urban Waste Expertise Programme (1995–2001).

Dumitrescu, I., Manciulea, I., Zaha, C., & Sauciuc, A. (2014). Recycling biomass waste to compost. In: I. Visa (ed.), *Sustainable energy in the built environment—Steps towards nZEB. Springer Proceedings in Energy* (pp. 229–241). doi:10.1007/978-3-319-09707-7_17.

Duncan, C. A. M. (1990). The centrality of agriculture: Between humankind and the rest of nature. *Dissertation Abstracts International. A, Humanities and Social Sciences, 50*(1).

ENT environment and management. (2006). *Estudi sobre el model de recollida de la fracció vegetal al Pla d'Urgell*. Vilanova i la Geltrú: Consell Comarcal del Pla d'Urgell.

European Commission. (2006). *Thematic Strategy for Soil Protection*. Communication from the Commission to the Council, the European Parliament, the European Economic and Social Committee and the Committee of the Region. COM/2006/0231 final

European Commission. (2015, December). Closing the loop—An EU action plan for the Circular Economy. COM (2015) 614 final.

European Commission. (2013a, December). *Report from the commission to the European Parliament and the Council on the Case for a Local Farming and Direct Sales Labelling Scheme*. COM(2013) 866 final.

European Commission. (2013b, December). *Report from the Commission to the European Parliament and the Council on the Case for a Local Farming and Direct Sales Labelling Scheme.* COM (2013) 866 final.

FAO. (2015a). *International year of Soils*. Available at http://www.fao.org/soils-2015/en/. Accessed June 1, 2016.

FAO. (2015b). *Soil is a non-renewable resource*. Online. Available at http://www.org/soils-2015/news/news-detail/en/c/275770/. Accessed June 1, 2016.

García, C., Hernández, T., Costa, F., & Ceccanti, B. (1994). Biochemical parameters in soils regenerated by the addition of organic wastes. *Waste Management and Research, 12*, 466–547.

Grimm, M., Jones, R., & Montanarella, L. (2002). *Soil erosion risk in Europe* (p. 40). Napoli: European Comunities.

Hawken, P., Lovins, A., & Lovins, H. (1999). *Natural capitalism: Creating the next industrial revolution* (p. 396). Little, Brown & Company.

Hiroki, S., Garnevska, E., & McLaren, S. (2016). consumer perceptions about local food in New Zealand, and the role of life cycle-based environmental sustainability. *Journal of Agricultural and Environmental Ethics, 29*(3), 479–505.

Holloway, L., Kneafsey, M., Venn, L., Cox, R., Dowler, E., & Tuomainen, H. (2007). Possible food economies: A methodological framework for exploring food production—Consumption relationships. *Sociologia Ruralis, 47*(1), 1–19.

Hottle, T. A., Bilec, M. M., Brown, N. R., & Landis, A. E. (2015). Toward zero waste: Composting and recycling for sustainable venue based events. *Waste Management*, 9 http://dx.doi.org/10.1016/j.wasman.2015.01.019.

Johnson, R. (2016, February 18). *The role of local and regional food systems in U.S. farm*. Specialist in Agricultural Policy.

Kennedy, C., et al. (2010). The study of urban metabolism and its applications to urban planning and design. *Environmental Pollution*. doi:10.1016/j.envpol.2010.10.022.

Kirwan, J., Ilbery, B., Naye, D., & Carey, J. (2013). Grassroots social innovations and food localisation: An investigation of the Local Food programme in England. *Global Environmental Change, 23*(5), 830–837.

Kneafsey, M., Venn, L., Schmutz, U., Balázs, B., Trenchard, L., Eyden-Wood, T. et al. (2013). *Short food supply chains and local food systems in the EU. A state of play of their socio-economic characteristics*. JRC Scientific and Policy Reports. Joint Research Centre Institute for Prospective Technological Studies, European Commission.

Langdale, G. W., West, L. T., Bruce, R. R., Miller, W. P., & Thomas, A. W., (1992). Restoration of eroded soil with conservation tillage. *Soil Technology, 5*, 81–90.

Llobera Sera, F., & Simon-Rojo, M. (2014). TERRAE municipal network: Boosting the local economy. *Urban Agriculture Magazine, 28*, 26–28.

Mäder, P., Fliessbach, A., Dubois, D., Gunst, L., Fried, P., & Niggli, U. (2002). Soil fertility and biodiversity in organic farming. *Science, 296*(5573), 1694–1697.

MAGRAMA. (2013). Estrategia “Más Alimento, Menos Desperdicio” (Strategy “More Food, Less Waste”).

Marmo, L. et al. (2004). Organic Matter and Biodiversity. Task Group 4 Exogenous Organic Matter. In L. Van-Camp, B. Bujarrabal, A. R. Gentile, R. J. Jones, L. Montanarella, C. Olazabal, & S. K. Selvaradjou (Eds.), *Reports of the Technical Working Groups*.

Marsden, T., Banks, J., & Bristow, G. (2000). Food supply chain approaches: Exploring their role in rural development. *Sociologia ruralis, 40*(4), 424–438.

Moragues, A., Morgan, K., Moschitz, H., Neimane, I., Nilsson, H., Pinto, M. et al. (2013). *Urban food strategies: The rough guide to sustainable food systems* (p. 24). Document developed in the framework of the FP7 project FOODLINKS.

Moran, N. (2015). *La dimensión territorial de los sistemas alimentarios locales. El caso de Madrid* (Ph.D. Dissertation). Technical University of Madrid.

Mundler, P., & Laughrea, S. (2016). The contribution of short supply chains to territorial development: A study of three Quebec territories. *Journal of Rural Studies, 45*, 218–229.

Nehls, T., Jiang, Y., Dennehy, C., Zhan, X., & Beesley, L. (2015). From waste to value: Urban agriculture enables cycling of resources in cities. In F. Lohrberg, L. Lička, L. Scazzosi, & A. Timpe (Eds.), *Urban agriculture Europe* (pp. 170–173). Berlin: Jovis.

Peña-Turruella, E., Carrión-Ramírez, M., Martínez, F., Rodríguez-Nodales, A., & Companioni-Concepción, N. (2002). *Manual para la producción de abonos orgánicos en la agricultura urbana* (pp. 58–69). Cuba: INIFAT-Grupo Nacional de Agricultura Urbana.

Plošek, L., Nsanganwimana, F., Pourrut, B., Elbl, J.,Hynšt, J., Kintl, A., et al. (2013). The effect of compost addition on chemical and nitrogen characteristics, respiration activity and biomass production in prepared reclamation substrates. *International Journal of Environmental Science and Engineering, 7*(11), 364–369 (World Academy of Science, Engineering and Technology).

Renting, H., Marsden, T. K., & Banks, J. (2003). Understanding alternative food networks: Exploring the role of short food supply chains in rural development. *Environment and Planning A, 35*, 393–411.

Rusco, E., Jones, R. J., & Bidoglio, G. (2001). *Organic matter in the soils of Europe: Present status and future trends* (p. 14). Institute for Environment and Sustainability, Joint Research Centre, European Commission.

Sage, C., Tornaghi, Ch., & Dehaene, M. (2015). Urban agriculture practices on the metabolic Frontier: Cases for Geneva and Rotterdam. In F. Lohrberg, L. Lička, L. Scazzosi, & A. Timpe (Eds.), *Urban agriculture Europe* (pp. 178–181). Berlin: Jovis.

Säumel, I., Kotsyuk, I., Hölscher, M., Lenkereit, C., Weber, F., & Kowarik, I. (2012). How healthy is urban horticulture in high traffic areas? Trace metal concentrations in vegetable crops from plantings within inner city neighbourhoods in Berlin, Germany. *Environmental Pollution, 165*, 124–132.

Schertz, D. L., Moldenhauer, W. C., Livingston, S. J., Weesies, G. A.., & Hintz, E. A. (1989). Effect of past soil erosion on crop productivity in Indiana. *Journal of Soil and Water Conservation, 44*, 604–608.
Schmitt, E., Keech, D., Maye, D., Barjolle, D., & Kirwan, J. (2016). Comparing the sustainability of local and global food chains: A case study of cheese products in Switzerland and the UK. *Sustainability, 8*(5), 419.
Schwarz, K., Cutts, B. B., London, J. K., & Cadenasso, M. L. (2016). Growing gardens in shrinking cities: A solution to the soil lead problem? *Sustainability, 8*, 141. doi:10.3390/su8020141.
Simon-Rojo, M., Llobera, F., Yacamán, C., Palmeri, F., Morán, N., Saralegui, P., et al. (2015). *Madrid Agroecológico: The power of civil society to foster food sovereignty in Local urban food policies in the global food sovereignty debate*. Bélgica: At Gante.
Simon-Rojo, M., Morales-Bernardos, I., & Sanz-Landaluze, J. (2017). Food movements swinging between autonomy and co-production of public policies in Madrid. *Nature & Culture* (in press).
Slater, R., Frederickson, J., & Yoxon, M. (2010). Unlocking the potential of community composting: Full project report.
Šarapatka, et al. (2011). *Agroekologie: východiska pro trvalé zemědělské hospodaření [Agroecology: Base for sustainable farming]* (p. 440). Bioinstitut: Praha.
Tornaghi, Ch., Sage, C., & Dehaene, M. (2015). Metabolism: Introduction. In F. Lohrberg, L. Lička, L. Scazzosi, & Timpe, A. (Eds.), *Urban agriculture Europe* (pp. 166–169). Berlin: Jovis.
Von Liebeg, J. (1863). In J. B. Foster (2000). *Marx's ecology: Materialism and nature* (p. 156). NYU Press.
Wittman, H., Desmarais A., & Wiebe, N. (2010). The origins and potential of food sovereignty. *Food sovereignty: Reconnecting food, nature and community*, 1–14.

## Author Biographies

**Marian Simon Rojo** is adjunct professor at the Department of Urban and Regional Planning (Faculty of Architecture, Universidad Politécnica de Madrid). She is engaged in several research projects on sustainable agri-food systems and urban-rural re-connection and is currently a member of the team in charge of designing the Food Strategy for the City of Madrid. She was a participant in the COST Action Urban Agriculture Europe, 2012–2016. Between 2012 and 2014, she coordinated a three year project, Peri-urban Agrarian Ecosystems in Spatial Planning, financed by the Spanish Ministry of Science and Innovation.

**Barbora Duží** works as a researcher at the Institute of Geonics AS CR, v.v.i., Department of Environmental Geography. She focuses on societal resilience and adaptation to climate change, the environmental and socio-economic aspects of food production and consumption, post-communist societal challenges, and environmental migration. Besides academia, she devotes time to environmental education and counselling, raising public awareness on environmental issues and tourism in the non-governmental sector. She cooperates with several NGOs, such as the Líska, Education and Information Centre of the White Carpathian Mountains and People in Need. She is also an active gardener and beekeeper.

# Chapter 12
# Conclusions: Promises and Challenges for Sustainable Agri-Food Systems

**Simron Jit Singh, Willi Haas and Eva Fraňková**

**Abstract** This chapter summarizes the main sustainability challenges (in terms of science and policy) of the current dominant agri-food system and presents insights derived from the cases in the volume. We return to the two main questions asked in the introductory chapter of the book. How useful is the socio-metabolic approach in studying the sustainability of local food systems (LFS)? To this, we identify three main methodological contributions: (1) That classic indicators (of material and energy flows) derived from the sociometabolic approach offers greater insights as well as lend power and rigour when combined with social, ecological, political and other dimensions; (2) Multi-dimensional and multi-scalar analyses can contribute not only to sustainability assessment of a particular LFS but also to broader theoretical and conceptual debates regarding sustainability and potential localisation of LFS; (3) Socio-metabolic studies on the local level provide detailed understanding of the particular LFS while revealing potential leverage points for intervention for improved system performance with respect to sustainability. Besides methodological insights, the chapter derives key lessons from the cases in the book, in particular the promising characteristics of both the historical and current local food system. We identify the following points as important: (1) A close proximity between the producers and consumers holds a very strong potential for systemic change of the current dominant agri-food system, but also the other way round, the growing distance obscures the sustainability challenges; (2) LFS proves

S. J. Singh (✉)
School of Environment, Enterprise and Development (SEED), University of Waterloo, Environment 3 Building, 200 University Avenue West, Waterloo, ON N2L 3G1, Canada
e-mail: Simron.Singh@uwaterloo.ca

W. Haas
Institute of Social Ecology, Faculty for Interdisciplinary Studies, Alpen-Adria University, Schottenfeldgasse 29, 1070 Vienna, Austria
e-mail: willi.haas@aau.at

E. Fraňková
Department of Environmental Studies, Faculty of Social Studies, Masaryk University, Joštova 10, 602 00 Brno, Czech Republic
e-mail: eva.slunicko@centrum.cz

 
E. Fraňková et al. (eds.), *Socio-metabolic Perspectives on the Sustainability of Local Food Systems*, Human-Environment Interactions 7,
https://doi.org/10.1007/978-3-319-69236-4_12

better in closing nutrient cycles on local and regional levels. This issue is also related to the importance of the multifunctionality of land use, and livestock use, in both the historical and the current LFS. As seen in our case studies, LFS cannot be seen as a panacea to address all sustainability challenges of the current dominant agri-food system, however, they hold great potential and therefore deserve further exploration.

**Keywords** Sustainability · Sustainable development goals (SDGs) · Agri-food system · Local food system · Social metabolism

## 12.1 The Challenge

Looking back at the valuable, both theoretical and empirical, contributions in this book, it is a humbling experience to crystallize these insights in just a few pages. Let's first begin by summarising the scientific and policy challenges we have addressed so far. Existing literature as well as the contributions in this book provide strong evidence that our current global agri-food system is unsustainable (Jurgilevich et al. 2016). In ecological terms, it consumes fossil fuel, water, and topsoil at unsustainable rates. It contributes to environmental degradation, via air and water pollution, soil depletion and diminishing biodiversity. Meat production contributes disproportionately to these problems (Horrigan et al. 2002). In terms of health, the world has been moving headlong towards an unhealthy pattern of food consumption for decades. In low and middle income countries, food intake from animal sources has grown rapidly with remarkable declines in physical activity for populations with income growth (Ng and Popkin 2012). As a result, it is estimated that over 12 million deaths in 2010 were attributable to unhealthy diets and physical inactivity around the globe (Lim et al. 2012). At the same time, nutritional deficiencies have stayed at very high levels, which caused 0.4 million deaths in 2015 (Wang et al. 2016) and food security is far from guaranteed by the present food system (Alder et al. 2012), as dramatically shown by the newly unfolding famines in Southern Sudan (Burns 2017).

That said, the challenge to influence dietary patterns in the modern world, in order to achieve more sustainable choices by consumers, is not a trivial one. In liberal democracies, especially when it comes to food issues, politicians have little interest in telling citizens what to buy and eat. At the same time, the obvious possible driver for obesity, categorised as an epidemic by health experts, can be found in the food system due to the increased supply of cheap, palatable, energy-dense foods, improved distribution systems to make food much more accessible and convenient, and more persuasive and pervasive food marketing (Cutler et al. 2003; Kitchen et al. 2004). Further, convenience food has enabled consumers to make choices without sparing much thought. No wonder then, that most consumers in the industrialised world are unaware, or lost in confusing information on the entire food chain and the poor social, environmental and animal welfare standards associated with their decisions (Jurgilevich et al. 2016).

In September 2015, representatives of all the 193 states of the United Nations (UN) embarked on an ambitious *Agenda 2030* with the launch of the Sustainable Development Goals (SDGs, UN 2015). The 17 goals are not mutually exclusive, and in the spirit of sustainability, none of the goals can be seen as stand-alone. For example, good health and well-being cannot be secured without climate action, or zero hunger. The 17 goals cut across a number of critical global environmental and social issues that mandate any sustainability agenda for cross-sector, stakeholder driven, inter-and-multidisciplinary approaches at multiple scales. The topic of this book is at the core of at least 8 of these goals: zero hunger (goal 2), good health and well-being (3), reduced inequalities (10), sustainable cities and communities (11), responsible consumption and production (12), climate action (13), life below water and on land (14 and 15 respectively).

However, as it stands, the Sustainable Development Goals (SDGs) have a long way to go, and the global food system has not provided its share in their accomplishment. Many sustainability scholars do not believe in small adjustments, they rather call for a fundamental change (Gliessman 2014). Public health experts also come to the conclusion that a fundamental change is urgently needed; they argue that increases in obesity seem to be driven mainly by the global food system (Swinburn et al. 2011) and, similarly, the key to reducing cardiovascular disease is making changes in the globalised food system (Anand et al. 2015a, b), and yet meeting the nutrition and NCD[1] targets in the SDGs will require concerted actions to take on the broader food system (Hawkes and Popkin 2015).

In dealing with the complexity of the problem, many studies adopt a rather reductionist perspective, focusing either on the social or natural part of agri-food systems, and very often only on a narrow aspect of either one. However, grasping something as simple and complex as "food" requires a whole system approach—beyond solely the agricultural fields, and often even beyond agroecology—to have an emphasis on Society-Nature interactions in general. These interactions are best analysed from a socio-ecological systems perspective, that focus on the inextricable connection between the biophysical as well as cultural dimensions. Moreover, attention needs to be given to both temporal and spatial scales.

In defining the "sustainability" of agri-food systems, Enric Tello and Manuel González de Molina (Chap. 2) argue that "agri-food systems are sustainable when they can meet human needs while maintaining the basic funds and ecosystem services of agroecosystems and cultural landscapes in both a reproducible way and a healthy ecological state, at local, regional and global scales." In pursuing this definition, the authors suggest the inclusion of societal, political, economic, and also historical perspectives of agri-food production. Such an approach implies a

[1]Noncommunicable diseases (NCDs), known also as chronic diseases, include for example cancer, cardiovascular diseases, diabetes, poor mental health and respiratory diseases. For nine global targets to reduce NCDs see https://ncdalliance.org/global-ncd-targets (2017-07-16).

large and complex research agenda to develop concrete criteria and indicators capable of capturing this complexity. This calls for large collaborative efforts, drawing on scholars versed in interdisciplinary research fields, such as political ecology, social/human ecology, industrial ecology, ecological economics, landscape ecology, agroecology and environmental economics, those with an ability to link and integrate a range of methods. This volume is a small step towards this goal.

## 12.2 Why Are Socio-Metabolic Perspectives on Local Food Systems Important?

We believe the socio-metabolic approach offers a very useful framework for capturing the complexity of this endeavour. The seven cases in this book (Chaps. 5–11) applied the concept of "social metabolism" (SM), tracking flows of matter and energy both within and through their systems' boundaries. The first significant strength of the contributions lies in the fact that most of the authors applied the social metabolism approach to gain insights beyond classic indicators of material and energy flows. For example, Chap. 5 looked at the long-term evolution of agroecological landscapes in terms of labour productivity vis a vis food and fuel requirements, diversity in land use and species richness, and the socio-economic and political implications of a growing dependency on external resources, including land grabbing and, more generally, ecological unequal exchange between Spain and countries in the Global South. These insights suggest not only a growing imbalance between available local resources, regional diets, and type of agricultural production, but also the breaking of regional nutrient cycles into one-way global flows that do not allow organic replenishment of the nutrients. Another historical study (Chap. 6) dealing with the issue of nutrient balance used the social metabolism approach to draw attention to another two "great challenges" of pre-industrial agriculture—social inequality in distribution of critical resources, and unstable provision of food for the local farming community.

Thus, the multi-dimensional and multi-scalar analysis of social metabolism cannot be underestimated. This approach allows understanding of the impacts of food related policy at different dimensions, such as diets, regulations, landscapes, land uses and international trade. Whether the cases focus on historical (long term) metabolism or contemporary, the SM approach underlines the importance of material closing loops locally to offset flows induced from elsewhere. Many of the SM indicators are simple and cost-effective and allow us to track the performance of the system over time. The system boundary in each case incorporates both the social and ecological dimensions, and allows one to derive a number of pressure indicators (focusing on the interaction between society and nature), and at times state indicators (focusing on either the social or the ecological dimension).

There are a number of sustainability indicators for agri-food systems (see FAO 2013), while many others serve as proxies. For example, Soil Organic Matter

(SOM), as argued by Tiziano Gomiero (Chap. 3) does not only relate to soil fertility, but also to soil biodiversity, its water-holding capacity, and resistance to erosion, thus providing plenty of information with relatively small effort. Several chapters present energy-related indicators. Chapter 2 introduces a very innovative adaptation of Energy Return On Investment (EROI) indicators to agroecosystems, enabling one to analyse not only external inputs and outputs (as in traditional EROI calculations), but also to capture internal energy flows significant both in traditional and industrialised agroecosystems. Chapter 3 introduces the specific concept of power expressing the rate or speed at which required energy (in terms of food but also in other required forms) is available to society. And Chap. 4 offers a conceptual and operational model of integrating energy loops and information into landscape and biodiversity modelling.

As such, Chaps. 2–4 stress the function of indicators to express not only the volume and speed of energy and material flows, but also their relation to concrete, agro-ecologically relevant funds, such as soil fertility, managed land, population/workforce etc. Also, they stress the need to understand the identity of the phenomena to be captured by concrete indicators. Whereas, for example, the significance of fossil fuel consumption might be adequately expressed in energy terms, and its further impact in terms of $CO_2$ emissions, for biocides the same units of analysis are possible (in socio-metabolic studies it is common to express their use in terms of embodied energy and $CO_2$ emissions), however, their sustainability relevant identity also includes their toxicity. Thus, adequate indicators should inform us not only about their energy demand and emission impact, but also their poisonous potential.

The wide variety of sustainability indicators covered in the case studies allow for the second important strength of the socio-metabolic approach, its potential to contribute to broader academic debates on sustainability and localisation of agri-food systems. One example is the recent debate on land sparing and land sharing strategies (as introduced in Chaps. 1, 2 and 4). Most chapters in this volume, both the theoretical and empirical ones, argue in favour of the latter, the land sharing approach. From the agroecosystem perspective, local scale matters as both agroecological processes and peasant knowledge have co-evolved, rooted in a specific place, and thus cannot be easily moved to other places where we might consider them more "efficient" as supposed by the land sparing approach. Some of our cases provide insights into household and neighbourhood dynamics and their decision-making processes, as well as the ways local communities respond to higher-level interventions (such as subsidies, markets, prices, climate, infrastructure, etc.). The cumulative effect of these decisions not only alters the local, but also scales up to the global.

Regarding scale, there is another example of theoretical argumentation where the socio-metabolic approach provides important insights. Within the critique of localisation, one line of argumentation pursues the notion that scale is socially constructed, and thus one cannot assume certain outcomes from applying a certain scale (Born and Purcell 2006).[2] The biophysical insight contradicts such a constructivist position; once considering closing the nutrient cycles and other biophysical flows within a certain agroecosystem, the physical scale does matter and is decisive for the sustainability outcome. As argued theoretically (Chap. 2) and in terms of agroecosystem modelling (Chap. 4), but also shown empirically (e.g. Chap. 5), the level of nutrient re-cycling on the local and regional level is directly dependent on the scale and related type of farming activities. In each of these chapters, either the cycles are integrated within the local agroecosystem (with implications for ecosystem functioning, biodiversity etc.), or are not, as in the case of current feedlots in Vallès county in Spain, where significant amounts of feed are imported from various countries worldwide implying global, opened flows of nutrients.

Clearly, this is not to say that local scale is always sustainable in every aspect. Chapter 6 warns us effectively against this pitfall, along with the related one of assuming traditional organic agriculture to be sustainable by definition. On the basis of detailed historical analysis, we learn lessons on the sometimes negative long-term nitrogen balance related to high pressure on land by the local farming community, often due to limited access to land and unequal social relations in the form of manorial tithes and taxes exerted on the peasant community by the landlords.

These insights lead us to the third strength of socio-metabolic studies—the fact that they are context specific, urging a greater insight into the system under consideration. Without this knowledge, it would be difficult to establish an effective systems boundary. For example, local-level analysis articulates how the biophysical interacts with the cultural (taboos, beliefs, institutions, practices, etc.). Clearly, materials and energy do not flow on their own, rather they are deliberately organised and reproduced by society through communication, contingent on their system of meaning. With site-specific studies, the peasants' know-how, narratives, identities, and tastes, as well as the biophysical attributes of landscapes, livestock, seeds, plants and irrigation systems become alive and relevant. Thus, it is through narratives that the characteristics of a farming system, and how it functions within given structural and socio-economic constraints and opportunities (including historically), become evident.

In several cases, the SM approach proved useful for identifying leverage points for intervention. One example is described in Chap. 10, focusing on the functional system, tracking materials in the supply-chain, revealing the different narratives, and thus different logics that shape the assessment of performance of three different organic tomato supply chains (local, mixed and global ones) in Catalonia, Spain. Despite all being certified as organic, the authors argue that their identity is so different that a straightforward comparison is not meaningful. Still, the supply-chain

[2]In their own words (Born and Purcell 2006: 195): "scale is socially produced: scales (and their interrelations) are not independent entities with inherent qualities but strategies pursued by social actors with a particular agenda. It is the content of that agenda, not the scales themselves, that produces outcomes such as sustainability or justice."

approach is useful to identify critical points for intervention (such as the efficiency of the distribution with the local system) and thereby opens the possibility to improve their sustainability performance.

A few of the cases reported bottom-up movements in the Global North (e.g. Greece and Spain) and South (Cambodia) where local practitioners and activists were able to identify leverage points for intervention using the SM approach. Depending on the context, these cases either serve as examples of "quiet sustainability" or "social innovation". The argument is that when it comes to a sustainability transition with respect to agri-food systems, we are not to expect a top-down transition driven by huge investments, national level policies and infrastructure. But, as some of our cases suggest, multi-level transitions based on compatibility with culture, markets, institutions, regulation and the local socio-ecological context will be required. New regimes are constantly being created in terms of local food systems, coupled with energy, waste management and livelihoods, and opportunities for new niches are abundant.

## 12.3 Further Insights from Our Case Studies

What conclusions, beyond methodological ones, can we derive from our cases on the sustainability of local, and above all localised agri-food systems? Several key issues come to the forefront in this respect. First, we argue that the close proximity between the producer and consumer could foster a transition to healthier dietary patterns in quantity and quality. We believe that this proximity offers a reset on how we perceive our food. As we have seen in Chaps. 5 and 9 with respect to the Mediterranean cases (the historical situation in Vallès County in Spain and current case of the island of Samothraki in Greece), a smaller distance between the producers and consumers resulted in a diet that is typically more "healthy" from today's perspective—less meat, more proteins from legumes, more fresh and less processed food. Here we find a "relatively high adherence to Mediterranean diet", with some deviations, such as—in the current case of Samothraki—higher consumption of white bread, sugar, coffee, alcohol and—to a lesser extent—meat. Even if the current economic situation is difficult, health and localness are more important factors in making food choices than price. Thus, food quality, tradition and meaning is still significant for people in large parts of the world. This meaning is preserved when consumers and producers are tied as a community, where one influences the other in terms of activities, rituals, and seasons.

On the other hand, as in the present situation in Vallès county, the producer-consumer disconnect affects food choices, as consumers remain oblivious of how they trigger unsustainable processes far away, such as land grabbing, social inequalities, and nutrient imbalance, to name a few. LFS would not only mean to "know" about these consequences and possibly include them in consumption choices, but to re-connect to local diets, to local farming communities and agroecosystems, to local biophysical conditions. In the case of Vallès, for example, this would mean that in 1999, meat production would have had to be 5.3 times lower if it had to be adjusted to the local biophysical capacity in terms of growing the required amount of animal feed.

Another, related issue of critical relevance is the vast nutrient losses associated with the increasing nutrient imbalance throughout the globe. While rich countries accumulate nutrients in their soil, poor countries lack access to nutrients and their soil suffers poor phosphorus access and low agricultural production (Schoumans et al. 2015). The livestock sector and the production of meat is one of the most nutrient-intensive agricultural sectors. The associated trade of inputs, feed and products shifts nutrients between continents and entails a vast virtual trade of land, and water, not embedded in the product but needed at the production site. Thus, the high Japanese import of meat implies that half of Japan's virtual nitrogen is lost in the US (Galloway et al. 2007). This global re-distribution of nutrients is inherent to industrialised agriculture in which a spatial division leads to an animal and plant production concentrated in separate areas. This causes excess manure in places where soils are already saturated with nutrients (Csatho and Radimszky 2009; Taghizadeh-Toosi et al. 2014).

Traditional systems followed a multifunction approach to livestock and farming (as discussed conceptually in Chaps. 2 and 4, and empirically in Chaps. 5, 6, and 7). The case studies clearly show that the more integrated LFS (heterogeneous land-uses, integration of cropland, pastures and woodland with a livestock system on a local scale) have more closed nutrient cycles—both in traditional organic systems, but also in the case for the current organic system discussed in Chap. 7. To ensure sufficient nutrient input for the soils, a significant internal biomass turnover within the farm system has to be ensured—the case studies describe both traditional and current concrete practices how to ensure this (Chaps. 5, 6, 7, and 8)—using green fertilisers (leguminous nitrogen fixing plants as part of the crop rotation scheme, buried biomass from post-harvest leftovers, compost, etc.), animal manure and possibly also humanure. Chapter 5 describes specific practices of biomass re-use in Vallès county, Spain called "formiguers"—small charcoal kilns where fresh biomass is buried and burnt to create charcoal that is incorporated into the soil.

From a biophysical perspective, for agri-food systems to be more sustainable, nutrient cycles have to be more closed on the local and regional level, and less dependent on fossil fuel-based inputs. From evidence, localised LFS fulfil this requirement better than the globalised ones. Not only this, there are additional benefits in retaining local nutrients. For example, Chap. 8 introduces small-scale community based use of modern technologies—a biogas system producing energy both for cooking and lightening. On a larger—city and regional—scale, the concrete institutional and economic possibilities of re-integration of biomass residues into the farm production process are explored and assessed in Chap. 11, suggesting a combination of small household and neighbourhood composting schemes, bigger composting facilities on the city level, and composting at decentralized network of peri-urban farms as optimal. The importance of closing nutrient cycles for the agroecological functioning of agroecosystems, heterogenous land uses and related farm-associated biodiversity, energy efficiency of farming systems, the value of site-specific peasant knowledge, potential for democratisation and more balanced power relations across the agri-food value chain are all critical.

Despite its strengths, we do not claim that the concept of a "local food system" is a panacea to solve all the problems related to the global agri-food system mentioned above. But it does provide promises for significant improvements, as well as hints at future pathways to be explored. Already, a number of initiatives and efforts can be seen around the globe, in different forms and at all scales, some more visible than others. On the conceptual level, a few key terms have become important for interpreting some of these efforts (also in this volume). For example, "leapfrogging" is discussed in Chaps. 6, 8 and 9. Combination of the local practices and models captured in these particular case studies opens up the possibility of avoiding the often one-way intensification and industrialisation path that sometimes seem (under socio-economic and demographic pressures) inevitable. Often, these locally embedded practices combine traditional knowledge with new methods of cultivation, distribution and institutional arrangements. Through the combination of the old and the new, not pursuing either extreme, such LFS are somewhere in between, or "halfway through" the intensification path, trying to balance and make use of both types of agri-food systems (Chaps. 5, 8, and 9). Possible advantages, and also trade-offs of such a position are discussed in Chap. 7. In seeking sustainable models of food production and consumption, we also stress the importance of existing practices that are not motivated by purposeful sustainability concerns, that we refer to as "quiet sustainability". For example, Chaps. 5, 7 and 8 describe existing practices and techniques that by their nature comply with sustainability, although not purposefully.

In conclusion, it is easier to say what is *not* sustainable. What a more sustainable agri-food system should look like, and above all, how to achieve this goal still remains a major challenge. The cases in this volume are indeed comforting, but too few to generalise or offer firm insights for science and policy. From what we've seen, LFS are promising, but in all the cases discussed in this book their sustainability depends on how exactly they are specifically implemented. Thus, they are in the first place not as comforting as large-scale technological solutions which are often sold as silver bullets but which in most circumstances like the green revolution do not live up to their promises. Against this backdrop, to engage in pathways based on LFSs is a very rich, but demanding option. In terms of policy, it needs a shift in principle, from policy directed towards international agro-markets and transnational companies to an adaptive reflexive governance approach dealing with a network of interrelated agri-food systems.

Such an approach requires social and societal learning that proceeds in a stepwise fashion, moving from single to double to triple-loop learning (Pahl-Wostl 2009). Inherent to this approach is a continous adaptation to new circumstances which makes it difficult to predict developments and to control processes, but which promises higher resilience as well as higher sustainability. It requires a governance structure with willingness to learn from LFS, to manage uncertainty and risk, as well as to understand and change structural constraints for promising cases of highly sustainable agri-food systems. In this, a major role needs to be attributed to non-state actors like local food initiatives. Scholars argue that such poly-centric systems are assumed to have a higher adaptive capacity and to be less vulnerable to disturbance (Pahl-Wostl 2009; Casado-Asensio and Steurer 2014; Jordan and Lenschow 2010; Mickwitz et al. 2009; Sonnino et al. 2014; Voß 2005). Thus, going along the path of a more localised food system means an explorative process of subsidiarity, making use of top-down, bottom-up, network and side-by-side governance elements for making all our food sustainable, after all.

## References

Alder, J., Barling, D., Dugan, P. et al. (2012). *Avoiding future famines: Strengthening the ecological foundation of food security through sustainable food systems*. A UNEP Synthesis Report.

Anand, S. S., Hawkes, C., De Souza, R. J., Mente, A., Dehghan, M., Nugent, R., et al. (2015a). Food consumption and its impact on cardiovascular disease: Importance of solutions focused on the globalized food system: a report from the workshop convened by the World Heart Federation. *Journal of the American College of Cardiology, 66*(14), 1590–1614.

Anand, S. S., Hawkes, C., de Souza, R. J., et al. (2015b). Food consumption and its impact on cardiovascular disease: Importance of solutions focused on the globalized food system. *Journal of the American College of Cardiology, 66*(14), 1590–1614.

Born, B., & Purcell, M. (2006). Avoiding the local trap: Scale and food systems in planning research. *Journal of planning education and research, 26*(2), 195–207.

Burns, J. J. (2017). *Preventing the World's next refugee crisis: Famine, conflict, and climate change in Nigeria, South Sudan, Somalian and Yemen*. American Security Project (ASP). https://www.americansecurityproject.org/wp-content/uploads/2017/05/Ref-0202-Preventing-the-Worlds-Next-Refugee-Crisis.pdf. Accessed July 20, 2017.

Casado-Asensio, J., & Steurer, R. (2014). Integrated strategies on sustainable development, climate change mitigation and adaptation in Western Europe: Communication rather than coordination. *Journal of Public Policy, 34*(3), 437–473.

Csatho, P., & Radimszky, L. (2009). Two worlds within EU27: Sharp contrasts in organic and mineral nitrogen–phosphorus use, nitrogen–phosphorus balances, and soil phosphorus status: Widening and deepening gap between Western and Central Europe. *Communications in Soil Science and Plant Analysis, 40*(1–6), 999–1019.

Cutler, D. M., Glaeser, E. L., & Shapiro, J. M. (2003). Why have Americans become more obese? *The Journal of Economic Perspectives, 17*(3), 93–118.

FAO. (2013). *SAFA—Sustainability assessment of food and agriculture systems indicators*. Rome: Food and Agriculture Organization of the United Nations. http://www.fao.org/fileadmin/templates/nr/sustainability_pathways/docs/SAFA_Indicators_final_19122013.pdf. Accessed July 15, 2017.

Galloway, J. N., Burke, M., Bradford, G. E., Naylor, R., Falcon, W., Chapagain, A. K., et al. (2007). International trade in meat: The tip of the pork chop. *AMBIO: A Journal of the Human Environment, 36*(8), 622–629.

Gliessman, S. R. (2014). Agroecology: The ecology of sustainable food systems. CRC press.

Hawkes, C., & Popkin, B. M. (2015). Can the sustainable development goals reduce the burden of nutrition-related non-communicable diseases without truly addressing major food system reforms?. *BMC Medicine, 13*(1), 143. http://bmcmedicine.biomedcentral.com/articles/10.1186/s12916-015-0383-7. Accessed July 15, 2017.

Horrigan, L., Lawrence, R. S., & Walker, P. (2002). How sustainable agriculture can address the environmental and human health harms of industrial agriculture. *Environmental health perspectives, 110*(5), 445. http://library.wur.nl/WebQuery/wurpubs/377942. Accessed July 20, 2017.

Jordan, A., & Lenschow, A. (2010). Environmental policy integration: A state of the art review. *Environmental Policy and Governance, 20*(3), 147–158.

Jurgilevich, A., Birge, T., Kentala-Lehtonen, J., Korhonen-Kurki, K., Pietikäinen, J., Saikku, L., et al. (2016). Transition towards circular economy in the food system. *Sustainability, 8*(1), 69.

Kitchen, P. J., Brignell, J., Li, T., & Jones, G. S. (2004). The emergence of IMC: A theoretical perspective. *Journal of advertising research, 44*(1), 19–30.

Lim, S. S., Vos, T., Flaxman, A. D., Danaei, G., Shibuya, K., Adair-Rohani, H., et al. (2012). A comparative risk assessment of burden of disease and injury attributable to 67 risk factors and risk factor clusters in 21 regions, 1990–2010: A systematic analysis for the Global Burden of Disease Study 2010. *The Lancet, 380*(9859), 2224–2260.

Mickwitz, P., Aix, F., Beck, S., Carss, D., Ferrand, N., Görg, C., et al. (2009). *Climate policy integration, coherence and governance* (No. 2). PEER.

Ng, S. W., & Popkin, B. M. (2012). Time use and physical activity: A shift away from movement across the globe: Declines in Movement across the Globe. *Obesity Reviews, 13*(8), 659–680.

Pahl-Wostl, C. (2009). A conceptual framework for analysing adaptive capacity and multi-level learning processes in resource governance regimes. *Global Environmental Change, 19*(3), 354–365.

Schoumans, O. F., Bouraoui, F., Kabbe, C., Oenema, O., & van Dijk, K. C. (2015). Phosphorus management in Europe in a changing world. *Ambio, 44*(2), 180–192.

Sonnino, R., Torres, C. L., & Schneider, S. (2014). Reflexive governance for food security: The example of school feeding in Brazil. *Journal of Rural Studies, 36,* 1–12.

Swinburn, B. A., Sacks, G., Hall, K. D., McPherson, K., Finegood, D. T., Moodie, et al. (2011). The global obesity pandemic: Shaped by global drivers and local environments. *The Lancet, 378*(9793), 804–814.

Taghizade-Toosi, A., Olesen, J. E., Kristensen, K., Elsgaard, L., Østergaard, H. S., & Lægdsmand, M. (2014). Changes in carbon stocks of Danish agricultural mineral soils between 1986 and 2009. *European Journal of Soil Science, 65*(5), 730–740.

UN, General Assembley, (2015). *Transforming our world: The 2030 agenda for sustainable development. General Assembley 70 Session.*

Voß, J.-P. (2005). Sustainability foresight. Methods for reflexive governance in the transformation of utility systems. *IHDP Update. Newsletter of the IHDP, 2005*(1), 18–20.

Wang, H., Naghavi, M., Allen, C., Barber, R. M., Bhutta, Z. A., Carter, A., et al. (2016). Global, regional, and national life expectancy, all-cause mortality, and cause-specific mortality for 249 causes of death, 1980–2015: A systematic analysis for the Global Burden of Disease Study 2015. *The Lancet, 388*(10053), 1459–1544.

## Author Biographies

**Simron Jit Singh** is Associate Professor in the School of Environment, Enterprise, and Development (SEED), and Associate Dean of the Faculty of Environment, University of Waterloo, Canada. Drawing on the concept of social metabolism, his research focuses on the systemic links between material and energy flows, time use and human wellbeing. His interest lies at the local and sub-national scales, and in particular small islands. He has conducted social metabolism studies in the Nicobar Islands (India), and Samothraki (Greece), and supervised work on the biomass metabolism of Jamaica (as part of the Canadian project Hungry Cities), the Region of Waterloo as well as Canada. As work-package leader, he led work on biomass flows and social conflicts in an EU project (EJOLT).

**Willi Haas** is senior researcher and lecturer at the Institute of Social Ecology in Vienna, an institute of the Faculty of Interdisciplinary Studies at the Alpen-Adria University. In the 90s, he became fascinated by social ecology, the study of society-nature relations across time and space. He is interested in the past transition from agrarian to industrial societies and the insights that can be drawn from these far-reaching changes for the next transition to a post- fossil society. In his view the question of how to overcome the system inertia and the unsustainable reproduction dynamics of the present fossil fuelled societies is crucial for fostering a sustainable future. He was chair of Greenpeace CEE for 9 years. Professionally, he was a public official at the Ministry of Social Affairs, director of the Institute of Applied Ecology (Vienna), acting director of the Environmental Monitoring Group (Cape Town) and researcher at the International Institute of Applied System Analysis (IIASA, based in Austria). At the Institute of Social Ecology he has headed numerous scientific projects and is the coordinator of the institute's team of thematic research coordinators.

**Eva Fraňková** works as an Assistant Professor at the Department of Environmental Studies, Faculty of Social Studies, Masaryk University in Brno, Czech Republic. Her long-term research interests and passions include the concept of eco-localisation, sustainable degrowth and various grass-root alternative economic practices including eco-social enterprises, local food initiatives etc. Recently she has been involved in the mapping of heterodox economic initiatives in the Czech Republic, and in research on the social metabolism of local food systems. She is also involved in several NGOs—the Association of Local Food Initiatives, the Society and Economy Trust, and NaZemi (OnEarth), a global education and Fair Trade organisation in the Czech Republic.

# Index





E. Fraňková et al. (eds.), *Socio-metabolic Perspectives on the Sustainability of Local Food Systems*, Human-Environment Interactions 7,
https://doi.org/10.1007/978-3-319-69236-4

The manufacturer's authorised representative in the EU is Springer
Nature Customer Service Centre GmbH, Europaplatz 3, 69115 Heidelberg,
Germany. If you have any concerns regarding our products, please
contact ProductSafety@springernature.com

Printed and bound by CPI Group (UK) Ltd, Croydon, CR0 4YY
05/10/2026
02228133-0004